Processing, Properties, and Design of Advanced Ceramics and Composites

Processing, Properties, and Design of Advanced Ceramics and Composites

Ceramic Transactions, Volume 259

Edited by

Gurpreet Singh

Amar Bhalla

Morsi M. Mahmoud

Ricardo H. R. Castro

Narottam P. Bansal

Dongming Zhu

J. P. Singh

Yiquan Wu

The American Ceramic Society

WILEY

Published by John Wiley & Sons, Inc., Hoboken, New Jersey.
Published simultaneously in Canada.

Library of Congress Cataloging-in-Publication Data is available.

ISBN: 978-1-119-32364-8
ISSN: 1042-1122

10 9 8 7 6 5 4 3 2 1

Contents

PROCESSING AND PERFORMANCE OF MATERIALS USING MICROWAVES, ELECTRIC, AND MAGNETIC FIELDS

ADVANCES IN COMPOSITES

Preface

This volume contains 33 papers presented during the Materials Science & Technology 2015 Conference (MS&T'15), held October 4–8, 2015 at The Columbus Convention Center, Columbus, Ohio. Papers from the following symposia are included in this volume: • Controlled Synthesis, Processing, and Applications of Structural and Functional Nanomaterials

- Advances in Dielectric Materials and Electronic Devices
- Innovative Processing and Synthesis of Ceramics, Glasses and Composites
- Advances in Ceramic Matrix Composites
- Rustum Roy Memorial Symposium on Processing and Performance of Materials Using Microwaves, Electric, and Magnetic Fields
- Sintering and Related Powder Processing Science and Technology
- Thermal Protection Materials and Systems
- Surface Protection for Enhanced Performance
- Ceramic Optical Materials, and
- Alumina at the Forefront of Technology

These conference symposia provided a forum for scientists, engineers, and technologists to discuss and exchange state-of-the-art ideas, information, and technology on advanced methods and approaches for processing, synthesis, characterization, and applications of ceramics, glasses, and composites.

Each manuscript was peer-reviewed using The American Ceramic Society's review process. The editors wish to extend their gratitude and appreciation to all the authors for their submissions and revisions of manuscripts, to all the participants and session chairs for their time and effort, and to all the reviewers for their valuable comments and suggestions.

We hope that this volume will serve as a useful reference for the professionals working in the field of synthesis and processing of ceramics and composites as well as their properties.

GURPREET SINGH
AMAR BHALLA

Morsi M. Mahmoud
Ricardo H. R. Castro
Narottam P. Bansal
Dongming Zhu
J. P. Singh
Yiquan Wu

Controlled Synthesis, Processing, and Applications of Structural and Functional Nanomaterials

ASSESSING THE LIMITS OF ACCURACY FOR THE TAUC METHOD FOR OPTICAL BAND GAP DETERMINATION

Dunbar P. Birnie, III
Rutgers University, Department of Materials Science and Engineering
New Brunswick, New Jersey, 08854-8065

ABSTRACT

Scientists and engineers working with nanotechnology and thin film optical devices often make use of "Tauc plots" to determine band gaps and evaluate the effect of processing conditions on the quality of coatings made for these applications. Broad-band optical data are easy to acquire and usually exhibit a region of reasonable transparency and then a sharp rise in absorption with increasing photon energy as the band-gap energy is exceeded. The shape of the onset of absorption is diagnostic of whether the band-gap is direct or indirect. Then, an appropriate linear regression can be used to extrapolate to the band gap value, though sometimes the extrapolation is quite far in absolute energy terms from the data used to make the extrapolation. This paper covers some of our recent work where we use known materials to standardize the fitting protocols and assess the accuracy of this simple method.

INTRODUCTION

In our earlier work with thin films (and for many studies in the literature that use the Tauc method) we've noticed that the distance of extrapolation in the fitting process may be relatively large, and the tail of sub-band-gap absorption can also be quite large. This raised the basic question about how accurate the Tauc method would be, and how to establish procedures that improve the accuracy of the fitting results[1]. We delved deeply into this problem by looking at ZnO thin films because they are an extremely well-studied material and ZnO is known to have a direct band gap. By looking closely at a population of over 120 thin film Tauc fits we found the band-gap results overall were consistent with a value of 3.27 +/- 0.05 eV, with evidence for two small outlier populations [1]. A subpopulation of higher gap values appeared to be caused by nanoparticle quantum confinement effects (not surprisingly), while a subpopulation of lower gap values appeared to be correlated with more defective samples. These were essentially cases that had stronger sub-band-gap absorption, which has the mathematical effect of shifting the intercept point somewhat to the left and making the confidence interval of the band-gap determination wider (less accurate). To quantify this effect and provide a figure of merit for identifying the more accurate samples, we introduced the "near-edge absorptivity ratio (NEAR)". And, when using the NEAR to focus on the more accurate data sets, we found that the Tauc method generally gave an experimental distribution of results with a standard deviation of only 0.033 eV, thus emphasizing the relatively high accuracy of the method in general.

We extend that work to the case of indirect band-gap materials and examine accuracy limits based on absorption coefficient values and coating thickness effects that can influence the signal-to-noise ratio of real optical absorption data. Indirect band-gaps are more difficult to characterize because their absorption intensities are characteristically weaker, which provides an added difficulty when most optical data are determined from thin film samples. We address this problem by working with single crystal data from silicon, probably the most well-characterized indirect band-gap material available.

BACKGROUND

The seminal work of Tauc, Grigorovici, and Vancu [2] presented a simple method that uses broad band absorption spectra and interpreted the shape of the absorption edge to arrive at a determination of the band gap, and its character. Their method was further developed in Davis and Mott's more general work on amorphous semiconductors [3, 4]. Together they've shown that the optical absorption strength depends on the difference between the photon energy and the band gap as shown in (Eq. 1):

$$(\alpha h v)^{1/n} = A(hv - E_g) \tag{1}$$

where h is Planck's constant, v is the photon's frequency, α is the absorption coefficient, Eg is the band gap and A is a proportionality constant. The value of the exponent denotes the nature of the electronic transition, whether allowed or forbidden and whether direct or indirect:

For direct allowed transitions	n=1/2
For direct forbidden transitions	n=3/2
For indirect allowed transitions	n=2
For indirect forbidden transitions	n=3

Typically, the allowed transitions dominate the basic absorption processes, giving either n=1/2 or n=2, for direct and indirect transitions, respectively.

Thus, the basic procedure for a Tauc analysis is to acquire optical absorbance data for the sample in question that spans a range of energies from below the band gap transition to above it. Then, plotting the (α h v) with various test exponents versus photon energy allows the researcher to decide which of the exponents gives the most linear plot. Finally, with this exponent, the line is extrapolated down to intersect the X-axis, which will be the band-gap value (as can be interpreted from Equation 1). Of the four exponent choices listed, it is usually found that either the ½ and 2 exponents are most frequently used (being associated with the allowed transitions).

ANALYSIS OF DIRECT-GAP MATERIALS

Zinc oxide was a good candidate for evaluating the Tauc method because it has been widely studied for a number of useful applications [5-13]. Among these applications the band-gap plays a central and fundamental role as it controls many absorption and conductivity phenomena. Single crystal optical studies have found a direct band gap of 3.3 eV[14-16], though many of the papers surveyed in our thin film analysis were collected from very well crystallized films or even epitaxially grown layers[1]. ZnO was also attractive as a reference material because of its high level of stoichiometry. While every stoichiometric compound must thermodynamically have point defects at some level (and therefore by definition be non-stoichiometric), the phase of ZnO has been experimentally studied and found to have very little deviation from the ideal 1:1 ratio. For example, the early work of Allsopp and Roberts found a slight zinc excess, but less than 50 ppm [17]. This is much more stoichiometric than many phases and thus provided a good calibration test-case for the Tauc method.

Figure 1 gives one example Tauc plot for ZnO where the absorption coefficient times the photon energy to the second power is plotted versus the incident photon energy[18]. The second power was used as zinc oxide is well known to have a direct allowed transition. The characteristic features of Tauc plots are evident: at low photon energies the absorption approaches zero – the material is transparent; near the band-gap value the absorption gets stronger and shows a region of linearity in this squared-exponent plot. This linear region has been used to extrapolate to the X-axis intercept to find the band gap value (here about 3.28 eV).

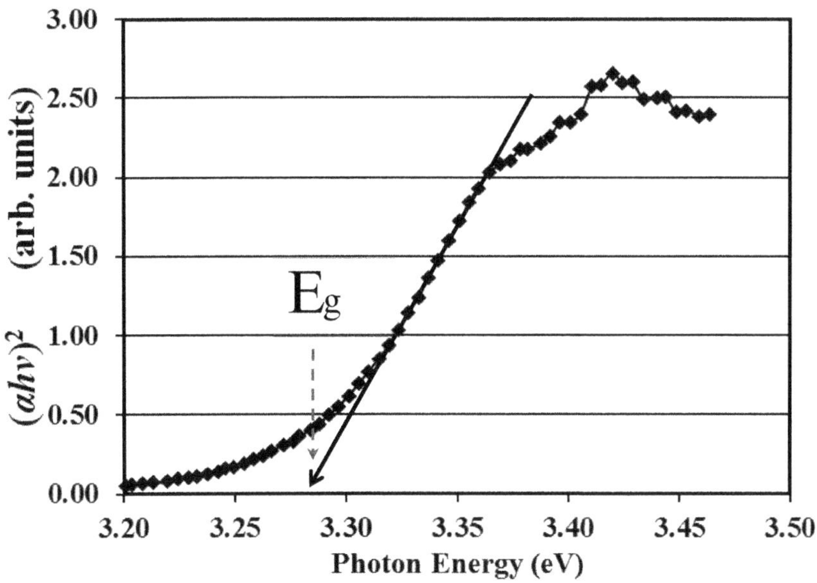

Figure 1: Example Tauc Plot from UV-Vis analysis of a ZnO thin film, illustrating the method of fitting the linear region to extrapolate the band-gap at the X-axis intercept, here about 3.28 eV. Data replotted from ref.[18].

At even higher energies the absorption processes saturate and the curve again deviates from linear. To select and justify a linear region for extrapolation one must understand the reasons for these lower and upper deviations from linear behavior. On the low energy end, the deviation from linearity can be associated with defect absorption states that are near the band edge. This phenomenon has been investigated by Urbach [19] and in subsequent years, therefore, identified as an "Urbach Tail." These states are usually described by an exponential function, corresponding to a typical distribution of density of states, evident in the absorption behavior seen in the example Tauc plot (Figure 1). On the high energy end, saturation of available transition states can be responsible for a leveling out of absorption strength in most collected spectra.

The absorption data are rooted in the possible optical transitions within the electronic structure of the material. Figure 2 (next page) shows the band diagram for ZnO[20], showing that the material is direct and that the band-gap derives from states at, Γ, the center of the Brillouin Zone. A representative direct optical transition is shown for a photon energy slightly larger than the band-gap energy.

Selecting the "right" points to use for fitting from Figure 1 is largely subjective, but could also have a profound effect on the extrapolated value for the band-gap. In our earlier work on ZnO, we tried to develop a completely unbiased method for picking the linear portion of the plot and finding the band gap value[1]. Digital data were processed in a spreadsheet to achieve a series of linear regressions corresponding to incremental portions of the data set. We typically fitted using an 11-data-point window for evaluating the local linear regression (using +/- 5 datapoints on either side for any given local fit), and then we slid this fitting-window along and tested the fit at every possible location. The impact of fitting window width can be illustrated in Figure 3 where we plot the R^2 value for each incremental linear regression fit for the data we extracted from the graph shown in Figure 1. Three different curves are presented that cover 5, 11, and 15 datapoint fitting windows, respectively. When fewer data points are used for fitting then better R^2 results are generally obtained (as a mathematical certainty). However, if the actual linear region is relatively short then using a bigger span of datapoints will force the inclusion of points that are clearly not part of the linear region and the R^2 value will be reduced. Or, similarly when fitting a line to a clearly curved part of the dataset, the same R^2 reduction will occur. Figure 3 illustrates this behavior with the general trend downward for the energy values between 3.2 and 3.3 eV (see arrow). Referring back to Figure 1, it can be seen that this is a

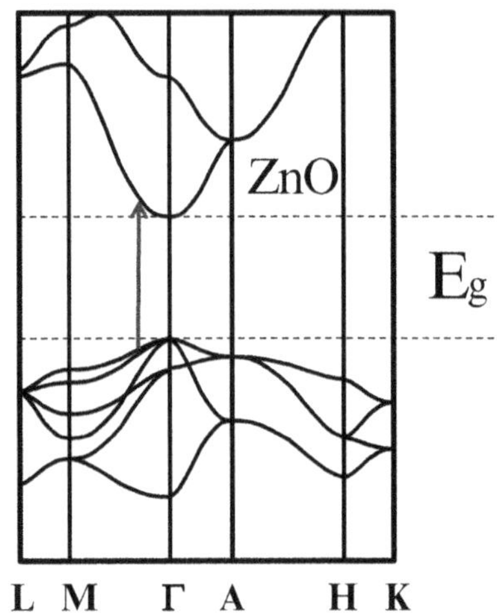

Figure 2: Band diagram for ZnO showing direct transitions at the central Γ point. One direct transition at slightly higher energy is illustrated.

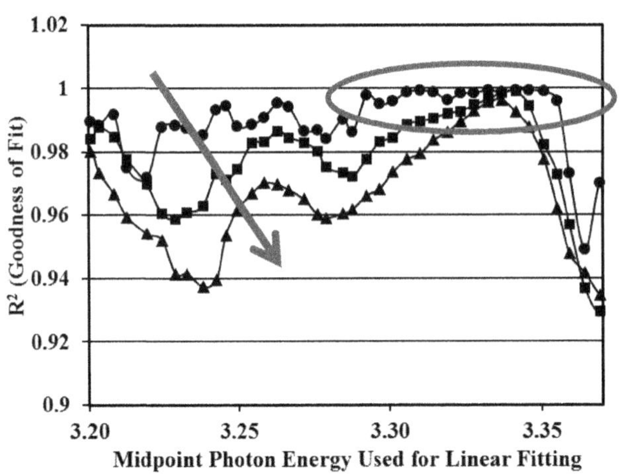

Figure 3: Linear regression fitting quality for different portions of the data as a method of choosing the best region for extrapolating the "Tauc gap": using (•) 5, (■) 11, or (▲) 15 sequential points.

region of upward curvature for the Tauc plot, so that extending the fit to larger ranges of data can force the inclusion of more curvature and therefore poorer fits (as shown). The best fit values are found in the mid-point photon energy region between 3.30 and 3.35 eV (as marked with the red ellipse in Figure 3). In our analysis, each linear regression can be evaluated to find the X-axis intercept (the band gap value). Figure 4 shows how this band gap/intercept value changes depending on which set of adjacent datapoints was used for linear regression fitting. The best R^2 values for fitting correspond to the band gap values highlighted with the ellipse, all around 3.28 eV. Note that the choice of the width of fitting window imposes only a slight change in extrapolated band gap value establishing a method-imposed precision of about +/-0.005 eV. Interestingly, the standard regression error from any specific fit can be used to calculate a confidence interval for specific fit's extrapolated bandgap value. For the data shown in Figure 1-4 the best-fit region is found to have 95% confidence intervals of +/-0.0025 eV, +/-0.0015 eV, and +/-0.0023 eV, for the 5, 11, and 15 point fitting windows, respectively. The smaller fitting window has a better R^2 value, but the extrapolation is poorer because it is based on a narrower range of energy values and fewer data points. The largest fitting window has a wider basis for making the extrapolation, but the R^2 value is a little lower and the confidence interval a little wider, too. In any case, these confidence intervals must be considered the best precision values for the technique, though when many measurements are considered and compared the accuracy is not as good as this.

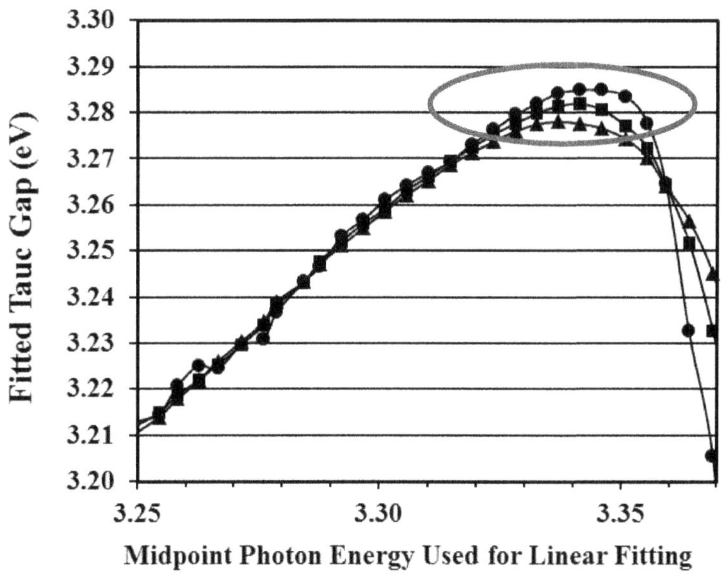

Figure 4: Fitted Tauc gap obtained from linear regression fits covered in Figure 1 and 3. The region where the best R^2 values resulted is also the region where the fitting slope is steepest and the largest Tauc gap intercept is found: (●) 5, (■) 11, and (▲) 15 sequential points used.

In view of the importance of the overlap between the linear region used for Tauc gap fitting and the lower energy Urbach tail absorption effects we also have tried to provide a quantitative measure for comparing different plots and data[1]. We noted that if there were no Urbach tail at all then the absorption would be zero up to the optical gap and then rise linearly according to equation (1). So we suggest that the $(\alpha h\nu)^2$ value measured at the Tauc gap should relate to the size of the Urbach tail, though perhaps not to its energy breadth or shape in detail. Often the Tauc plots are put on an arbitrary units scale, so we normalize this value by comparing it to a $(\alpha h\nu)^2$ value at slightly higher energy. To make it generalizable we suggested normalizing using a value taken at 2% higher energy than the Tauc gap that has been determined by the fitting process, ie. @ hv = 1.02Eg. To generalize this further and make the concept applicable to indirect materials, also, we take the square root and correct for the 2% difference in photon energy to arrive at a factor we call the "Near-Edge Absorptivity Ratio", or NEAR, which is essentially the ratio of the absorption coefficients at those two energy values.

$$NEAR = 1.02\left\{\frac{(\alpha h\nu)^2|_{h\nu=E_{gap}}}{(\alpha h\nu)^2|_{h\nu=1.02E_{gap}}}\right\}^{1/2} = \frac{\alpha(E_{gap})}{\alpha(1.02E_{gap})} \qquad (2)$$

Note that this ratio is dimensionless and can thus be evaluated from $(\alpha h\nu)^2$ graphs even when arbitrary units are used in the plots. The +2% offset is arbitrary and merely intended to probe how steep the curve is close to the Tauc gap. Similarly, when the NEAR factor would be applied to an indirect-gap material (where the $(\alpha h\nu)^n$ would have been plotted with a ½ power, then the ratio would need to be squared to yield a dimensionless absorption coefficient ratio.

This NEAR factor was shown to be correlated with the accuracy of the band-gap determined using the Tauc method [1]. Certainly, sharper absorption data that come closer to the X-axis before exhibiting their Urbach tail will be mathematically more likely to have a more accurate band-gap value.

ANALYSIS OF INDIRECT-GAP MATERIALS

The case of indirect band-gap materials is quite a bit more complicated since every indirect band-gap material will also, eventually, have direct band optical transitions that start to come into play at photon energies above the indirect gap value. In this regard, many publications provide both the n=1/2 and n=2 plots and use these to extract the indirect edge as well as the lowest of the direct band transitions. And, in many of these analyses, the transition from indirect to direct is not so starkly delineated; there is often a rather broad energy range where the data might be contributing to the linear regression fit for both values – even though each line would have to be built from the assumption of having a dominant contribution from one type or the other.

We address this situation by using the best optical data possible for silicon, an indirect-band-gap material, as a proxy for well crystallized thin film indirect-gap materials. This allows us, eventually, to assess the limits that might be found for distinguishing the indirect and direct band edges and key data signatures that help quantify these assessments. Figure 5 (next page) shows a band-diagram for silicon [21-23]. Silicon is an indirect band-gap material where the indirect transition goes from the center of the Brillouin Zone (Γ) and requires the simultaneous absorption or emission of a phonon in one of the <100> directions of the lattice (denoted X), resulting in the slanted vector shown. Direct optical transitions happen vertically, but will require larger energy for silicon because the electron must be excited from a filled state into an available state at higher energy. One such direct transition is shown. Interestingly, the earliest direct transitions for silicon happen up near 3 eV, which is quite a bit larger than the room temperature band-gap of around 1.12 eV. Figure 6 (next page) shows the rapid rise in absorption coefficient for energies above 3 eV and much lower values between 1.1 and 3 eV (associated with indirect optical transitions).

In the basic understanding of how the energy levels vary with electron momentum (k) result in shapes that often have parabolic shapes at symmetry points (for example at the valence band edge and other high symmetry points. And, this parabolic shape is the mathematical reason that the Tauc exponents were found to be the values listed above in equation (1).

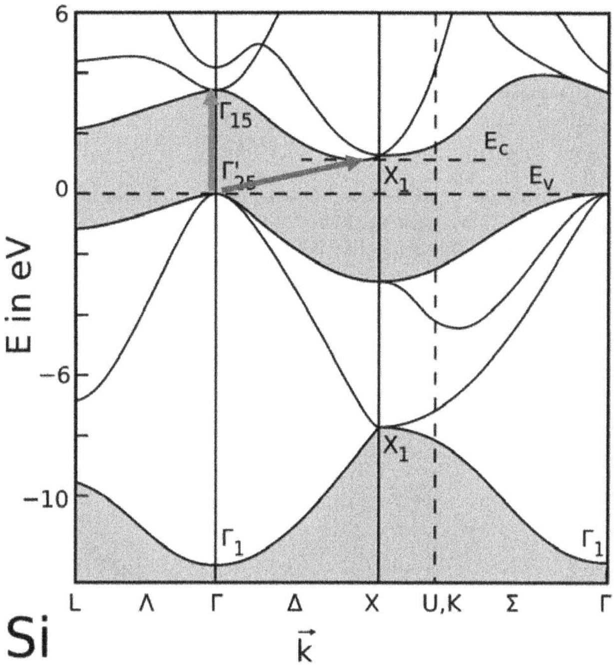

Figure 5: Band structure for silicon showing the indirect transition that is responsible for the onset of absorption and a direct transition (blue arrow) that occurs at certain higher energy values (figure adapted from Chelikowsky [21]).

Knowing that silicon is an indirect material we immediately move to replotting the data using the ½-exponent (the n=2 case from Equation 1, above). Figure 7 (next page) shows this plot for photon energies up to 3 eV (before the sharp rise caused by direct transitions). It can be seen that a large segment of the data can be well represented by a moderately straight light, supporting Tauc's formalism for an indirect material. The data deviate upward from the nominally-linear part for energies above about 2.2 eV, which could be expected for the following reasons. First, the E-vs-k energy level lines must be parabolic to yield the exponents derived by

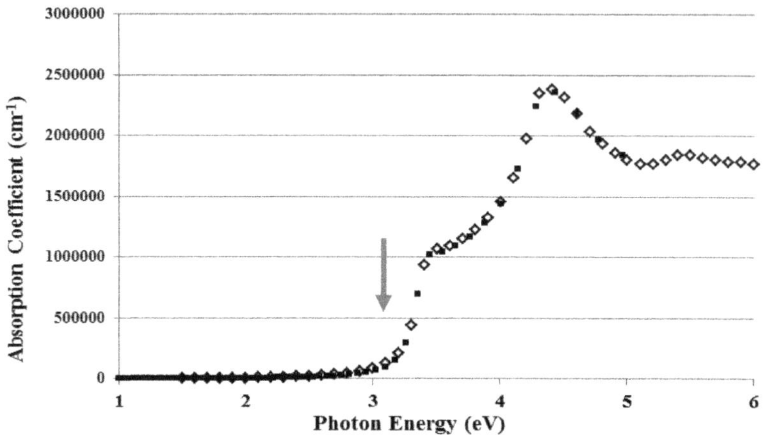

Figure 6: Dispersion of absorption coefficient at 300°K for intrinsic silicon. Open symbols are data from Aspnes and Studna[24]; filled symbols are from Green and Keevers[25]. The rapid rise above 3 eV (indicated by the arrow) comes from the onset of direct transitions.

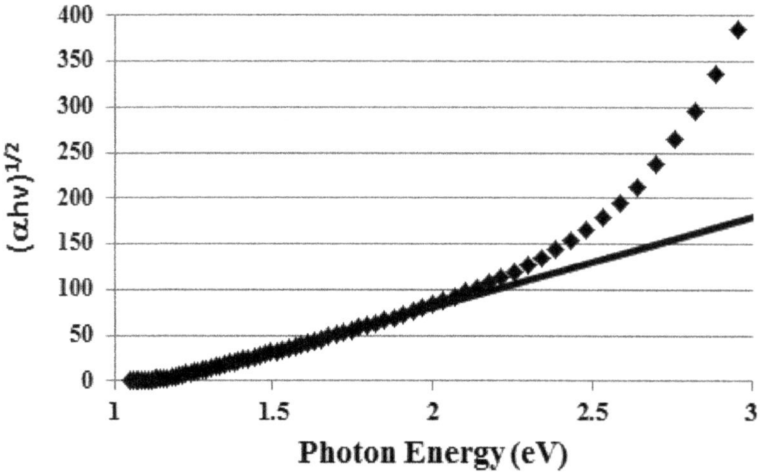

Figure 7: Overview Tauc plot examination of the single crystal silicon absorption data of Green and Keevers[25]. Data are plotted up to 3 eV showing the long region of good linearity (from the band edge up to about 2.2 eV) and including the higher energy range where there is strong upward deviation from the ideal ½-power indirect correspondence.

Tauc. However, as seen in Figure 5, there may be a central region of nice parabolic shape, but there is usually inflection and leveling of the lines at some point. This indicates a higher density of states, which will translate into a larger differential absorption effect as we evaluate at increasingly large photon energies. It is also possible that we might start to have the early contributions from direct transitions which will be added to the basic starting indirect behavior. Further, we might have other secondary indirect transitions that might begin to add into the main transition that was illustrated as the mechanism in Figure 5. All of these factors will lead to an upward deviation of the Tauc plot at higher energy values. For the present discussion we confine our assessment to the region that is nicely linear.

Since we are especially interested in using the Tauc method to extrapolate back to the X-axis to solve for the indirect optical band-gap, we turn our attention to the lower left part of this curve, as shown in Figure 8. Interestingly, the indirect plot displays two linear regions indicating two different indirect transitions. These correspond to the cases where a phonon needs to be either (A) absorbed from the lattice or (B) is emitted into the lattice[26]. Case A has the shallower line, which intersects at (Egap – Eph), where Eph is the energy of the phonon vibration. Case B, with the steeper line, intersects at (Egap + Eph). The intersection points found here were 1.0490 ± 0.0094 and 1.1351 ± 0.0012, resulting in a nominal phonon energy of 0.043 ± 0.010 eV, which is quite consistent with experimental and modeling studies of the phonon dispersion in silicon[27-30].

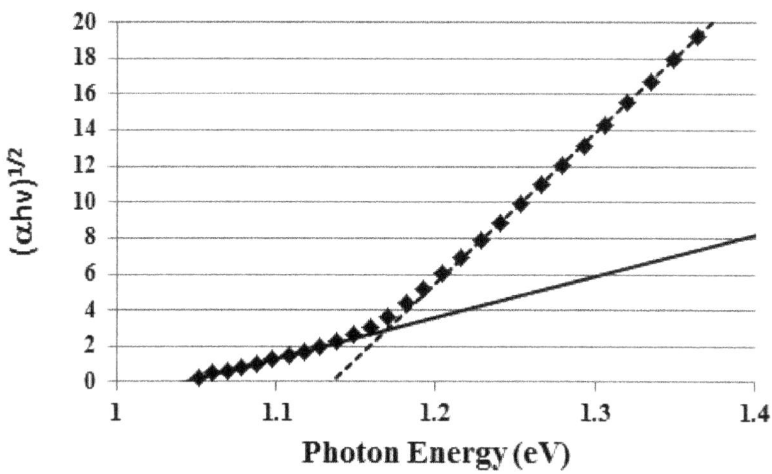

Figure 8: Close-up of the near-edge region of the indirect Tauc plot for silicon [25]. Two different indirect regions are found: the solid line includes phonon absorption and the dashed line is for phonon emission.

While the single crystal data for silicon can be resolved into both phonon absorption and emission processes, it would be rather unusual for thin film optical absorption data to extend down this low and thus band-gap values derived by the Tauc method are usually reported without acknowledging the phonon processes inherent in these indirect transitions. Notably, then, experimental Tauc plots are derived from the steeper sloped region (case B) that extends to higher absorption coefficient values.

Now, we focus our attention more broadly to assess our ability to extrapolate from more limited optical absorption data. This is required because thin film samples may not be thick enough to accurately measure the relatively small absorption coefficient values covered in Figure 8. For example, if we had a 200 nm film that had α = 1000 cm^{-1} at a certain wavelength, then still 98% of the light intensity would be transmitted, so depending on the film quality and the spectrometer used it might be difficult to measure that α value. Figure 9 signifies this limit showing that the larger α values will necessarily be more accurate (for a given film thickness

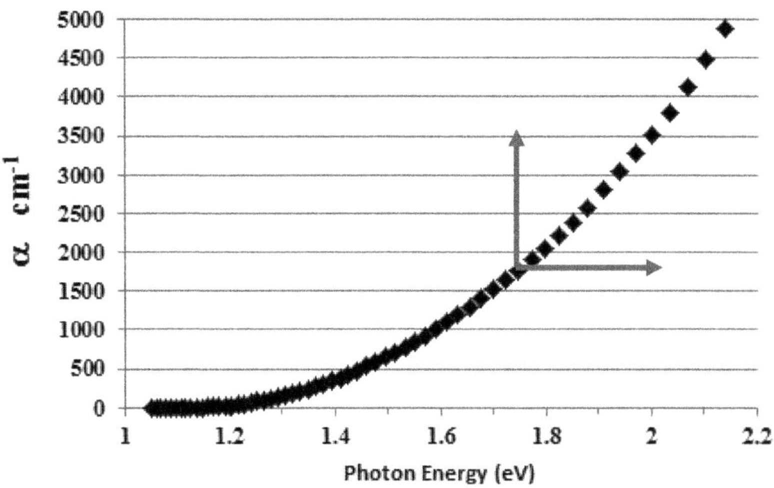

Figure 9: Absorption coefficient values for silicon in the region from 1.05 through 2.20 eV [25]. The parabolic shape associated with an indirect transition is clear.

and spectrometer) and that the linear regression of the type shown in Figures 7 and 8 would most sensibly be done for the photon energies above some threshold. This issue is further complicated by the fact that many thin films may be defective to some degree and exhibit sub-band-gap absorption or band tailing, the "Urbach tail mentioned above [19]. These states are usually described with an exponential function having absorption coefficient values that remain well above zero quite further into the gap.

Now, referring back to Figure 7, we test the degree of linearity that would be found (and the extrapolated band-gap) if we were fitting a Tauc plot using data only above some selected lower limit absorption coefficient value and working upwards with the more confident data. For simplicity we take a linear regression that is based on 10 adjacent data points (though one would normally select that fitting window to be as large as reasonable for the existing data set, as discussed before in our direct-gap work[1]). Figure 10 (next page) gives the band-gaps and extrapolated confidence interval values found when we do each of the possible 10-adjacent-point linear regressions for the silicon data in Figure 7. The X-axis in Figure 10 is draw to correspond to the photon energy of the left-most data point used in the linear regression. The typical width of energy values thus used for these test fits spanned about 0.1 eV at the lower energies and grew to about 0.3 eV when fitting at the higher energies. For example, referring back to Figure 4, the 10-point-regression that starts at 1.1 eV (on the flatter line) will extend well into the area on the steeper line, resulting in a somewhat poorer fit.

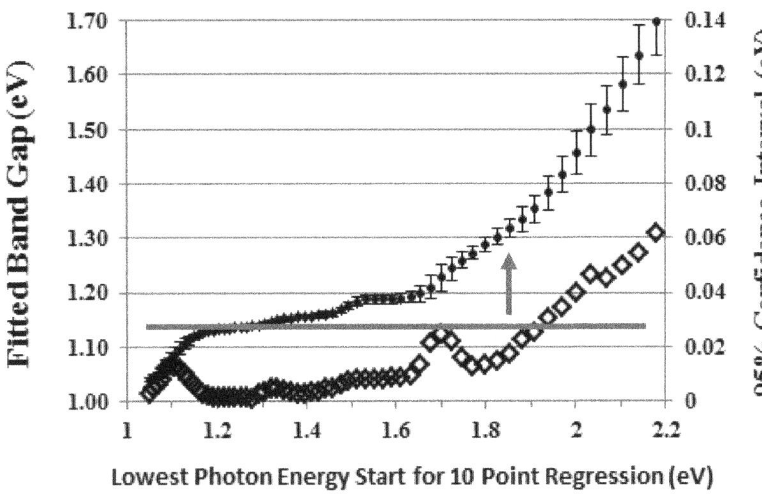

Figure 10: Fitted indirect band-gap values for each 10-point linear regression window (small circles) including their mathematical confidence interval based on the variability within each 10-point window. The specific values of mathematical confidence interval are referenced to the right side scale (open diamonds).

So, if we were working with a thin coating of single crystal silicon we might have trouble measuring the smaller α values and could be forced to fit a linear regression through a part of the data that appeared linear but might already be including some of the data from incipient direct transitions, for example, or a wider range of indirect transitions. It is no surprise that if we use a fitting window that is at higher energies then the confidence interval gets gradually wider (because we are extrapolating further). Overlaid onto Figure 10 is a flat line at the indirect band-gap value (determined using the steeper line from Figure 8, above). The higher-photon-energy-base fitted lines (those using the larger absorption coefficient values) find band-gap intercept values that are much higher above the Eg value than would have been predicted by the linear-regression's projected confidence interval. As noted above, this is partly attributable to the change in curvature for the E-vs-k energy levels in Figure 5, which provides for slightly more absorption at higher photon energies and slight upward curvature of the α(hv) curve, compared with what would have been predicted by the lowest energy threshold responses. Thus, we find that systematic error might be relatively large when associated with fitting using a range of energy values somewhat far from the lowest absorption coefficient values. The R^2 values and calculated confidence intervals must be considered the precision of the measurement, not the accuracy.

DISCUSSION

Application of the Tauc method to thin films is then especially strongly influenced by the thickness of the coating being measured. If the coating is especially thin then only the larger α values are sampled leading to a likely systematic bias that over-estimates the band-gap value. And, since the α values associated with those indirect transitions can be rather small it is questionable whether indirect materials can be fairly evaluated using this method. For example, as shown with silicon, there will always be some form of direct transition that eventually overshadows the indirect transition.

CONCLUSIONS

Direct and indirect band-gap model material data have been examined to help understand in greater detail the limitations of the Tauc method when applied to thin film materials. Direct-gap materials have been shown to have band-gap accuracy in the 1% range when considering the whole range of error contributions that might be important for real thin film sample preparation and characterization. A figure-of-merit has been applied that can help differentiate datasets that might have more mathematical likelihood of providing high accuracy band-gap values. This figure of merit is also applicable to indirect materials, but the absorption coefficient values found near the indirect-gap are frequently too low to be measured with typical samples, raising the question of whether good indirect-gap determinations can be made with the Tauc method.

ACKNOWLEDGEMENTS

Funding for this research was generously provided by the Malcolm G. McLaren fund at Rutgers University.

REFERENCES

1. Viezbicke, B.D., S. Patel, B.E. Davis and D.P. Birnie, *Evaluation of the Tauc Method for Optical Absorption Edge Determination: ZnO Thin Films as a Model System.* Physica Status Solidi, B, 2015. **252**(8): p. 1700-1710.
2. Tauc, J., R. Grigorovici and A. Vancu, *Optical properties and electronic structure of amorphous germanium.* Physica Status Solidi, 1966. **15**: p. 627-637.
3. Davis, E.A. and N.F. Mott, *Conduction in non-crystalline systems V. Conductivity, optical absorption and photoconductivity in amorphous semiconductors.* Philosophical Magazine, 1970. **22**: p. 903.
4. Mott, N.F. and E.A. Davis, *Electronic processes in non-crystalline materials.* 2nd ed. 1979: Clarendon Press (Oxford and New York).
5. Ozgur, U., Y.I. Alivov, C. Liu, A. Teke, M.A. Reshchikov, S. Dogan, V. Avrutin, S.J. Cho and H. Morkoc, *A comprehensive review of ZnO materials and devices.* Journal of Applied Physics, 2005. **98**(4).
6. Ozgur, U., D. Hofstetter and H. Morkoc, *ZnO Devices and Applications: A Review of Current Status and Future Prospects.* Proceedings of the Ieee, 2010. **98**(7): p. 1255-1268.
7. Rodnyi, P.A. and I.V. Khodyuk, *Optical and luminescence properties of zinc oxide (Review).* Optics and Spectroscopy, 2011. **111**(5): p. 776-785.
8. Gupta, T.K., *APPLICATION OF ZINC-OXIDE VARISTORS.* Journal of the American Ceramic Society, 1990. **73**(7): p. 1817-1840.
9. Bagnall, D.M., Y.F. Chen, Z. Zhu, T. Yao, S. Koyama, M.Y. Shen and T. Goto, *Optically pumped lasing of ZnO at room temperature.* Applied Physics Letters, 1997. **70**(17): p. 2230-2232.
10. Mitra, P., A.P. Chatterjee and H.S. Maiti, *Chemical deposition of ZnO films for gas sensors.* Journal of Materials Science-Materials in Electronics, 1998. **9**(6): p. 441-445.
11. Mitra, P., A.P. Chatterjee and H.S. Maiti, *ZnO thin film sensor.* Materials Letters, 1998. **35**(1-2): p. 33-38.

12. Platzer-Bjorkman, C., T. Torndahl, D. Abou-Ras, J. Malmstrom, J. Kessler and L. Stolt, *Zn(O,S) buffer layers by atomic layer deposition in Cu(In,Ga)Se-2 based thin film solar cells: Band alignment and sulfur gradient.* Journal of Applied Physics, 2006. **100**(4).
13. Abb, M., B. Sepulveda, H.M.H. Chong and O.L. Muskens, *Transparent conducting oxides for active hybrid metamaterial devices.* Journal of Optics, 2012. **14**(11).
14. Thomas, D.G., *The exciton spectrum of zinc oxide.* Journal of Physics and Chemistry of Solids, 1960. **15**(1-2): p. 86-96.
15. Srikant, V. and D.R. Clarke, *On the optical band gap of zinc oxide.* Journal of Applied Physics, 1998. **83**(10): p. 5447-5451.
16. Jellison, G.E. and L.A. Boatner, *Optical functions of uniaxial ZnO determined by generalized ellipsometry.* Physical Review B, 1998. **58**(7): p. 3586-3589.
17. Allsopp, H.J. and J.P. Roberts, *Non-stoichiometry of zinc oxide and its relation to sintering: Part 1 - Determination of non-stoichiometry in zinc oxide.* Transactions of the Faraday Society, 1959. **55**: p. 1386-1393.
18. Wang, M.D., D.Y. Zhu, Y. Liu, L. Zhang, C.X. Zheng, Z.H. He, D.H. Chen and L.S. Wen, *Determination of thickness and optical constants of ZnO thin films prepared by filtered cathode vacuum arc deposition.* Chinese Physics Letters, 2008. **25**(2): p. 743-746.
19. Urbach, F., *The Long-Wavelength Edge of Photographic Sensitivity and of the Electronic Absorption of Solids.* Physical Review 1953. **92**: p. 1324.
20. Vogel, D., P. Krueger and J. Pollmann, *Self-interaction and relaxation-corrected pseudopotentials for II-VI semiconductors.* Physical Review, B, 1996. **54**(8): p. 5495-5511.
21. Chelikowsky, J., *Silicon in all its forms.* MRS Bulletin, 2002. **27**(12): p. 951-960.
22. Chelikowsky, J.R. and M.L. Cohen, *Electronic structure of silicon.* Physical Review B, 1974. **10**(12): p. 5095-5107.
23. Chelikowsky, J.R. and M.L. Cohen, *Nonlocal pseudopotential calculations for the electronic structure of eleven diamond and zinc-blende semiconductors.* Physical Review B, 1976. **14**(2): p. 556-582.
24. Aspnes, D.E. and A.A. Studna, *Dielectric Functions and Optical-Parameters of Si, Ge, GaP, GaAs, GaSb, InP, InAs, and InSb from 1.5 to 6.0 eV.* Physical Review B, 1983. **27**(2): p. 985-1009.
25. Green, M.A. and M.J. Keevers, *Optical-Properties of Intrinsic Silicon at 300 K.* Progress in Photovoltaics, 1995. **3**(3): p. 189-192.
26. Kasap, S. and P. Capper, *3. Optical Properties of Electronic Materials: Fundamentals and Characterization*, in *Springer Handbook of Electronic and Photonic Materials*, S. Kasap and P. Capper, Editors. 2006, Springer: New York, NY. p. 47-77.
27. Yin, M.T. and M.L. Cohen, *Abinitio Calculation of the Phonon-Dispersion Relation - Application to Si.* Physical Review B, 1982. **25**(6): p. 4317-4320.
28. Nilsson, G. and G. Nelin, *Study of the Homology between Silicon and Germanium by Thermal-Neutron Spectrometry.* Physical Review B, 1972. **6**(10): p. 3777-3786.
29. Tubino, R., L. Piseri and G. Zerbi, *Lattice Dynamics and Spectroscopic Properties by a Valence Force Potential of Diamondlike Crystals: C, Si, Ge, and Sn.* Journal of Chemical Physics, 1972. **56**: p. 1022.
30. Sui, Z.F. and I.P. Herman, *Effect of Strain on Phonons in Si, Ge, and Si/Ge Heterostructures.* Physical Review B, 1993. **48**(24): p. 17938-17953.

INVESTIGATION OF PYROAURITE-TYPE ANIONIC CLAY-DERIVED MIXED OXIDES WITH VARIOUS COMPOSITIONS

Jonathan Gabriel[1], Aarti Patel[1], Ewul Ebenezer[2], Andrei Jitianu[2,3*], Mihaela Jitianu[1*]

[1]William Paterson University, Department of Chemistry, 300 Pompton Road, Wayne, NJ 07470
[2]Lehman College – City University of New York, Department of Chemistry, Davis Hall, 250 Bedford Boulevard West, Bronx, NY 10468
[3]PhD. Program in Chemistry and Biochemistry, The Graduate Center of the City University of New York, New York, NY 10016, USA
* Corresponding author

ABSTRACT

Pyroaurite, a natural mineral with the formula $Mg_6Fe_2(OH)_{16}(CO_3){\bullet}4H_2O$ belongs to the large class of anionic clays and contains positively charged Mg(II) and Fe(III) layers alternating with layers containing carbonate ions and water molecules. Mesoporous pyroaurite-type anionic clays with different Mg/Fe molar ratios (1:3-5:2) have been successfully synthesized by coprecipitation method. The corresponding oxides were obtained by thermal treatment at 550°C and 900°C. The effect of samples' composition on the structural and textural characteristics of starting pyroaurite and corresponding oxides has been investigated. The crystallite size, the phase composition and BET surface area values were found to strongly depend on the Mg/Fe ratios. TEM micrographs revealed that sheet-like morphology of the high Mg(II) content samples was preserved at 550°C, due to a special decomposition mechanism of pyroaurite.

INTRODUCTION

Pyroaurite and sjögrenite compounds are two closely similar minerals that are two stacking modifications of typical composition $Mg_6Fe_2(OH)_{16}(CO_3){\bullet}4H_2O$[1]. Their structure is based on $Mg(OH)_2$ – brucite-like layers, in which some of the Mg(II) ions are randomly substituted by Fe(III) leading to positively charged layers, their charge being compensated by carbonate (CO_3^{2-}) ions, located in the interlayer region along with water molecules. Those two minerals are polytypes that differ only in layer stacking, the rhombohedral pyroaurite (3R polytype) and the hexagonal sjögrenite (2H polytype) creating two subgroups that belong to the much larger class of layered compounds called anionic clays[1,2]. The anionic clays have been largely investigated due to their applications as anion exchangers, catalysts and catalysts precursors, adsorbents and UV stabilizers, just to mention a few[3]. Cationic compositions along with specific characteristics of the interlayer play an important role in numerous applications[4].

Lately, anionic clays derived mixed oxides have been proved to be effective in desulfurization processes[5,6]. A high level of sulfur in fuels is not desired due to the formation of SO_x from the combustion of sulfur-containing compounds. SO_x causes acid rain and this in turn causes damage to buildings and affects dramatically the ecosystem balances[5]. Moreover, sulfur compounds poison the noble metal catalysts used in automobile catalytic converters and this leads to fuel to be incompletely combusted and consequently to emissions of incompletely combusted hydrocarbons, carbon monoxide, nitrogen oxides[7]. An interesting promising potential application of derived oxides is as absorbents for sulfur based compounds, due to their basic properties[7]. So far, different types of absorbents have been used, such as supported metal oxides (support is typically alumina or carbon), mixed metal oxides, metal ion-exchanged zeolites. While all those materials are quite effective, the main difference that anionic clays derived oxides bring is that they show very good promise to work at ambient temperature, as opposed to elevated temperatures required for the other absorbents[7]. Based on the particularities of the layered structure and decomposition mechanism, the metal oxides have increased basicity relative to the clay precursor[7]. Due to these properties, anionic clays derived mixed oxides have been top candidates for absorbent applications for acidic species, such as CO_2, NO_x and SO_x, mercaptans, organosulfides, CS_2 and thiophene[7].

Surface properties and morphology of the oxides, and subsequently their absorption capacities, catalytic activity and efficiency as supporting material is strongly dependent on their chemical composition. The mixed oxides obtained after such a calcination process have very interesting properties, such as: high surface area, basic properties, formation of homogeneous mixture of oxides with very small crystallite size[8]. The key to these special characteristics resides in the distinctiveness of anionic clays decomposition that occurs in two steps: first, dehydration followed by the collapse of the lamellar structure[9]. The intermediate structure preserves the layered characteristics of anionic clays, and the overall topology is unchanged. Second, the decomposition of the anions and collapse of the structure leads to formation of a new 3-dimensional network[8]. The increase of surface area occurs together with this contraction of the layers and is possibly related to the associated strain development in the material. The calcination temperature is the key parameter to be controlled in this process.

This present work describes synthesis and characterization of a series of mixed oxides derived from synthetic phases structurally related to pyroaurite minerals. Pyroaurite-like materials with three different Mg/Fe molar ratios have been synthesized by precipitation and then calcined at two different temperatures. The structural and morphological characterization of the mixed oxides and their precursors indicated differences in structure, thermal behavior and surface characteristics as a function of the cation composition.

EXPERIMENTAL

Materials

All materials (Reagent grade) were used as received, without further purification. All solutions have been prepared with deionized water. Magnesium nitrate hexahydrate, >99% and iron nitrate nonahydrate, >98% (Sigma Aldrich) have been used for pyroaurite synthesis, along with sodium carbonate >99.5% and sodium hydroxide >98% (Sigma Aldrich).

Sample preparation

The pyroaurite-type samples with Mg/Fe molar ratios of 1:3, 2:1 and 5:2 were synthesized by the coprecipitation method at a constant pH=8. An 0.5M aqueous solution of $Mg(NO_3)_2 \cdot 6H_2O$ and $Fe(NO_3)_3 \cdot 9H_2O$ containing Mg(II) and Fe(III) in the required ratios was added dropwise under mechanical stirring (600 rpm) simultaneously with a basic solution containing 0.5M Na_2CO_3 and 0.5M NaOH in a reaction vessel at a constant pH=8. The reaction vessel was equipped with a heating mantle that maintained a constant temperature of 60°C during the reaction. Once the reaction was completed, the samples were aged for an hour at 60°C, then filtered, washed at room temperature with deionized water, and subsequently dried at 110°C overnight. Thermal treatment was carried out at 550°C and 900°C, respectively, at a heating rate of 5°C/min. Samples were kept at each temperature for 1h.

Sample characterization

Powder X-Ray diffraction patterns were recorded using a Rigaku Ultima IV Theta-Theta with a Cu Kα radiation (1.54056 Å) in the 2θ range 5-80°, step size 0.02 at a speed of 0.1°/min. For the determination of the average crystallite size, Williamson-Hall method has been employed:

$$\beta_L = K\lambda / L\cos\theta$$

where λ is the wavelength (CuKα), β_L is the size broadening, θ the Bragg angle.

The FTIR Spectra were recorded in the ATR mode in the range 4000-400 cm^{-1} using a Nicolet FTIR IS10 spectrometer equipped with the Smart Orbit HATR attachment (Thermo Scientific) at a resolution of 4 cm^{-1}.

Combined thermal analysis measurements (thermogravimetry TG and differential thermal analysis DTA) were carried out with a Seiko EXTAR TGA/DTA 6200 thermal analyzer using platinum crucibles. The measurements were carried out under flowing air at a heating rate of 5°/min up to 1000°C.

The surface area and porous structure of the samples was determined by absorption/desorption of nitrogen at 77K using a Tristar II 3020 BET surface analyzer. Samples were previously outgassed at 110°C (pyroaurite samples) and 300°C (oxide samples), respectively, in a nitrogen flow overnight. Surface areas were determined using the Brunauer-Emmet-Teller (BET) method and the pore size distribution was evaluated by nitrogen/helium mixture desorption using Barrett, Joyner & Halenda (BJH) method.

Transmission electron microscopy was carried out by means of a Hitachi Model 7700 under high resolution at an acceleration voltage of 100kV and a current of 10 μa. Samples were prepared on Formvar-coated thick carbon coated 200 mesh copper grid. Micrographs were taken at 50,000x magnification.

RESULTS AND DISCUSSION

The observed peaks in the X-Ray diffraction patterns (Figure 1) indicate well-crystallized samples for Mg/Fe ratios of 2:1 and 5:2 (JCPDS card 01-086-0181). A rather amorphous nature of the sample with Mg/Fe ratio of 1:3 is indicated by very broad and low intensity peaks (Figure 1 insert), almost not noticeable when overlaying all the patterns. For all samples, the diffraction peaks can be indexed with the space group *R3m*, revealing that the samples are crystallized in a rhombohedral structure characteristic to the pyroaurite polytype[1]. No secondary crystalline phases such as $Mg(OH)_2$ or $Fe(OH)_3$ have been identified in the diffractograms of the prepared samples.

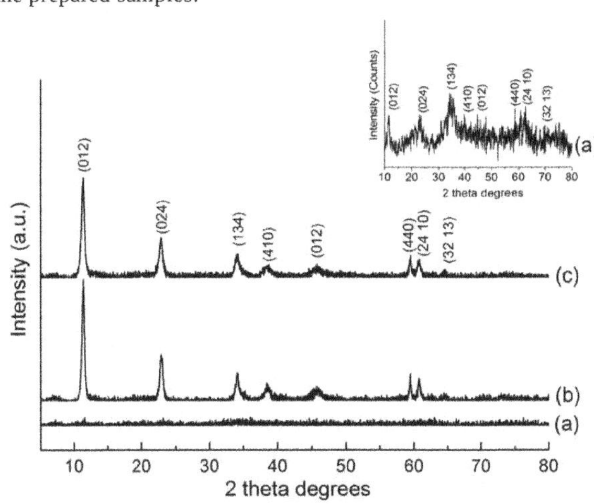

Figure 1. X-Ray diffraction patterns of pyroaurite samples:
a. Mg/Fe 1:3 (Insert), b. Mg/Fe 2:1, c. Mg/Fe 5:2;

Noteworthy changes are observed for the X-Ray diffraction patterns of the samples calcined at 550°C (Figure 2). The well-defined pyroaurite peaks for the samples with Mg/Fe molar ratios 2:1 and 5:2 (Figure 1b, c) are replaced by less intense broader peaks (Figure 2b, c), suggesting a nanocrystalline material with very small nanoparticles. This strong change in the X-Ray patterns is an indication of the fact that thermal decomposition of pyroaurite occurs topotactically via a disordered structure[9]. Conversely, the sample with Mg/Fe ratio of 1:3 displays high-intensity narrow peaks, suggesting larger well crystallized particles (Figure 2a). The periclase Mg(Fe)O (JCPDS card 01-076-2585) was mainly detected for samples with Mg/Fe ratios 2:1 and 5:2 along with a small amount of magnesioferrite, $MgFe_2O_4$ spinel structure $MgFe_2O_4$ (JCPDS card 01-074-8054). These results reveal that the mixed oxides may primarily have a brucite-type structure with Fe(III) randomly distributed in the octahedral sites. Also, it is possible to consider the

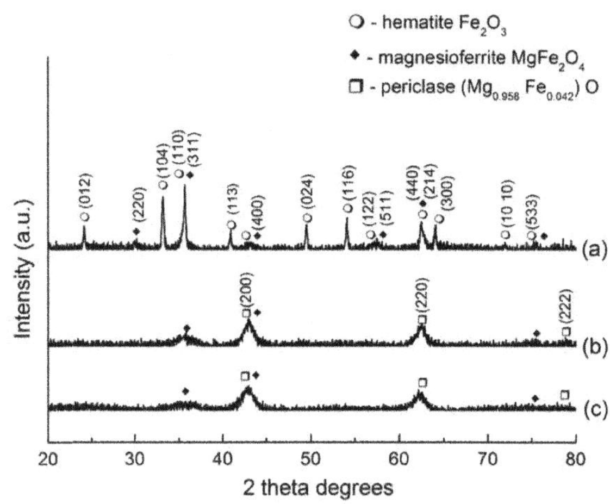

Figure 2. X-Ray diffraction patterns for the oxides obtained at 550°C:
a. Mg/Fe 1:3, b. Mg/Fe 2:1, c. Mg/Fe 5:2.

formation of Mg(Fe)O solid solutions, based on the Fe(III) ion being smaller (0.49 Å) than Mg(II) (0.72 Å)[10]. The FTIR spectra support these results and will be discussed in detail later in the paper. The diffraction pattern of the sample with Mg/Fe ratio of 1:3 is very different from the ones of the samples with Mg/Fe ratios of 2:1 and 5:2. The lines reveal formation of mainly hematite Fe_2O_3 phase (JCPDS card 01-071-5088) along with a small quantity of magnesioferrite. This sample has the highest Fe(III) content and it appears that by decomposition it develops mainly a Fe(III)-only containing phase. With a further increase of calcination temperature to 900°C, the broad peaks for samples with Mg/Fe ratios of 2:1 and 5:2 narrow and become more intense, indicating the presence of crystalline samples (Figure 3b,c). Peaks for the Fe(III)-rich sample (Figure 3a) with Mg/Fe 1:3 become sharper and more intense, as well, suggesting that crystallinity increases with temperature for this sample. No additional oxide phases have been identified at 900°C compared to the ones at 550°C.

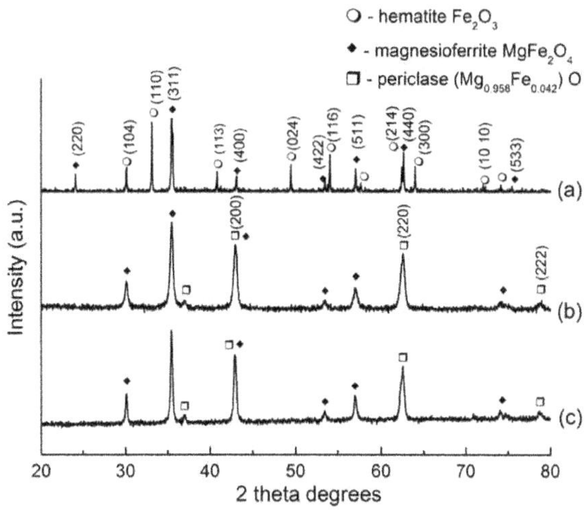

Figure 3. X-Ray diffraction patterns for the oxides obtained at 900°C: a. Mg/Fe 1:3, b. Mg/Fe 2:1, c. Mg/Fe 5:2.

Table I. Crystallite size for the Mg/Fe 1:3 sample and derived oxide phases obtained at, 550° and 900°C.

Sample	Crystallite size (Å)				
	As prepared	Oxides at 550°C		Oxides at 900°C	
	Pyroaurite	Hematite Fe_2O_3	Magnesioferrite $MgFe_2O_4$	Hematite Fe_2O_3	Magnesioferrite $MgFe_2O_4$
Mg/Fe = 1:3	3.1	102	865	563	900

Table II. Crystallite size for the Mg/Fe 2:1 and Mg/Fe 5:2 samples and derived oxide phases obtained at 550° and 900°C.

Sample	Crystallite size (Å)				
	As prepared	Oxides at 550°C		Oxides at 900°C	
	Pyroaurite	Periclase $(Mg_{0.958}Fe_{0.042})O$	Magnesioferrite $MgFe_2O_4$	Periclase $(Mg_{0.958}Fe_{0.042})O$	Magnesioferrite $MgFe_2O_4$
Mg/Fe = 2:1	141	50	23	75	31
Mg/Fe = 5:2	155	29	5.92	81	36

The crystallite sizes as determined from XRD measurements are reported in Tables I and II, for the sample with Mg/Fe ratio of 1:3 and samples with Mg/Fe ratios of 2:1 and 5:2, respectively. The evolution of crystallite sizes is shown for all phases formed. An interesting behavior is found for the sample with Mg/Fe 1:3 and the derived two oxide samples. A very small crystallite size (0.3 nm) was found in the starting pyroaurite sample, while slightly larger, within the small nanoparticles domain, was identified for the samples with higher Mg/Fe ratio (14-15 nm). The evolution of the crystallite size with temperature is very different, depending on the Mg/Fe ratio, a sharp increase for both phases at 550°C and further at 900°C is displayed by the Fe(III)-rich sample. For the samples with Mg/Fe ratios of 2:1 and 5:2, an actual decrease in crystallite size was found for the oxide phases at 550°C, followed by a moderate increase at 900°C. This leads to the suggestion that the decomposition mechanism for samples is different, depending on the Mg/Fe ratio, as it will be described in the next section.

The thermal decomposition was investigated by TGA / DTA experiments (Figure 4). All samples start to lose weight from room temperature up to 550°C. The thermal decomposition consists of two clear steps for samples with Mg/Fe ratios of 2:1 and 5:2 (Figure 4a,b). The first step corresponds to the loss of physisorbed and interlayer water[9] and is accompanied by an endothermal effect on the DTA curve at 170 and 180°C, respectively. The second decomposition step relates to the loss of carbonate ions and the hydroxyl groups in the brucite-type layer[9], with two clear endothermal effects on the DTA curve at 290-305°C and 354-357°C, respectively. The thermal decomposition of the sample with Mg/Fe ratio of 1:3 is different (Figure 4c). Decomposition occurs in one step, basically a continuous weight loss was observed from room temperature till around 550°C. The DTA curve for this sample only shows two endothermal effects in the temperature range 100-360°C, one at 160°C, and the second one at 345°C. The absence of the peak at 290-300°C suggests a disordered interlayer structure in those samples, leading to a gradual weight loss during heating. The amorphous nature of this sample leads to different decomposition features. Two more additional endothermal effects were observed for this sample with Mg/Fe ratio of 1:3, at 468°C and at 588°C. Those effects are not accompanied by weight loss, thus they may be attributed to phase transformation with temperature. Since the latter effects were not observed for the other two pyroaurite samples, it is an indication that the Fe(III)-rich sample with Mg/Fe ratio of 1:3 contains another phase besides the pyroaurite that could not be detected by X-Ray diffraction due to lack of crystallinity. Those findings for the Fe(III)-rich sample correlate well with the different phase composition identified by X-Ray diffraction at 550°C and beyond.

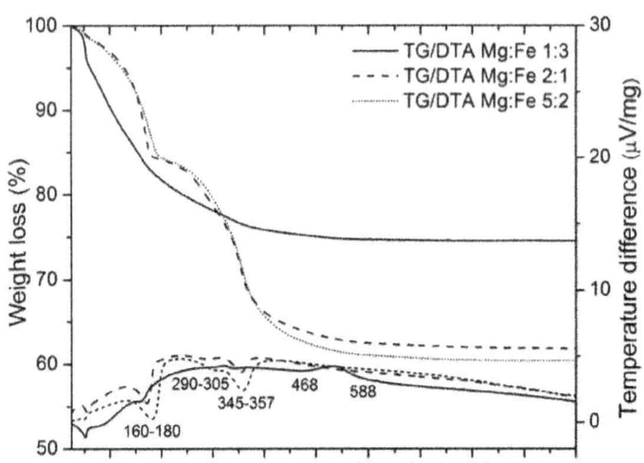

Figure 4. Thermal analysis of the pyroaurite samples.

FTIR analysis (Figure 5) reveals characteristic absorptions for carbonate-containing pyroaurite. All spectra show absorption bands at ~ 3380 cm^{-1}, attributed to hydroxyl stretching bands (υOH)[11], and located at lower wavenumbers than in simple hydroxides with no hydrogen bonds (> 3600 cm^{-1}). The band at 1640 cm^{-1} is attributed to the deformation of (δHOH) angle of water molecule[3]. Vibrational spectroscopy is widely applied in investigating the structure of the interlayer anion, in this present case, carbonate. As reported in the literature for anionic clays[12,13], carbonate anions in the interlayer are generally in a symmetric environment, characterized by a D_{3h} planar symmetry, with three IR active absorption bands as in the case of free carbonate anion, as follows: 1360 cm^{-1} (υ_3), 880 cm^{-1} (υ_2), 680 cm^{-1} (υ_4). Our spectra show an intense band situated at 1350 cm^{-1} (υ_3) along with the one at 680 cm^{-1} (υ_4), thus showing the presence of highly symmetrical D_{3h} carbonate anions. Yet, the presence of the band at 1470 cm^{-1} in the spectra of our samples with Mg/Fe ratios of 1:3 and 2:1 is attributed to a lowering in the symmetry of the carbonate anion to C_{2v} or C_s, as well as to a disordered nature of the interlayer[14,15]. The lowering symmetry of the carbonate ions is due to the restricted environment into which the anions are placed and to various intensity of the hydrogen bonds[3]. It appears that the intensity of the band at 1470 cm^{-1} decreases from the Fe(III)-rich sample with Mg/Fe ratio of 1:3 to the sample with Mg/Fe ratio of 2:1, suggesting that the lower symmetry of carbonate ions and the interlayer disorder becomes less significant in samples with higher Mg to Fe ratios. For the sample with Mg/Fe ratio of 5:2, only highly symmetric carbonate ions were identified (only the band at 1350 cm^{-1} is present). The absorption band at 550 cm^{-1} in the spectra of samples with Mg/Fe ratios of 2:1 and 5:2 is attributed to Mg-O vibrations of the brucitic layer[1]. This band is not very visible for the Fe(III)-rich sample, due to a very low Mg(II) content.

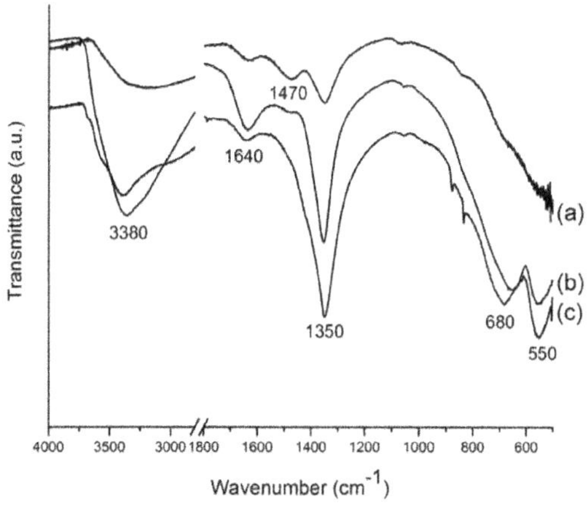

Figure 5. FTIR spectra of pyroaurite samples. a. Mg/Fe 1:3, b. Mg/Fe 2:1, c. Mg/Fe 5:2.

The FTIR spectra at 550°C and 900°C (Figure 6) show conversion of the pyroaurite structures into the corresponding oxides. The interpretation of the spectra is carried out in the light of structures of spinel, Fe_2O_3 and MgO species that have been identified by X-Ray diffraction. "Normal" spinel structures consist in a close-packed arrangement of oxide ions, in which the tetrahedral holes are occupied by the divalent cation (MgO_4 tetrahedra) and the octahedral holes are occupied by the trivalent cation (FeO_6 octahedra)[16]. In some cases, total or partially "inverse" spinel structure can be formed. In this latter case, all or a part of the divalent cation will become octahedral, substituting part or all of the trivalent cation in the octahedral sites[16]. Hematite Fe_2O_3 consists of a close-packed oxide ions lattice with the Fe(III) in the octahedral holes, creating FeO_6 octahedra[16]. Magnesium oxide MgO is a highly ionic compound that has Mg(II) occupying the octahedral holes of a close-packed oxide ions lattice, creating MgO_6 octahedra[16].

The observed FTIR vibration bands for spinels have been labelled υ_1, υ_2, υ_3 and υ_4 in the succession of decreasing wavenumbers[17]. According to Tarte et al[17]. υ_1, υ_2 have been assigned to vibrations of the lattice of condensed octahedra. υ_3 seems to be related to a complex vibrations involving the participation of both types of cations, tetrahedral and octahedral, while υ_4 has been assigned to some type of vibration involving the tetrahedral cation[17]. FTIR spectra recorded for the oxides at 550°C and 900°C are very similar, bands become more defined at the higher calcination temperature, as the oxides crystallinity increases, as found by X-Ray diffraction. Due to the limitation of our instrument, the band due to υ_4 vibration could not be visualized, since it is located below 300 cm^{-1}, below our spectral range of our instrument. Moreover, some spinels do not always display all the bands. For the sample with Mg/Fe ratio of 1:3 calcined at 550°C, the band at 520 cm^{-1} can be assigned to vibrations of (FeO_6) octahedra in Fe_2O_3[19], while the band at 425 cm^{-1} is assigned to υ_3 vibrations of (MgO_6) octahedra[18]. The shoulder at 473 cm^{-1} can be related to the formation of spinel structure, and assigned to the υ_1 vibrations of (FeO_6) octahedra[18,19] within the spinel structure.

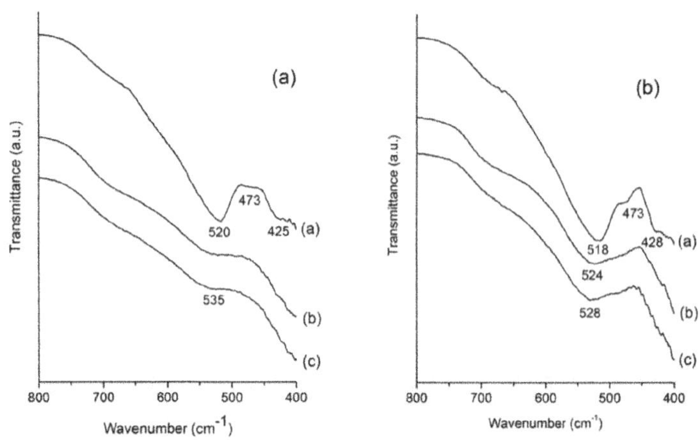

Figure 6. FTIR spectra of a. oxides at 550°C, b. oxides at 900°C
a. Mg/Fe 1:3, b. Mg/Fe 2:1, c. Mg/Fe 5:2.

The assignment of these bands is true to the evolution of the sample with Mg/Fe ratio of 1:3 towards 900°C, the main change being that the shoulder at 473 cm^{-1} intensifies as the spinel structure becomes more crystallized at higher temperature. The band at 518-520 cm^{-1} in the spectra of the oxides corresponding to the sample with Mg/Fe ratio of 1:3 can be due also to the isolated tetrahedral (MgO_4) groups in the spinel structure[18]. Thus, this band may be very well an overlap between vibrations of (FeO_6) octadedra in Fe_2O_3 structure and vibrations of isolated tetrahedral (MgO_4) groups in the spinel structure. The presence of the υ_3 band due to (MgO_6) octahedra suggests a degree of inversion of the $MgFe_2O_4$ spinel, according to the above mentioned spinel features[16]. Samples with higher Mg/Fe ratios 2:1 and 5:2 mainly exhibit a band at 535 cm^{-1} in the spectra of the oxides obtained at 550°C that shifts to slightly lower wavenumbers for the oxides at 900°C. Given the samples composition at 550°C, which is mainly MgO with small inclusions of Fe(III), this band can be attributed to the highly ionic MgO, for which two bands, one near 550 cm^{-1} and one near 400 cm^{-1} corresponding to the longitudinal optic and transversal optic frequencies, respectively have been reported[17]. Due to the spectral range of our instrument, only the band near 550 cm^{-1}, present at 535 cm^{-1} for the oxides at 550°C could be identified. As the spinel phase crystallizes better at 900°C, it is evident that the shift of this band towards lower wavenumbers is related to its overlap with the band due to the isolated tetrahedral (MgO_4) groups in the spinel structure, as mentioned earlier. As the Mg content of the samples increases however from sample with Mg/Fe ratio 2:1 to 5:2, the band shifts towards higher wavenumbers, towards the value for pure MgO.

Table III. BET values for pyroaurite samples and corresponding oxides.

Sample	BET (m^2/g)		
	Pyroaurite	Oxides at 550°C	Oxides at 900°C
Mg/Fe = 1:3	262.04	25.96	2.09
Mg/Fe = 2:1	81.21	85.21	28.25
Mg/Fe = 5:2	49.87	94.26	13.97

Different appearances of the nitrogen absorption isotherms as a function of both Mg/Fe ratios and temperature suggest modifications of the porosity characteristics (Figures 7-10). The absorption isotherms of the uncalcined pyroaurite samples are very dissimilar, suggesting modifications of porosity with the Mg/Fe molar ratio (Figure 7). The Fe(III)-rich sample with Mg/Fe molar ratio 1:3 displays an isotherm with particular characteristics. For $p/p_o < 0.35$, multilayer adsorption is implied, and reveals microporous characteristics for this sample. For $0.35 < p/p_o < 0.6$, the isotherm shape suggests the presence of pores with a wide size range, and non-uniform in size and/or shape, as per the irregular form of the hysteresis loop. For $p/p_o > 0.6$, the absorption follows more or less a type IV isotherm[20].

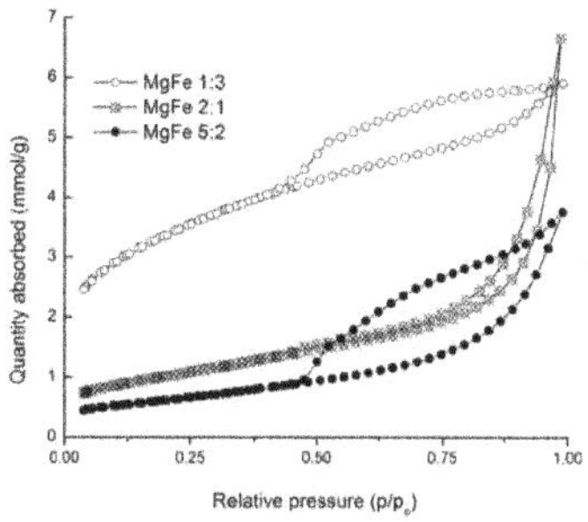

Figure 7. Adsorption isotherms for all pyroaurite samples

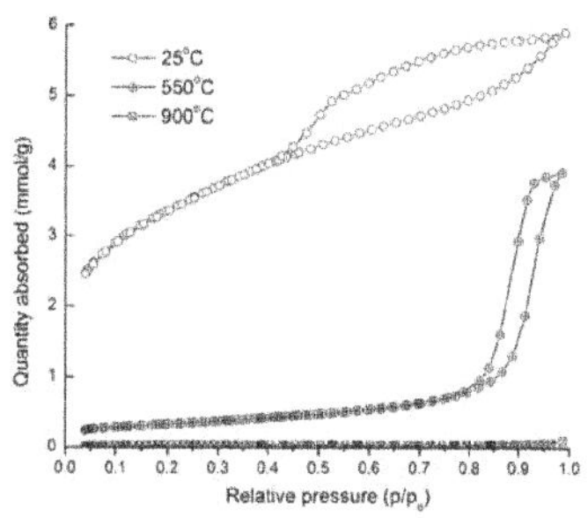

Figure 8. Adsorption isotherms for Mg/Fe 1:3 at all studied temperatures.

Samples with molar ratios Mg/Fe 2:1 and 5:2 show type IV absorption isotherms, with a H3 type hysteresis. Sample with Mg/Fe 2:1 displays however a very narrow hysteresis loop, suggesting more uniformly shaped pores. Conversely, the sample with Mg/Fe 5:2 molar ratio shows a broad H3 hysteresis loop, in the middle range of relative pressure. For this latter sample, for the values of pressure $p/p_o < 0.5$, the expected adsorption process is by formation of a monolayer, while the multilayer adsorption is occurring at high p/p_o. The loop type suggests that aggregates of plate-like particles may form non-uniform slit shaped pores[20].

The evolution of the adsorption isotherms with the temperature up to 900°C illustrates that samples with Mg/Fe ratios 2:1 and 5:2 maintain the same adsorption characteristics with evolving temperature (Figures 9 and 10), while remarkable changes are occurring for the sample with Mg/Fe 1:3 ratio (Figure 8). For this latter sample, at 550°C, the shape of the curve is typical for a type IV isotherm with a H1 hysteresis. The isotherm form and the narrow hysteresis loop with almost parallel branches indicate the presence of mesopores with almost regular geometry[20]. The adsorption is very low at 900°C, the adsorption and desorption branches are close to zero adsorbate value.

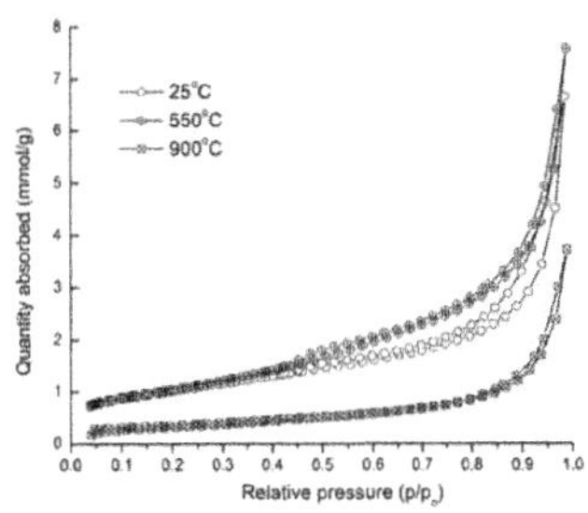

Figure 9. Adsorption isotherms for Mg/Fe 2:1 at all temperatures

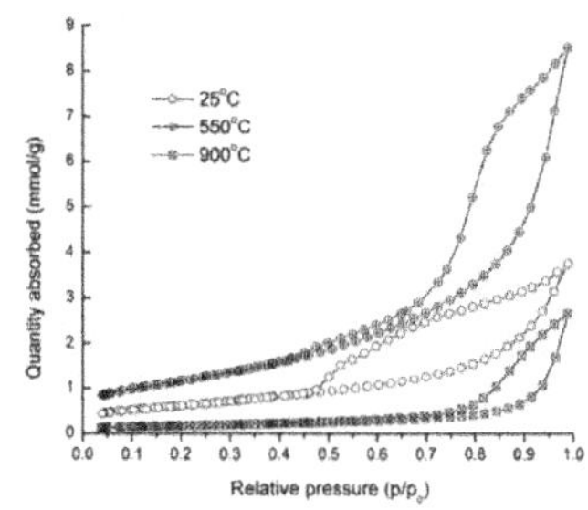

Figure 10. Adsorption isotherms for Mg/Fe 5:2 at all temperatures.

The calculated BET values (Table III) show the highest BET value for the Fe(III)-rich sample before calcination, due to the amorphous nature and very small crystallites, however the BET values for the oxides drops significantly till almost non-porosity is found in the sample at 900°C. The BET values correlate well with the significant increase in the crystallite size with temperature for this Fe(III)-rich sample. For the other two samples, BET values increased after heat treatment at 550°C (sample with Mg/Fe ratio 5:2) or remained about the same for the sample with Mg/Fe ratio 2:1 (Table III). As expected, all BET values decreased for the oxides at 900°C, due to the increase in crystallite size.

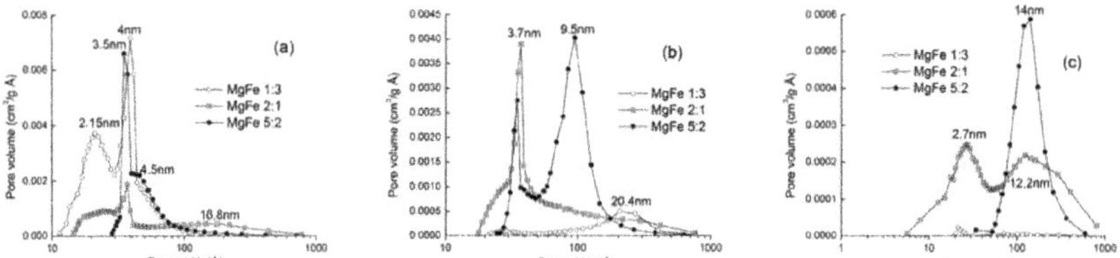

Figure 11. Particle size distribution for all samples:
a. pyroaurite, b. oxides at 550°C, c.oxides at 900°C.

One of the most interesting aspects of the anionic clays, including pyroaurite, is the formation of oxides with small crystallite size and high surface area[8] due to their special decomposition mechanism[9], which is the best illustrated by sample with Mg/Fe ratio 5:2. Sample with Mg/Fe ratio 2:1 shows a small increase in BET value at 550°C, while sample with Mg/Fe ratio 1:3 definitely leads to oxides with low BET values and high crystallite size for 550°C. It becomes thus clear that the main phase in this sample it is not pyroaurite, but an amorphous, most likely a Fe(III)-rich phase that does decompose by a different mechanism, leading to an accelerated crystallite size. This is supported also by the X-Ray diffraction results that show that the main oxide phase obtained by decomposition of sample with Mg/Fe ratio 1:3 is hematite Fe_2O_3 (Figures 2 and 3).

Pore size distribution (PSD) is mostly bimodal for starting pyroaurite samples (Figure 11a), samples with Mg/Fe 1:3 and 5:2 display a sharp peak with a maximum at 3.5 and 4 nm, respectively. Sample with Mg/Fe ratio 2:1 has a different pore distribution, over a large range, with maxima at 4 nm and 16.8 nm. At 550°C (Figure 11b), the sample with Mg/Fe ratio 2:1 shows a more narrow pore size distribution around 3.7nm, while the one with Mg/Fe 5:2 displays a two high intensity peaks for a bimodal pore distribution centered around 3.7 nm and 9.5 nm. The PSD curve for the sample with Mg/Fe 1:3 at 550°C is broad and shifted towards a larger size pores, with a maximum centered at 20.4 nm. At 900°C (Figure 11c), only samples with Mg/Fe ratios of 2:1 and 5:2 display PSD curves, the one Fe(III)-rich sample is almost non-porous, as suggested by the very low BET surface area (Table III). The PSD for the Mg/Fe 2:1 sample is bimodal, with two maxima at 2.7 nm and 12.2 nm, while the PSD for the Mg/Fe 5:2 is centered around a single maximum at 14 nm. Consequently, at all temperatures, samples are mesoporous, no pores larger than 21 nm were found in samples at all temperatures.

TEM micrographs of the pyroaurite sample with Mg/Fe 1:3 illustrate that this sample has an amorphous nature (Figure 12a), while sheet-like particles are observed for the other two samples with higher Mg(II) content (Figure 12b,c). At 550°C, particles for the sample Mg/Fe ratio 1:3 become more defined in shape (Figure 13 a). The sheet-like structure for samples with Mg/Fe 2:1 and 5:2 is preserved at this temperature (Figure 13 b,c), even though the pyroaurite has decomposed in the corresponding oxides. This preservation of shape is most likely due to the topotactic decomposition mechanism[9] of pyroaurite. At 900°C, rhombohedral particles are predominant for all samples, owing to rhombohedral shape of either hematite or spinel structures (Figure 14).

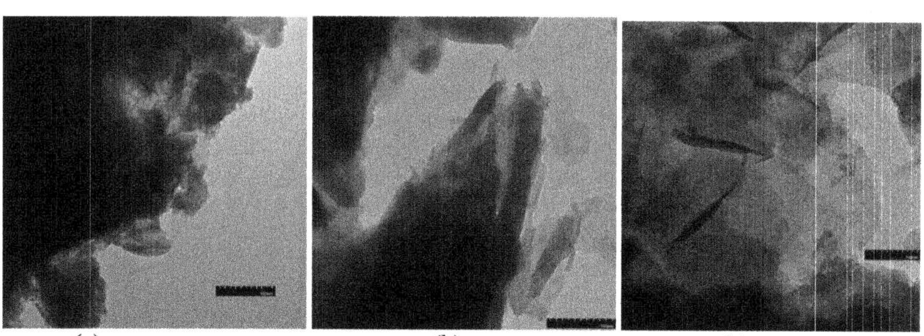

(a) (b) (c)

Figure 12. TEM micrographs for pyroaurite samples: a. Mg/Fe=1:3, b. Mg/Fe= 2:1, c. Mg/Fe 5:2.

(a) (b) (c)

Figure 13. TEM micrographs for oxides at 550°C: a. Mg/Fe=1:3, b. Mg/Fe= 2:1, c. Mg/Fe 5:2.

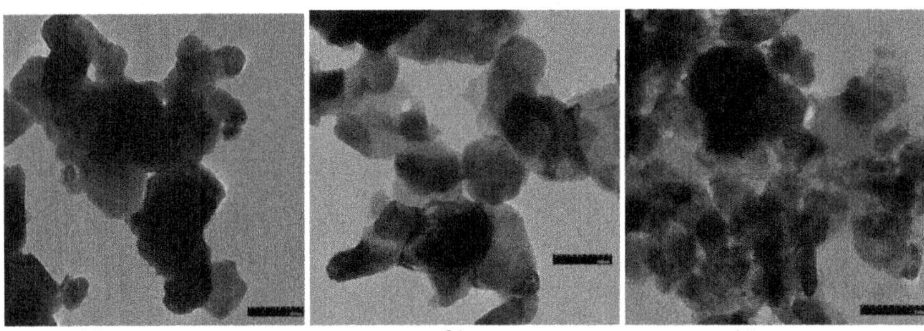

(a) (b) (c)

Figure 14. TEM micrographs for oxides at 900°C: a. Mg/Fe=1:3, b. Mg/Fe= 2:1, c. Mg/Fe 5:2.

CONCLUSIONS

Pyroaurite-like structures with various Mg/Fe molar ratios have been obtained by coprecipitation. Structural, morphological and surface properties were found to be very dependent of the cation ratios. The Fe(III)-rich sample was mainly consisting of an amorphous Fe(III)-phase that led to oxides with very low surface area, while the samples with higher Mg(II) content consisted of crystalline pyroaurite lead to oxides with high surface area. High Fe(III) content is not desirable, in spite of an initial high surface area and small crystallite size. These desired characteristics are not preserved after calcinations. The special structure and decomposition mechanism of samples with Mg/Fe ratios 2:1 and 5:2 led to oxides with high surface area at 550°C, along with a smaller crystallite size compared to the starting material. These make those compositions good candidates for catalytic reactions and/or adsorption of acidic species.

Acknowledgements

We acknowledge the financial support of the William Paterson University through Center for Research summer grant 2015 and ART grant 2014-2016.

REFERENCES

[1]P.G.Rouxhit and H.F.W. Taylor, Thermal decomposition of Sjögrenite and Pyroaurite, *Chimia*, **23**, 480-85 (1969).

[2]H.F.W. Taylor, Crystal structures of some double hydroxide minerals, *Mineralogical Magazine*, **39(304)**, 377-89 (1973).

[3]O.P. Ferreira, O. L. Alves, D. X. Gouveia, A. G. Souza Filho, J.A.C. de Paiva, and J.Mendes Filho, *J. Solid State Chem.* **177**, 3058-69 (2004).

[4]H.C.B. Hansen and C. Bender Koch, Synthesis and characterization of pyroaurite, *Applied Clay Sci.*, **10**, 5-19 (1995).

[5]Novel process for removing sulfur from fuels, US Patent, US 2010/0025301 A1, WO 2005/116169 A1.

[6]S. Dhir, R. Uppaluri and M.K. Purkait, Oxidative desulfurization: Kinetic modelling, *Journal of Hazardous Materials*, **161,** 1360-68 (2009).

[7]D. E. Sparks, T. Morgan, P.M. Patterson, S. A. Tackett, E. Morris, and M. Crocker, New sulfur adsorbents derived from layered double hydroxides I: Synthesis and COS adsorption, *Applied Catal. B: Env* **82**, 190-98 (2008).

[8]F. Cavani, F. Trifiro, and A. Vaccari, Hydrotalcite-type anionic clays: preparation, properties and applications, *Catal. Today* **11**, 173-301 (1991).

[9]M. Bellotto, B. Rebours, O. Clause, J. Lynch, D. Bazin, and E. Elkaim, Hydrotalcite Decomposition Mechanism: A Clue to the Structure and Reactivity of Spinel-like Mixed Oxides, *J. Phys. Chem.* **100**, 8535-42 (1996).

[10]Y. Ohishi, T. Kawabata, T. Shishido, K. Takaki, Q. Zhang, Y. Wang, K. Nomura, and K. Takehira, Mg-Fe-Al mixed oxides with mesoporous properties prepared from hydrotalcite as precursors: Catalytic behavior in ethylbenzene dehydrogenation, *Appl. Catal. A: Gen* **288**, 220-31 (2005).

[11]V. Rives, Mater. Characterisation of layered double hydroxides and their decomposition products, *Mater. Chem. Phys.* **75**, 19-25 (2002).

[12]M.J.H. Hernandez-Moreno, M.A. Ulibarri, J.L. Rendon, and C.J. Serna, IR Characteristics of Hydrotalcite-like compounds, *Phys. Chem. Min.* **12**, 34-38 (1985).

[13]C.J. Serna, J.L. Rendon, J.E. Iglesias, Crystal chemical study of layered $[Al_2Li(OH)_6]^+H^-$ •nH_2O, *Clays Clay Min* **30**, 180-82 (1982).

[14]D.G. Evans, and R.C.T. Slade, Structural Aspects of Layered Double Hydroxides, *Struct. Bond.* **119**, 1-87 (2006).

[15]D.L. Bish, Deviations from the ideal disordered structure in minerals of the pyroaurite group, *Program and Abstracts, 6th International Clay Conference.* Oxford (1978).

[16]A.F. Wells, Structural Inorganic Chemistry, Fifth Edition, Clarendon Press, Oxford (1984).

[17]J. Preudhomme and P. Tarte, Infrared studies of spinels-III. The normal II-III spinels, Spectrochimica Acta, 27A, 1817-35 (1971).

[18]J. Preudhomme and P. Tarte, Infrared studies of spinels-I. A critical discussion of the actual interpretations, Spectrochimica Acta, 27A, 961-68 (1971)

[19]F. Kovanda, V. Balek, V. Dornicak, P. Martinec, M. Maslan, L. Bilkova, D. Kolousek, and I.M. Bountsewa, Thermal behaviour of synthetic pyroaurite-like anionic clay, Journal of Thermal Analysis and Calorimetry, 71, 727-37 (2003).

[20]G. Carja, R. Nakamura, T. Aida, and J. Niiyama, Textural properties of layered double hydroxides: effect of magnesium substitution by copper or iron, *Microporous and Mesoporous Materials*, **47**, 275-84 (2001).

FORMATION AND CHARACTERIZATION OF NANO-SCALE TITANIUM CARBIDES IN A TITANIUM TRIALUMINIDE INTERMETALLIC

Edward A. Laitila
Donald E. Mikkola
Michigan Technological University
Houghton, Michigan, USA

ABSTRACT

Nano-scale cubic titanium carbide particles can be produced within the powder particles of a chromium-modified titanium trialuminide intermetallic by mechanonanosynthesis; that is, milling the intermetallic powder in the presence of an organic process control agent. The amount of the carbide in the microstructure can be very large, actually exceeding the amount of intermetallic, and is controlled simply by the milling time and addition of titanium powder. The small carbide particles produced in the heavily cold worked intermetallic powder are of the order of 2 nm, which is also the scale of the dislocation substructure, posing a challenge to the characterization of the milled material. Despite the severe peak broadening and overlap in the x-ray diffraction patterns it has been possible through profile fitting and synthesis to define and model the microstructure, as well as to follow the changes caused by consolidation giving insight not only into the microstructure formation, but the unusually high strengths of the consolidated composite material.

INTRODUCTION

Contrary to the binary titanium trialuminide which is very brittle, the chromium-modified titanium trialuminide intermetallic $(Al,Cr)_3Ti$ which forms the cubic $L1_2$ structure[1] has shown some compressive ductility along with excellent oxidation resistance and good high-temperature properties.[2] Given the moderate but good density-compensated strength, the original goal of this research was to strengthen this cubic $L1_2$ chromium-modified titanium trialuminide intermetallic by incorporating dispersoids of TiB_2 reduced in size by the "mechanical alloying" process. This process first developed by Benjamin[3] is often used to describe any severe mechanical deformation action by a milling process. More specific definitions have been applied such as mechanical alloying relating to the occurrence of solid-state alloying, whereas mechanical milling refers only to a size reduction.[4] In the current work attempts at mechanical alloying/milling combinations of elemental and/or prealloyed intermetallic powders with TiB_2 powders in a dry environment were unsuccessful and suggested the use of a process control agent (PCA) to facilitate milling. Historically there have been concerns about contamination by any added liquid with these types of milling processes.[5] Investigation of this issue of PCA contamination during mechanical alloying/milling of the trialuminide powder in an initial long time milling process monitored with x-ray powder diffraction (XRD) gave the unexpected results shown in Figure 1. That is, the domination of the XRD pattern by a new titanium carbo-nitride phase comprised of the isomorphous phases TiC and TiN identified by the * and + symbols and representative stick patterns respectively superimposed on the XRD pattern. Originally it was thought the TiB_2 phase shown would be the dispersoid. However, with the formation of the Ti(C,N) phase with a crystallite size an order of magnitude smaller and with this phase also being a hard particle the original approach was abandoned for the more favorable reaction product. This result changed the direction of the research to focus on the *in situ* reactions particularly those between the titanium in the intermetallic powder and the carbon from the PCA. As will be shown, this technique produces a hard, very fine, dispersoid of the carbide phase, the amount of which can be easily controlled by simple processing parameters such as processing time.

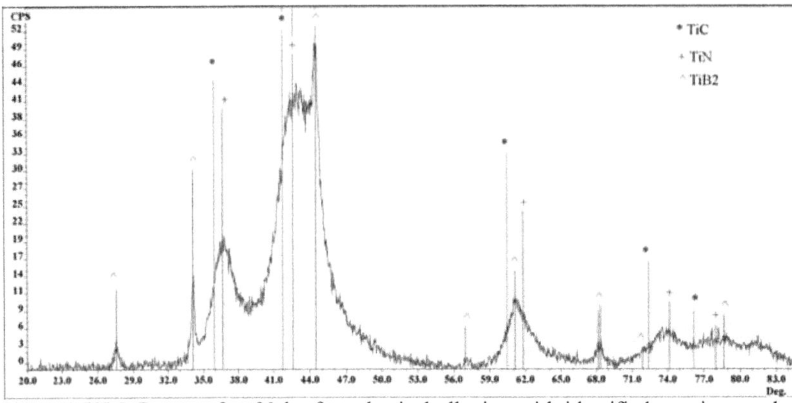

Figure 1 XRD Pattern after 90 h of mechanical alloying with identified reaction product phase Ti(C,N) with TiC identified by the * symbol and TiN identified by the + symbol with the representative stick patterns for each with the admixed TiB_2 identified by the ^ symbol.

Traditionally, the term *nanometer* region relates to features with sizes from 1 to 100 nm; however, in the current work sizes in the single-digit nanometer range are of interest and will be distinguished from the larger sizes with the term *nano-scale*. Unfortunately, when particle sizes, dislocation substructures, and grain sizes of metal matrix composite materials are in the nano-scale region characterization of the material by electron beam microscopy or x-ray diffraction (XRD) analysis becomes difficult. With electron beam imaging, scanning microscopy resolution has size limits near that of the lower nanometer regime, and with transmission microscopy the nano-scale features are often much smaller than the thickness of thin sections. In addition, the free surfaces created by thinning can change the energetics dramatically and hence the equilibria of the system. X-ray diffraction studies provide atomic-scale resolution in a natural three-dimensional environment, but the multiphase scattering effects complicate the patterns and while the structures can be determined, separating these effects has been difficult. Specifically, multiphase nano-scale crystallite sizes produce diffraction patterns comprised of many superimposed extremely broad overlapping diffraction peaks resulting in "amorphous-like" diffraction patterns thus masking important features of the diffraction, especially for phases with similar lattice parameters. Despite these difficulties, some simple procedures will be detailed here that lead to a much improved understanding of such patterns.

EXPERIMENTAL PROCEDURES

The current experiments will show that the systematic uptake of elements from the PCA can produce beneficial hard particle reaction products which ultimately yield composite materials with nano-scale features. The original goal to mechanically alloy the $L1_2$ intermetallic $Al_{66}Cr_9Ti_{25}$ chromium-modified titanium trialuminide with a hard particle dispersoid was therefore accomplished with TiC as a reaction product. This process is a variation of the mechanical alloying process with adequate amounts of the PCA (hexanes C_6H_{14}) providing the carbon. This new process which can control the amount of the nano-scale hard particle reaction products that develop *in situ* is given the name Mechanonanosynthesis (MNS).

All MNS processing here was accomplished with a SPEX™ mill using a 6 to 1 charge ratio, about 90 g of mill media, 15 g of powder, and 18 mL of hexanes, in an argon environment. The post-processed powders are referred to as "as-milled" powder. In the first experiment a detailed characterization was carried out on powders having undergone the same MNS processing conditions of 13.65 g of prealloyed powders ($Al_{66}Cr_9Ti_{25}$) with 1.35 g of pure titanium sponge, an amount that was determined based on the amount of titanium to completely react with carbon to form TiC, while the 20 h processing time and amount of titanium in the experiments were based on an estimated carbon content of 2 wt. % when titanium trialuminide elemental powders were mechanically alloyed for 20 h.[6] Two prealloyed powders were used: a Reading Alloying (RA), which contained a small amount of carbon (0.25 wt. %), and a Master Alloy (MA). The variable in this case was the MNS processing time systematically varied for times of 0.5, 1, 5, 10, 15, 20, 25, 30, and 40 h. In the second experiment, powders with larger amounts of carbide were formed based on the same 13.65 g of $L1_2$ intermetallic alloy with 1.35 g of elemental titanium processed for a target period of 20 h[6] as just described. However, in this case the amount of titanium and MNS processing time were systematically increased by factors of 1, 2, 3, 5, and 6 (with corresponding decreases in $L1_2$ intermetallic) with these powders used for consolidation into bulk samples.

Half of the powders produced were degassed by vacuum annealing at 500°C to remove hydrogen with these powders then encapsulated in quartz ampoule in an argon environment. These encapsulated powders were annealed at 1000°C for 2 h to produce powders to be used as annealed reference powders.

The powders processed by MNS were examined by XRD with a Scintag (Division of Thermo ARL, Dearborn, MI) XDS-2000 instrument outfitted with 1 and 2 mm beam slits with 0.5 and 0.3 mm receiving slits at a 45 kV and 35 mA power setting for the copper target tube using dwell times of 20 sec, or more, with step sizes of 0.03 to 0.05° and a 2θ range from 10 to 120°. Surface topography of the as-milled (post MNS processing) powders was imaged with a Hitachi (Pleasanton, CA) S-4700 field emission scanning electron microscope. Carbon analysis was accomplished with a LECO (St. Joseph, MI) model C/S 224 inert gas fusion-chromatograph.

RESULTS

Carbon analysis of alloys created from the second experiment in which case the elemental titanium content in the mixture and the MNS time were multiplied by factors of 1, 2, 3, 5, and 6 resulting in a near linear uptake of carbon as a function of processing time as shown in Figure 2. The base composition for this series of alloys was 13.65 g of prealloyed trialuminide and 1.35 g of titanium for a 20 h MNS time with two alloys created at 20 h; denoted RA containing a small initial amount of carbon from the synthesis process, with the MA having no initial carbon present.

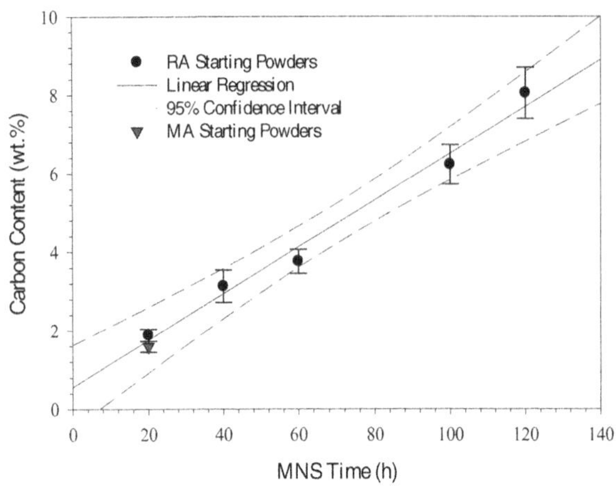

Figure 2 Dependence of carbon uptake on MNS processing time with elemental titanium amounts increased by the same factor for each test increment.

Characterization of the as-milled powders with XRD was accomplished by a combination of "intelligent" profile-fitting and synthesis of the individual Pearson VII peak profile models for each phase using the DMSNT (Scintag, Inc.) peak profile fitting software. The intelligent profile fitting simply requires detailed knowledge of the phases present and based on this information subsequent placement of peaks of proper intensity, approximate width, and in the proper position, provided the best simulation of the experimental information. Difficulties can arise in this process by not providing adequate initial models for the "first-guess" of the summation of each model; so that the required peaks may not be present or may be in the wrong positions with incorrect shapes, but may mathematically satisfy the convergence despite a lack of a physical description. The initial long-term milling experiment made possible the identification of the new phases formed providing the needed information for proper location to place each of the diffraction peaks from all phases identified. Several iterations (including background removal) were then carried out to help validate the process and provide statistical information with all data reported as an average of all individual profile peak analyses for each diffraction peak in these complex patterns. Examples of the XRD patterns for MNS times of 5 to 40 h are shown in Figure 3. All patterns, including the "amorphous-like" nature of the 40 h pattern with the very broad maximum peak, could be broken down by proper angular placement of the individual diffraction peaks for each phase identified with convergence providing detailed peak parameters.

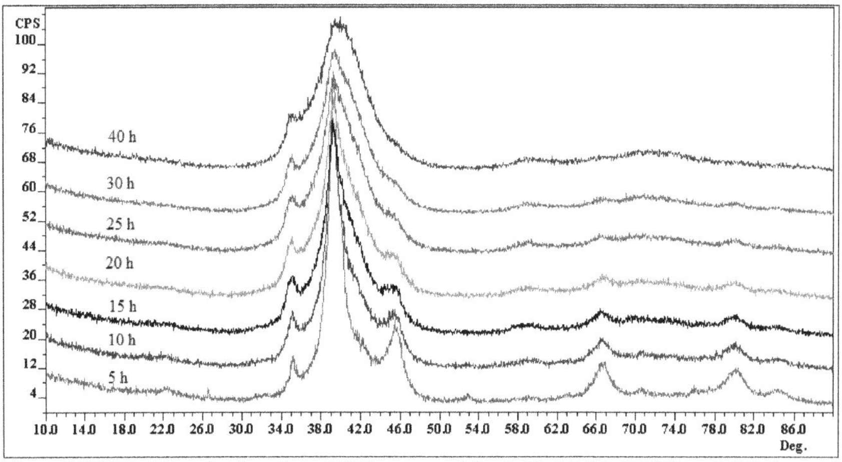

Figure 3 XRD patterns for the alloys MNS processed for 5, 10, 15, 20, 25, 30, and 40 h times all with similar starting chemistries.

An example of a deconvoluted diffraction pattern using this method is shown in Figure 4 with all identified phases accounted for, the $L1_2$ intermetallic, TiC, and TiH_2 (the hydrogen is removed later by vacuum annealing at 500°C). Once the individual diffraction peaks are profile fit to the model, by minimization of the error between the sums of individual profiles with the raw data, individual peak parameters can be used for subsequent XRD analysis. Lattice parameter values determined for each phase helped in validation of the analysis.

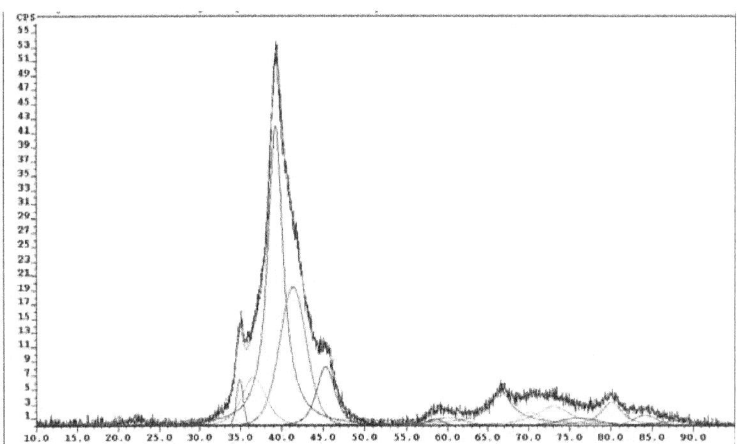

Figure 4 Profile fit example for phases identified in the XRD pattern for the 20 h MNS processing time.

Comparison of the XRD pattern of the as-milled material with that of annealed powder both with the same intensity scale, Figure 5, illustrates the easily observable increase in the background scattering over the whole 2Θ range. It should be noted the powder packing densities were essentially identical for each sample. The increased background scattering has been attributed to the random static displacements of atoms at the grain boundaries and this boundary phase will be given the designation of nano-grain boundary (NGB). This measurable scattering, increasing regularly with MNS processing time, makes it possible to estimate the volume fractions of the NGB phase from the intensity difference between the annealed and as-milled patterns.[7] Using this method in combination with the direct comparison method the total volume fractions of both crystalline and non-crystalline regions can be determined.

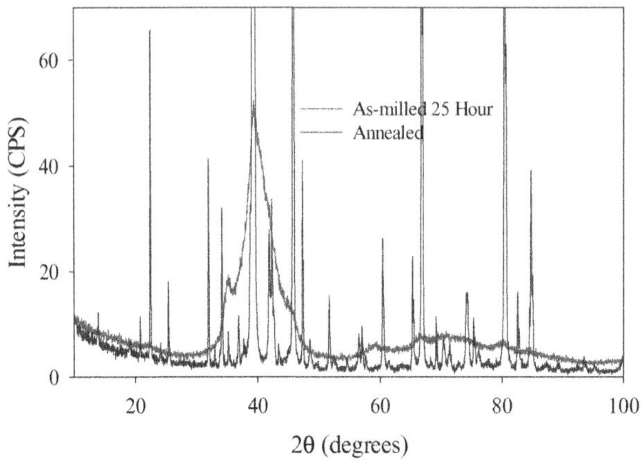

Figure 5 Comparison of as-milled powder to annealed powder with the same intensity scale.

Applying this overall approach to the experiment in which MNS processing time was increased with the same initial powder chemistry gives the results shown graphically in Figure 6. The NGB component increases rapidly as the crystallite size decreases with little change after MNS processing times of 10 - 15 h. As expected a rapid increase in TiC is observed with MNS time that appears to increase at a lesser rate as time progresses. However, there is no elemental titanium (diamond-shaped data points in Figure 6) present after the 15 h MNS time despite the still increasing TiC content. The $L1_2$ intermetallic decreases in a smooth systematic fashion as expected; prior to the correction for the background scattering the $L1_2$ intermetallic phase unrealistically increased initially then decreased. Another reaction product is the TiH_2 phase which increases at a much lower rate than the TiC. As noted, the hydrogen was easily removed as a post-MNS processing step prior to consolidation.

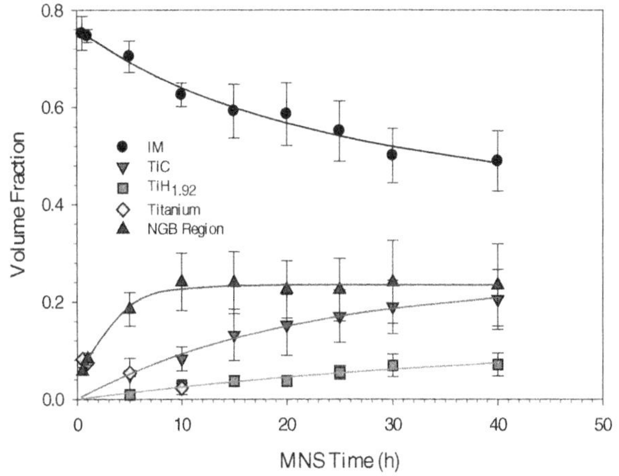

Figure 6 Phase volume fractions for the various MNS processing times all with the same starting composition.

The determined individual peak profiles represented with the common XRD Pearson VII peak model can be synthesized and then used as input to a Warren-Averbach analysis to determine crystallite size and microstrains. Each determination reported is the mean of at least three individual profile fit iterations. The Warren-Averbach method requires an annealed pattern for the Stokes correction of instrumental effects in the typical manner, this was accomplished with a well-annealed -400 mesh powder pressed and structurally detailed.[8] This process of developing the Warren-Averbach crystallite size and microstrain data for the series of alloys resulting from MNS processing of the same composition powders has been described earlier in the Powder Diffraction Journal.[7] Results of this analysis are illustrated in Figure 7 showing similar responses for both the <111> and <200> directions. The microstrain approaches a large value of 1 % elastic strain; however, it is a realistic strain considering the consolidated materials have been shown to be able to support these elastic strains. Conversely the crystallite size shows a rapid decrease asymptotically to a value of about 2 nm a size similar to that suggested as a theoretical size limit of a crystalline material.[9] The error bars are 95 % confidence intervals developed based on multiple background fit/profile fit synthesis followed by Warren-Averbach analysis. Note three of the data points show no error bars as these all converged with very similar models in multiple profile fit attempts.

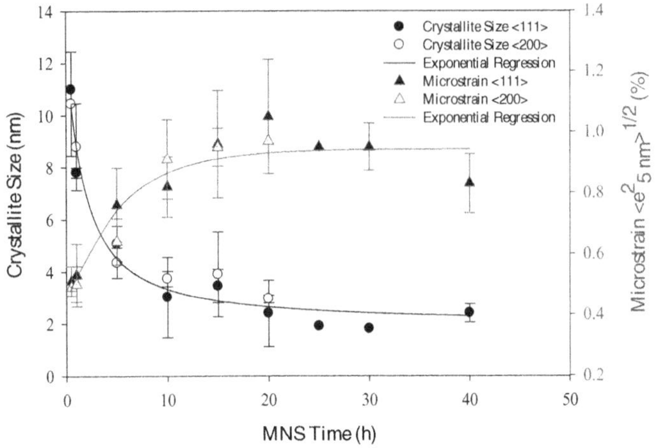

Figure 7 Warren-Averbach results of synthesized diffraction peaks for <111> and <200> directions in the $L1_2$ intermetallic phase.

Traditionally, dislocation densities have been estimated from the crystallite size, D, microstrain, $\langle \varepsilon^2 \rangle^{1/2}$, and Burgers vector according to $\rho = 2\sqrt{3}\langle \varepsilon \rangle^{1/2} / bD$. While the Burgers vector could not be determined a common room temperature dislocation observed in this material is the a/3<112> partial. Milligan[10] suggested that in general there is a different deformation mechanism for each of four regions of grains sizes; grain sizes above 1 μm traditional deformation by full dislocations nucleated within the grains; 30 nm to 1 μm where plastic deformation occurs by dislocations nucleated from grain boundary sources; 20 to 30 nm deformation dominated by partial dislocations that only pass through the grains; and those below about 10 nm where deformation occurs by grain boundary-mediated processes. It can be noted that the extremely small size crystallites observed here suggests this latter case will apply. The estimated dislocation densities determined for these MNS processed samples range from 4.6×10^9 mm^{-2} at 0.5 h to constant values of about 5.4×10^{10} mm^{-2} for 20 h or beyond. These latter values are extremely large, about an order of magnitude larger than dislocation densities of ~ 5×10^9 mm^{-2} in heavily cold rolled metals.[11]

The second experiment produced powders for consolidation that were processed by MNS with the base elemental titanium amount of 1.35 g with a 20 h processing time, both parameters were incremented by 1, 2, 3, 4, and 6 times (with a corresponding decrease in the amount of $L1_2$ intermetallic). As shown in Figure 8 the XRD peaks for the TiC phase at MNS processing time of 120 h dominated the XRD pattern indicating that very large amounts of reaction product can be easily produced in the as-milled state.

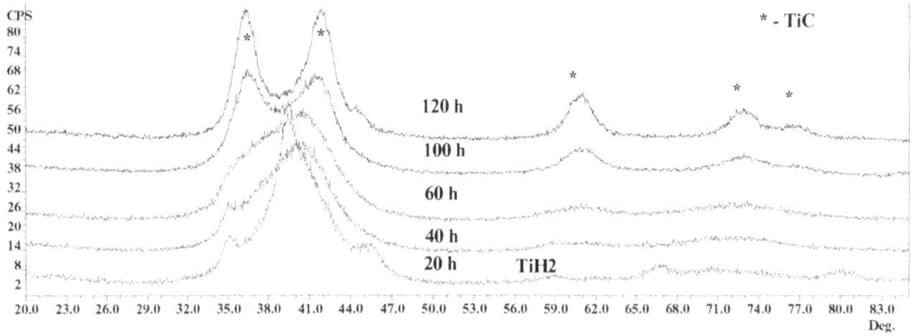

Figure 8 XRD patterns of as-milled powder for longer MNS processing times and similarly larger amounts of elemental titanium.

Microscopic study of the as-milled particle surfaces revealed heavy deformation of the surface that appeared to be made up of highly plastically deformed or "smeared" particles on larger particle surfaces such as the particle shown in Figure 9 highlighted by the black arrow. Assuming an original spherical shape for the smeared particle and considering the small thickness of the smeared particle on the surface, the estimated plastic deformation undergone by the particle is quite large, certainly exceeding typical ductilities observed in the large-grained $L1_2$ intermetallic. This is unusual given the mechanical properties of the $L1_2$ intermetallic. Close examination shows that the entire surface is littered with these smaller particles smeared and bonded to the surface of larger particles. Little attention has been given to this behavior in the literature, despite the likelihood that these surface features reflect fundamental aspects of the "mechanical alloying" process.

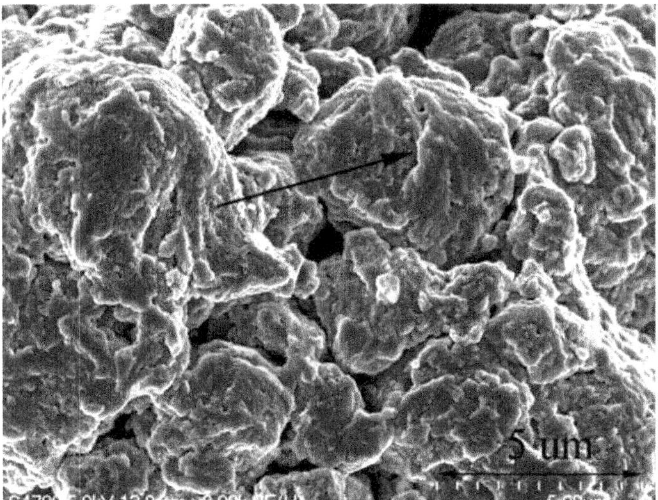

Figure 9 Field emission electron microscope micrograph of powder particle surface from MNS processing for 10 h.

DISCUSSION

MNS processing produces crystallite sizes and reaction products with microstructural features that have the smallest expected size for a polycrystalline material. The small size of the crystallites and measurable increase in the background scattering suggest microstructural features in agreement with observations by Gleiter et al.[12] for a 6 nm grain size sample indicating the background scattering was from atoms comprising the grain boundary region. Recall that the volume fraction of this NGB phase was 0.24 of all the phases (Figure 6). However, it represents 0.48 of the $L1_2$ intermetallic material as these would only include grain boundaries within that phase. In earlier work by Gleiter et al.[13] on a 6 nm grain size iron sample the grain boundary atoms were estimated to be close to 50 % in number and volume, having a value similar to that experimentally determined here, with these atoms on the grain boundaries suggested to have a "gas-like" structural arrangement.

This severe deformation processing causes significant defect generation with extremely large dislocation densities. However, the detailed nature of these defects/dislocations has not been determined. A detailed study of dislocations and defects is an important focus for future work in analyzing as-milled powders by transmission electron microscopy. The small intermetallic crystallites are expected to be defect free; therefore the defects must reside on the surface of these crystallites providing the origin for the observed large microstrains. In large-grained polycrystals the grain boundary surface is atomically thin and often has a superstructure consisting of coincident lattice sites[14] that are typically several to tens of atoms apart. Since these distances can easily exceed the expected size of the individual grain facets on the 2-3 nm grains assuming a tetrakaidecahedral shape, this type of large-grain boundary structure cannot exist. Grain boundary structural changes have also been observed in work by Valiev et al.[15] who suggested a new grain boundary structure emerging in sub-micrometer/nanometer grain sized materials (i.e. sizes closer to the 100 nm upper limit of nanometer range) termed a non-equilibrium grain boundary (NEGB). These consist of large numbers of non-equilibrium extrinsic grain boundary dislocations and show changes in fundamental properties that are typically microstructure-independent, such as Debye temperature and elastic modulus.[16] This can be taken as support for the hypothesis that the grain boundary structure changes in these nano-scale materials are necessary to accommodate the large increase in surface area, concurrent with a decrease in each grain facet surface area becoming a small number of unit cells.

Based on the detailed characterization of the as-milled powder an average microstructural model can be suggested considering only the $L1_2$ intermetallic and the TiC phase. Combining the small crystallite size, extremely high dislocation (or defect) density, with the boundary structure change to an NGB structure leads to the model presented in Figure 10. It is derived from the mean dimensions and volume fractions experimentally determined for a tetrakaidecahedron cut through the largest diameters of space-filling polygons yielding the classic hexagonal shape for the average grains. The grain interior modeled by the crystallite size is represented by regular close-packed atomic spacing (no attempt was made to represent exact orientation differences in grains) with the more loosely defined atoms comprising the defect region surrounding the crystallites and random displacements represented by the regions between grains (these atom placements only visually highlight the atomic density differences). The defects must accumulate on the periphery of the crystallites, which are dislocation-free by definition, with the additional change in the grain boundary structure to that of the NGB structure. The small crystallites would be expected to have a very high surface energy thus increasing the internal energy; however, it might be that this crystalline-defect-NGB atomic structural change lessens the surface energy by changing the structural requirements for compatibility. Therefore, the defects on the surface of these crystallites may well act as a structural accommodation gradient to the random static displacements of the grain boundary atoms thereby lowering the internal energy of the system; i.e., a natural response to the rapidly developed very large increase in surface area. Thus, the grain size can be taken as the center to center distance of the NGB phase regions. These expected and observed changes in the grain boundary structure, support the work of Gleiter et al.[17], who have also shown orders of magnitude increases in diffusion at the grain boundaries and large diffusivities at very low temperatures. Specifically, they reported that the diffusivity for 8 nm grain size copper at 353 °K was fourteen orders of magnitude greater than bulk self-diffusion and about three orders of magnitude greater than with GB self-diffusion in large-grained polycrystalline copper. This three-dimensional contiguous structural arrangement of the NGB's may well be the key to understanding this behavior.

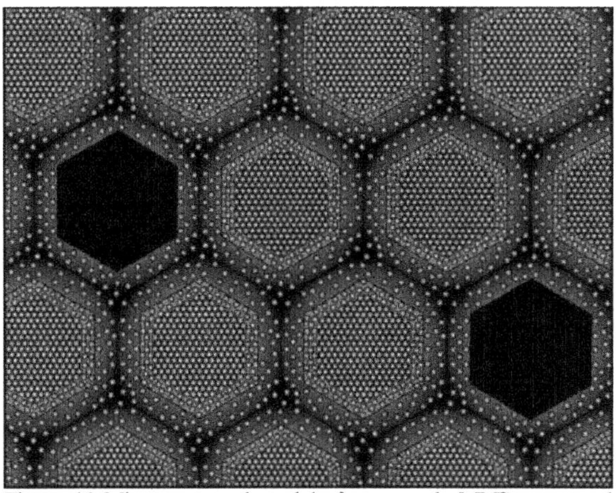

Figure 10 Microstructural model of nano-scale MNS processed as-milled material considering mean sizes of the $L1_2$ intermetallic and TiC phases and data developed in this analysis with the area representing the volume fraction calculated for these regions.

Rigney et al.[18] have shown experimentally the microstructural similarities between the transfer (or tribo) layer from sliding wear of two metals and the deformed materials in the mechanical alloying process developed by Benjamin.[3] In that work it was determined that the tribo layer had grain sizes of less than 10 nm with similar grain sizes also found internally in the mechanically alloyed powder. However, most of the grains in the mechanically alloyed powder were larger and it was concluded that while there are similarities the deformation in the mechanically alloyed material was inhomogeneous compared to the tribo layer. Also it is important to note that the nature of the surface of the mechanically alloyed powder was not investigated. Both these wear surfaces and as-milled powders show alloying, thus mass transport occurs, but plausible detailed explanations of these alloying events are sparse. Considering the structural arrangements developed during MNS processing, along with the Gleiter et al.[17] work on increased diffusivity of nano-scale grain boundaries, it might be expected that this type of NGB structure is developed at the points of contact from the severe deformation in both environments providing a mechanism of enhanced short-circuit diffusion. Importantly, the NGB structure can be considered a contiguous phase creating a three-dimensional highway for rapid diffusivity and hence the immediate and rapid alloying that occurs in both processes. Therefore, these surface events in "mechanical alloying" processing may play a dominant role in the alloying and grain size reduction as the deformation reaches the limit the material can accommodate under the given conditions. These structures can also play an important role in accomplishing mass transport not only with the alloying but also with the reactions that occur. Further, the triple junctions are possible nucleation sites for the TiC in these contiguous NGB regions. This NGB structure may also play a role in other deformation processes such as cold welding.

Fundamentally important in this study is the discovery of the *in situ* reactions occurring during the MNS processing and their simple dependence on an adequate supply of reactants. For the case of the chromium-modified titanium trialuminide milled with elemental titanium (in the experiment where both the titanium content and MNS processing time were increased) a linear uptake of carbon and production of TiC was observed. When the excess titanium content is fixed the TiC content begins to saturate with all elemental titanium completely reacted by the 15 h MNS processing time. Thus, it is quite clear that titanium is extracted from the base chromium-modified titanium trialuminide alloy but at a much lower rate. Of course, the carbon was always available.

A similar behavior was observed in the MNS processing of iron powder using hexanes as a PCA with *in situ* reaction to produce the Fe_3C phase in amounts dependent on MNS processing time. This suggests that there are a multitude of systems capable of producing a variety of matrix and reaction products in the form of a powder with an extremely high density of non-equilibrium nuclei able to produce unique microstructures utilizing conventional powder metallurgy techniques. The $L1_2$ intermetallic with at least 0.15 volume fraction of carbide has an ultra-high yield strength above 2 GPa, and was produced by conventional powder consolidation with cold isostatic pressing followed by hot isostatic pressing into a fully dense material.[19]

CONCLUSIONS

The MNS process can create composites with engineered volume fractions of particles having nano-scale dimensions by *in situ* reactions. In the system studied here the size of the crystallites in the $L1_2$ intermetallic and size of TiC formed *in situ* during the MNS processing approach the smallest known dimensions for crystalline materials of about 2 nm. At these sizes the grain boundary facet area is of the order of ten unit cells making a coincident type lattice typical of large-grained materials difficult to form. These observations suggest a random static atomic displacement of the atoms in the NGB's and that this disordered structure may well be a response to the lack of lattice sites on these grain boundary facets coupled with the large number of defects on the periphery of the crystallites buffering this structural change and thereby providing thermodynamic stability. These contiguous networks of NGB regions are expected to result in very high diffusivities providing short-circuit diffusion paths for rapid mass transport in low temperature processes such as mechanical alloying.

The origins of the commonly-observed surface smeared particles on the surfaces of larger as-milled particles appear to be point of contact severe deformation of localized material which creates the NGB structure as the material experiences the limits of deformation, in this case reaching a steady state of small size about 2 nm. These structural changes make possible grain boundary mediated plasticity in this near room temperature processing, a process that is similar to superplastic behavior, which although observed at higher temperatures with larger grain sizes, is also dependent on grain boundary diffusion processes. Large deformation related to small crystallites has been reported by Rigney et al.[18] in their wear studies and they suggested a smallest size of about 3 nm near that developed during MNS processing. Such deformation may be important in other severe deformation point of contact processes such as cold welding.[20]

REFERENCES

[1] S. Zhang, J.P. Nic, and D.E. Mikkola, New Cubic Phases Formed by Alloying Al_3Ti with Manganese and Chromium, *Scripta Met.*, Vol. 24, 57-62, 1990

[2] D.E. Mikkola, J.P. Nic, S. Zhang, and W.W. Milligan, Alloying of Al_3Ti to Form Cubic Phases, *ISIJ International (Japan)*, 31, 10, 1076-1079, Oct. 1991

[3] J.S. Benjamin, Dispersion-Strengthened Superalloys by Mechanical Alloying, *Metallurgical Transactions*, Vol. 1, 2943-2951, October, 1970

[4] C. Suryanarayana, Mechanical Alloying and Milling, *Progress in Materials Science*, Vol. 46, 1-184, 2001

[5] Alan Arias, Chemical Reactions of Metal Powders with Organic and Inorganic Liquids during Ball Milling, *NASA TN D-8015*, NASA, Washington D.C., September 1975

[6] S. Srinivasan, P.B. Desch, and R.B. Schwarz, Metastable Phases in the Al_3X (X = Titanium, Zirconium, Hafnium) Intermetallic System, *Scripta Metallurgica*, Vol. 25, 2513-2516, 1991

[7] E.A. Laitila, D.E. Mikkola, Employing X-ray Scattering to Characterize Materials with Grain Sizes in the Nano-Regime, *Powder Diffraction*, Vol. 23, Issue 2, 96-100, June 2008

[8] J.P. Nic, Structure/Property/Composition Relationships in Cubic Al_3Ti-Base Intermetallic Alloys, *MS Thesis*, Michigan Technological University, 1992

[9] C. Suryanarayana, Nanocrystalline Materials, *International Materials Reviews*, Vol. 40, No. 2, 41-64, 1995

[10] W. W. Milligan, Mechanical Behavior of Bulk Nanocrystalline and Ultrafine-Grain Metals, *Comprehensive Structural Integrity*. Elsevier, Amsterdam 529 (2003)

[11] D. Hull and D.J. Bacon, *Introduction to Dislocations*, Pergamon Press, 1984

[12] R. Birringer, U. Herr, X. Zhu, and H. Gleiter, X-Ray Diffraction Studies of the Structure of Nanometer-Sized Crystalline Materials, *Physical Review B*, Vol. 35, No. 17, 9085-9090, 1987

[13] R. Birringer, H.P. Klein, P. Marquardt, and H. Gleiter, Nanocrystalline Materials an Approach to a Novel Solid Structure with Gas-Like Disorder?, *Physics Letters*, Vol. 102A, No. 8, 365-369, 1984

[14] W. Bollmann, *Crystal Defects and Crystalline Interfaces*, Springer-Verlag Berlin Heidelberg, 1970

[15] R.Z. Valiev, V.Yu. Gertsman, and O.A. Kaibyshev, Grain Boundary Structure and Properties Under External Influences, *Physica Status Solidi*. A, 97, 11-56, 1986

[16] R.K. Islamgaliev, I.V. Alexandrov, and R.Z. Valiev, Bulk Nanostructured Materials from Severe Plastic Deformation, *Progress in Materials Science*, Vol. 45, 103-189, 2000

[17] J. Horvath, R. Birringer, and H. Gleiter, Diffusion in Nanocrystalline Material, *Solid State Communications*, Vol. 62, No. 5, 319-322, 1987

[18] L.H. Chen, M.G.S. Naylor, A.R. Rosenfield, and D.A. Rigney, Wear in Sliding Systems, *Wear*, Vol. 100, 195-219, 1984

[19] E.A. Laitila, and D.E. Mikkola, Ultra-High Strength Composite Materials Developed by Mechanonanosynthesis Processing, *to be published in the MS&T2015 Proceedings*

[20] N. Bay, Cold Welding. Part 1: Characteristics, Bonding Mechanisms, Bond Strength, *Metal Construction*, Vol. 18, No. 6, 369-372, 1986

GROWTH KINETICS OF LANTHANUM PHOSPHATE CORE/SHELL NANOPARTICLES DOPED WITH Ce-Tb AND Eu

M.C Molina Higgins [1], J. V. Rojas [1]

[1] Mechanical and Nuclear Engineering, Virginia Commonwealth University, 401 West Main Street, Richmond, Virginia 23284-3067301

Keywords: Lanthanide phosphate, nucleation, luminescent nanoparticles

Abstract

Lanthanide phosphate ($LnPO_4$) nanoparticles have been used for biomedical and luminescent devices, in applications ranging from drugs carriers to fluorescent lamps, because of properties such as relatively low toxicity, light absorption and luminescence. This properties are generally affected by size and shape of the nanoparticles. In this work we studied the growth and size evolution of $LnPO_4$ nanoparticles with core and 4 shells-core structure. Lanthanide phosphate nanoparticles were prepared by mixing aqueous solutions of lanthanum chloride ($LaCl_3$), containing a fraction of Cerium-Terbium (Ce-Tb) and Europium (Eu), and Sodium Tripolyphosphate (NaTPP) in a molar ratio of Ln/NaTPP=1. Each solution was aged for 3 hours at~90°C. During the heating process aliquots were collected periodically and analyzed with UV-Vis spectroscopy. Also, XRD, TEM, HRTEM and SAED were used to characterize the crystalline structure and morphology of the nanoparticles. The crystalline structure of the nanoparticles was found to be rhabdophane structure ($LnPO_4 \bullet xH_2O$) and an average particle size of ~4.2 nm for the core and ~5.9 nm for the core-4 shells structure was obtained. Additionally the HRTEM and SAED analysis confirmed the rhabdophane structure observed with XRD.

1. Introduction

The luminescence in phosphors has been an interesting topic for scientist during the last 100 years. The first phosphorus used was synthesized in 1866 by Theodore Sidot and its first use was in television screens [1]. In the recent years, phosphorus at the nanoscale level have been used successfully in applications such as drug carriers [12], fluorescents lamps, display devices, cathode ray tubes, scintillation detection, drug and luminous painting [1]. The luminescent properties of nanophosphors often depend on the shape and size distribution of the nanoparticles. Therefore studies regarding the growth kinetics of this type of nanomaterials, as well as they mechanism of energy transfer has been reported by several authors [2, 3, 4].

The growth of luminescent nanoparticles, and nanoparticles in general, follows the LaMer theory. In this model, the reactive material begins to collide and form clusters, subsequently the clusters become stable and start to grow. These two stages are called nucleation and growth [8]. In order to stabilize the clusters and prevent aggregation, capping agents such as NaTPP [5], oleic acid, [6] cetyltrimethylammonium bromide (CTAB) [7], etc. are added to the synthesis providing a better control of the growth process [8]. The LaMer mechanism is useful to explain the stages of nucleation and growth of several types nanocrystals, however it does not account for the kinetics of the process [9]. In the interest of studying the growth kinetic of the nanoparticles, ions with luminescent properties are placed inside the material, and the changes in the luminescence are followed in time [5].

$LnPO_4$ materials has been successfully used as phosphors in fluorescent lamps and in scintillation detection [17]. Commercially, inorganic macrocrystals are preferred for the applications aforementioned, however the manufacturing of these crystals is expensive. Consequently fabrication of $LnPO_4$ nanocrystals, offers a low cost alternative for new phosphor technologies. Interesting enough, $LaPO_4$ doped with ions such as Ce-Tb and Eu are a promising materials for biomedical applications such as biolabeling, phototherapy and optical imaging [14, 17]. $LaPO_4$ nanomaterials when doped with Ce-Tb or Eu under UV- excitation at a wavelength of 254 nm become green or red emitters, respectively. In the first case, the energy transfer process occurs between the Ce and the Tb. The Tb acts as an impurity, allowing forbidden transitions in the nanocrystal. When a Ce electron reaches a Tb activation center, a relaxation takes places between the energy bands $^5D_4 \rightarrow {}^7F_J$ and a photon is emitted [17]. On the other hand, when a $LaPO_4$ matrix is doped with Eu an electron transfer between the Oxygen (O) 2p shell and the 4f shell of Eu occurs. When the electron return to the valence band, a photon is emitted [14].

In this work, an aqueous colloidal synthesis of Lanthanide phosphate ($LnPO_4$) core-shell structures will be presented. Two types of nanoparticles were made. The first one is $LaPO_4$ doped with Cerium (Ce) and Terbium (Tb), and the second one is $LaPO_4$ doped with Europium (Eu). Furthermore, growth of the nanoparticles with core and core with four shells, was investigated through Ultraviolet-Visible Spectroscopy (UV-Vis). The final product was characterized with X-Ray Diffraction (XRD), transmission electron microscopy (TEM), high resolution transmission electron microscopy (HRTEM), and selected area electron diffraction (SAED).

2. Experimental Procedure

2.1 Materials

Lanthanum chloride ($LaCl_3$) heptahydrate, Cerium Chloride ($CeCl_3$) heptahydrate, Terbium Chloride ($TbCl_3$) hexahydrate, Europium Chloride ($EuCl_3$) hexahydrate and Sodium triphosphate pentabasic ($Na_5P_3O_{10}$ or NaTPP), were obtained from Sigma Aldrich. Additionally, deionized water (DI) (18MΩ) collected from an EMD Millipore Direct QTM 3UV was used for the preparation of the aqueous solutions.

2.2 Aqueous Colloidal Synthesis

Aqueous colloidal synthesis of the core and core-shell $La_{40\%}Ce_{45\%}Tb_{15\%}PO_4$ nanoparticles was carried out with the aforementioned materials. Firstly, a 0.1 M solution of $LnCl_3$ (Ln=La, Ce, Tb) was prepared by mixing $LaCl_3$, $CeCl_3$ and $TbCl_3$ with volume percentages of 40%, 45%, and 15%, respectively. The solution was then added to a 0.1 M NaTPP with a volume ratio of 1:1 and the mixture was vortexed until a colorless liquid was obtained. Subsequently the clear solution was stirred for 3 hours at 90 °C in a 100SH Fisher Scientific Stirring Hotplate. Then, the suspension was allowed to cool at room temperature, transferred to a dialysis membrane and dialyzed against DI water for 24 h. Finally, the clear suspension of core $La_{40\%}Ce_{45\%}Tb_{15\%}PO_4$ was collected. In order to add the first shell, a solution of 0.1M NaTPP: 0.1 $LnCl_3$ (40% $LaCl_3$, 45% $CeCl_3$, 15% $TbCl_3$) with 2:1 v:v ratio was added to the aqueous core solution. For the shell deposition, the ratio between the Ln Core and Ln shell was 1:1. Subsequently, the mixture was vortexed and placed in the hotplate. The mixture was kept under stirring and heating for 3 h at ~ 90°C. Also, the core-1 shell samples was dialyzed overnight. The procedure above for the addition of shells was repeated 3 more times until 4 shells were added on top of the core. Dialysis of the nanoparticles was carried out between the additions of each shell in order to remove unreacted species. In addition, $La_{75\%}Eu_{25\%}PO_4$ core-shells nanoparticles were also synthesized, following the same procedure mentioned above.

2.3 Sample Characterization

The time evolution absorbance of the resulting nanoparticles, during its synthesis process, as well as their particle size, particle size distribution and crystallinity were evaluated. The evolution of the UV-Vis absorbance of the NaTPP-$LnCl_3$ mixture with time throughout the synthesis process of the nanoparticles analyzed through a GENESYS 10S UV-Vis spectrophotometer. An aliquot from the solution during the heating process was taken every 0.5 hours. Afterwards, the aliquots were diluted in deionized water. Furthermore, the particle size and particle size distribution of the core and core- 4 shells were evaluated by TEM using a Zeiss Libra 120 Plus operating at a voltage of 120kV. For TEM, HRTEM and SAED analysis, the samples were collected after the dialysis a small portion was diluted 100 times in acetone. Next, the samples were sonicated for ~20 minutes to break up any existing aggregates. Finally, a formvar carbon coated copper grid was immersed in the diluted solution and allowed to dry at room temperature. The particle size analysis was performed with the software Image J. the crystalline structure of the nanoparticles was evaluated with XRD. The analysis was carried out using a PANalytical X'Pert Pro diffractometer with a Copper anode (CuK) at a voltage of 45kV and a current of 40 mA. The samples for XRD were prepared by drying the aqueous suspension under vacuum at ~ 35°C. Afterwards the sample was ground to a fine powder. Finally, the fine powder was deposited on the spinning sample stage of the diffractometer. Evaluation of the XRD patterns was performed with the software X'Pert Highscore Plus.

3. Results and Discussion

The aqueous synthesis of $LnPO_4$ nanoparticles is based on the reaction between solutions containing rare earth ions and polyphosphate salts [5]. The heating process implemented in the synthesis of the nanomaterial, activates the hydrolysis of the tripolyphosphate (TPP) group [12]. The hydrolysis promotes a first order reaction in the polyphosphate chain. As a result, the chain splits progressively until the TPP transforms into orthophosphate groups (PO_4^{-3}) [11]. The anion of the PO_4^{-3} reacts with the rare-earth salts cations producing stabilization and size control during the nucleation and crystallization of the $LnPO_4$. As the heating time increases the initial colorless solution becomes turbid due to the coalescence and precipitation of the nanoparticles [12]. The removal of unreacted species through dialysis leads to a translucent suspension. Nonetheless, the addition of shells on the core nanoparticles results in white suspension.

The UV-Vis time evolution spectra for the core structure for both, $La_{40\%}Ce_{45\%}Tb_{15\%}PO_4$ and $La_{75\%}Eu_{25\%}PO_4$ nanoparticles, are presented in figures 1a) and 1 b) respectively. The absorption spectra of $LnPO_4$ nanostructures doped with Ce and Tb, shown in Figure 1.a), evidence two characteristic peaks at 248 nm and the second one at 302 nm at a time equals 0 h, before the starting of the heating process. These features are related to Ce-TPP polymeric bond complexes [10]. The disappearance of the peak at 305nm as the synthesis time increase is due the TPP transforming into PO_4^{-3} (orthophosphate) and forming covalent bonds with the Ln species. This leads to the formation of $LnPO_4$ crystalline nanostructures. The absorbance peak at 273 nm is related with the amount of crystalline $LnPO_4$ phase [5].In addition this peak is continuously growing during the synthesis process. The peak at 273 nm is associated with transitions between the 4f-4f5d of the Ce in the hydrated $LaPO_4$ matrix. The energy transfer within the nanophosphor is due to the interactions between Ce and Tb. The energy travels from a Ce^{+3} to Ce^{+3} until a Tb^{+3} is reached, subsequently green light is emitted due to Tb^{3+} relaxation and energy transfer between the $^5D_4$ level and the ^7F_J state. It is important to notice, that the relaxation that takes places in Tb is under investigation [17].

The results regarding the core structure of $LaPO_4$ nanoparticles doped with Eu are observed in figure 1.b. The peak at 230 nm represents the interaction of the O^{-2} ligand with the rare earth metal (Eu^{+3}). In this case the absorption and emission of the matrix and the dopant depend on the Eu concentration [13]. The energy transfer mechanism of the nanostructures doped with Eu is due the electron movement between the filled 2p shell of oxygen (which acts as the valence band) and a partially filled 4f shell of Eu^{3+}. This transition occurs at ~ 230 nm. Afterwards, the electron return to the valence state, a photon is emitted and a red light is produced. The band gap of $LaPO_4$ doped with Eu, diminished until a Eu concentration of 50wt% is added to the matrix. Experimentally no decrease in the bad gap is observed from this concentration value [14]. Equation (1) was used to calculate the energy transitions of the undoped and doped material. Where E is the energy transition (J), h is the Planck constant (6.62 x10^{-34} J/s), c is the velocity of a photon (3x108 m/s) and λ is the wavelength for each case.

$$E = \frac{hc}{\lambda} \quad (1)$$

The energy transitions calculation by this method leads to values of 5.96 eV, 5.39 eV and 4.56 eV for the $LaPO_4$, $LaPO_4$ doped with Eu and $LaPO_4$ doped with Ce and Tb respectively. The energy transition corresponding to $LaPO_4$ matches with the value presented in the literature for hexagonal $LaPO_4$. In monazite structures this value has been reported to be ~8 eV. [18]. The separation between the conduction and the valence band in various materials doped with Ce^{+3} ions such as $LaSi_3N_5$ and $Sr_3V_2O_8$ has been reported as 4.6 eV [19, 20]. $LaPO_4$ bulk when doped with Ce and Tb present five different energy transitions, due to the different split levels of the excited state ^2D ($5d^1$). These energies are 4.46 eV, 4.76 eV, 5.2 eV, 5.8 eV and 6.05 eV [21]. However at the nanoscale level, for the Tb^{+3} emission due to f-d transition above 5.6 eV are suppressed [21].Similarly, for materials doped with Eu transition energy values between 5.4 and 5.5 eV has been found in compounds such as Y_2O_3 and $InBO_3$ [22, 23]. Most of the energy transfer in $LaPO_4$ nanoparticles doped with Eu is due to the interaction between Eu and O. Another three additional transitions can be observed at $^7F_{0,1} \rightarrow {}^5D_4$, $^7F_{0,1} \rightarrow {}^5G_3$ and $^7F_0 \rightarrow {}^5L_6$, but these peaks have a low intensity because they are forbidden transitions and they cannot be observed with UV-Vis [14].

Changes in absorbance of the matrix during the addition of shells were analyzed and the results are presented in figures 1.c and 1.d. The figures show the time evolution of the absorbance for the most intense peaks, 273 nm and 230 nm, in the $LnPO_4$ doped with Ce-Tb and Eu respectively. The growth of the nanoparticles is a heterogeneous process were the $LnPO_4$ shells grows on top of the core structure. Consequently, the core acts as a nuclei template [10]. The absorbance in the UV-Vis spectra is related to the absorption cross section of the material.

Therefore, an increase in the absorbance will indicate an increase in the amount of crystalline $LnPO_4$ nanoparticles synthesized since there is more material absorbing light [16].

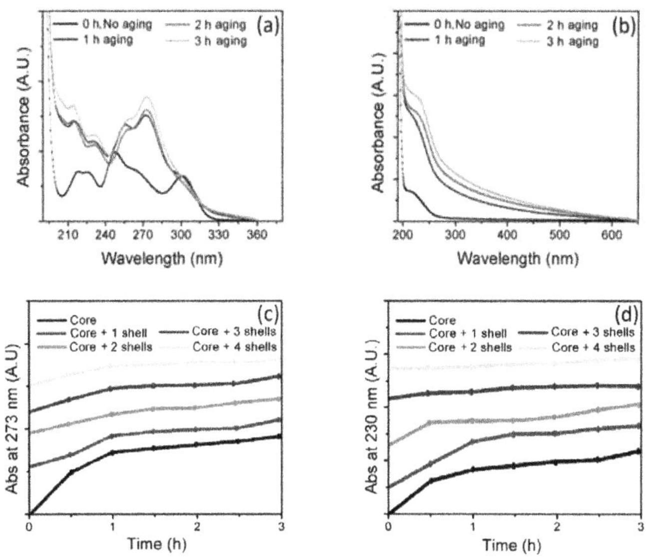

Figure 1. a) And **b)** are the UV-Vis absorption spectra at different time of synthesis from the $La_{40\%}Ce_{45\%}Tb_{15\%}PO_4$ and $La_{75\%}Eu_{25\%}PO_4$ nanoparticles, respectively. **c)** And **d)** Time evolution of the absorption bands at 273 nm and 230 nm for the $La_{40\%}Ce_{45\%}Tb_{15\%}PO_4$ and $La_{75\%}Eu_{25\%}PO_4$ nanoparticles respectively.

The morphology of the nanoparticles was studied by TEM. The TEM images showed an average particle size of 4.3 ± 0.7 nm and 5.9 ± 0.8 nm for the core and core-4 shells of the $La_{40\%}Ce_{45\%}Tb_{15\%}PO_4$ nanoparticles. Analysis of the average particle size in the $La_{75\%}Eu_{25\%}PO_4$ nanophosphor was 4.0 ± 0.6 nm and 5.8 ± 0.8 nm for the core and core-4 shells respectively. In figure 2, TEM micrographs and size distribution of the nanoparticles are shown. Although the results indicates a narrow size distribution, particles of several sizes are observed in the TEM images for core-4 shells. This suggest that part of the precursor NaTPP: Ln do not contribute to the growing of the shells on the nanoparticles but to the formation of new core $LaPO_4$ nanoparticles. Also aggregation of the nanoparticles is observed, as well as a spherical morphology. The dark shade surrounding the nanoparticles is related with the amount of surfactant still present after the 4 dialysis processes.

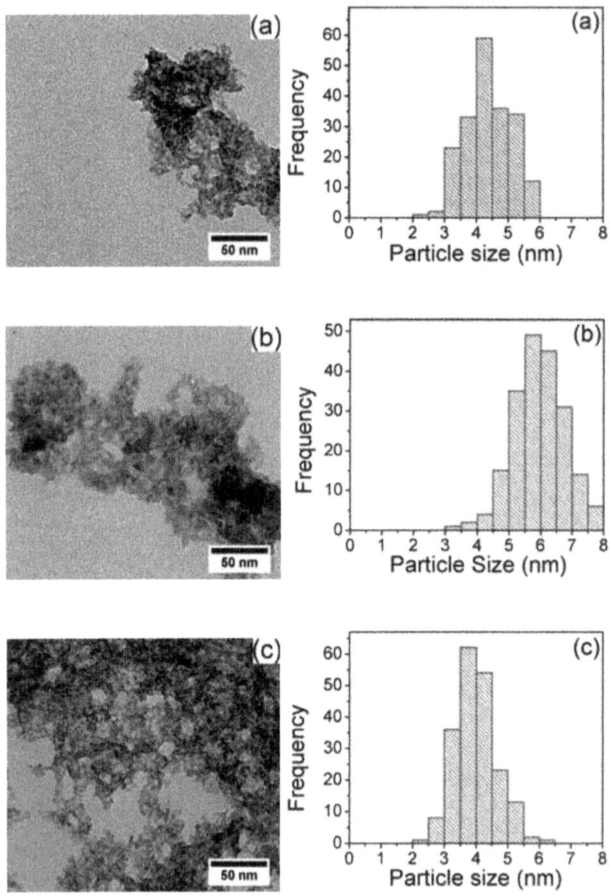
(a)
50 nm
(a)
Frequency
Particle size (nm)
(b)
50 nm
(b)
Frequency
Particle Size (nm)
(c)
50 nm
(c)
Frequency
Particle size (nm)

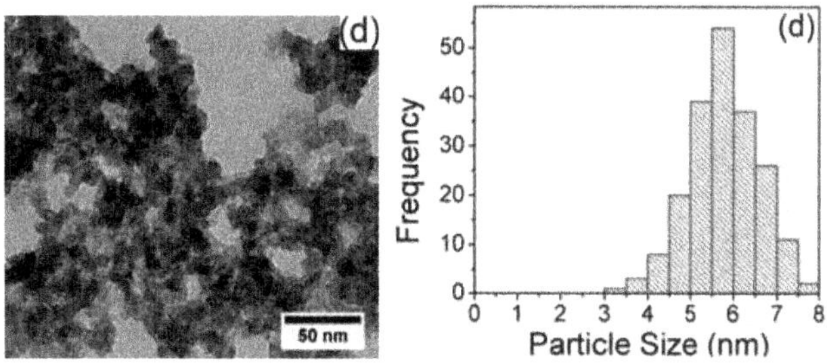

Figure 2. TEM Micrographs and their respective size distribution of **a)**. Core $La_{40\%}Ce_{45\%}Tb_{15\%}PO_4$ **b)**. Core-4 shells $La_{40\%}Ce_{45\%}Tb_{15\%}PO_4$ **c)**. Core $La_{75\%}Eu_{25\%}PO_4$ d). Core-4 shells $La_{75\%}Eu_{25\%}PO_4$.

Evaluation of the crystalline structure of the nanoparticles was carried out through XRD analysis. The XRD patterns shown in figure 3 revealed that the aqueous synthesis of $LnPO_4$ nanoparticles produce a rhabdophane structure (PDF 000-46-1439, $LnPO_4$ • xH2O), that correspond to a hydrated form of hexagonal $LnPO_4$ [10]. In the $LnPO_4$ patterns there is no indication of additional peaks regarding the $LaPO_4$ spectra. Absence of these peaks confirms the complete solubility of the Tb, Ce and Eu in the $LaPO_4$ matrix. The peaks of $LnPO_4$ are shifted to higher 2θ angles with respect to the $LaPO_4$ nanoparticles. The shifting is expected due the different cell parameters of the attained nanoparticles. The atomic radii of La (~187 pm) is slightly larger than the atomic radii of the dopants Ce, Tb and Eu (181 pm, 180 pm and 185 pm respectively). Consequently, a decrease in the unit cell parameters takes places when a dopant replaces a La atom in the lattice [24]. In addition, the XRD patterns were used to calculate the particle size using the Scherrer equation assuming a spherical nanoparticle shape. The results for the $LnPO_4$ nanoparticles doped with Ce and Tb were 4.4 nm, 5.3 nm, 5.6 nm, 5.9 nm, and 6.2 nm for the core, core-1 shell, core-2 shells, core-3 shells, and core-4 shells respectively. Furthermore, the size of the $LnPO_4$ nanoparticles doped with Eu were 4.3 nm, 5.2 nm, 5.5 nm, 5.9 nm, and 6.2 nm for the core, core-1 shell, core-2 shells, core-3 shells and core-4 shells respectively. The particle size results encountered for the $LaPO_4$ doped with Ce and Tb revealed an average thickness of the shells of 0.23 nm, while the nanoparticles doped with Eu had an average thickness of the shell of 0.24 nm Therefore the thickness of the shell corresponds approximately to one atomic layer. Interestingly enough, the attain value of the thicknesses of the second, third and fourth shells for the nanoparticles doped with Ce-Tb and Eu were approximately 0.15 nm, which is less than the atomic radius of La, Ce and Tb. This is an evidence during the process for the addition of shells, new nuclei are formed. Those nuclei grow with a smaller particle size and at a low nucleation rate [10]. As a result the mean particle size measured by the different characterization techniques shift towards smaller particle size values.

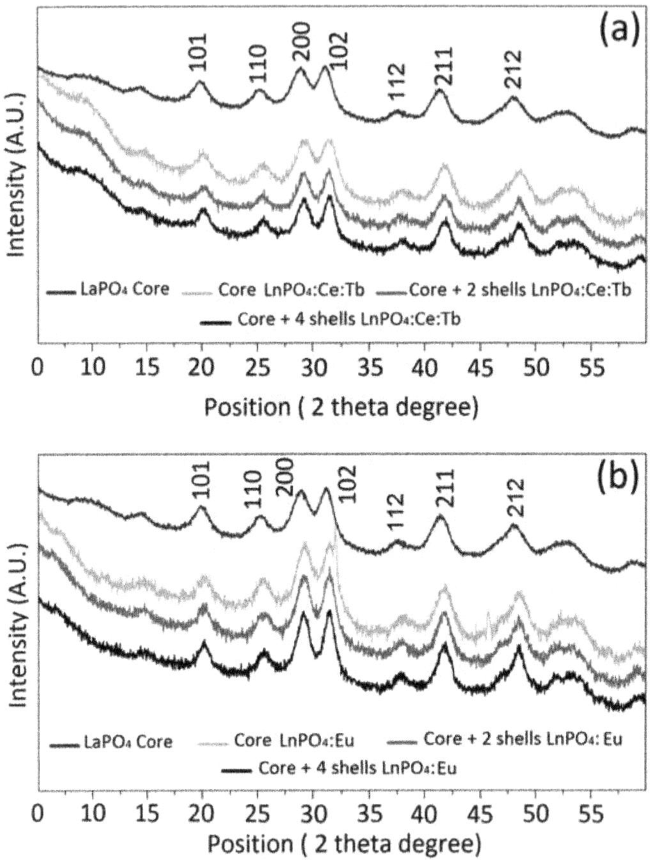

Figure 3. XRD Pattern of **a)**. Core shell Structures $La_{40\%}Ce_{45\%}Tb_{15\%}PO_4$ and **b)** Core shell structures $La_{75\%}Eu_{25\%}PO_4$

The crystalline structure of the nanoparticles was also evaluated with HRTEM and SAED and the results are shown in figure 4. The HRTEM image of the core-4 shells nanoparticles doped with Eu revealed a 0.32 nm d_{200} that is in closely agreement to the 0.31nm d_{200} found in the XRD pattern. It is important to notice that, in the HRTEM image, the core is not distinguishable from the shells. Given that the chemical composition of the core and shells is identical, no diffraction contrast was observed. Furthermore, the lattice fringes identified in the core-4 shells images correspond to a single crystalline domain [25]. On the other hand, the d-spacing values found through the SAED pattern for the core-4 shells nanoparticles doped with Ce and Tb match with the ones obtained with XRD. The XRD d_{200} was 0.31 nm, while the SAED d-pacing in the plane (200) was 0.32 nm.

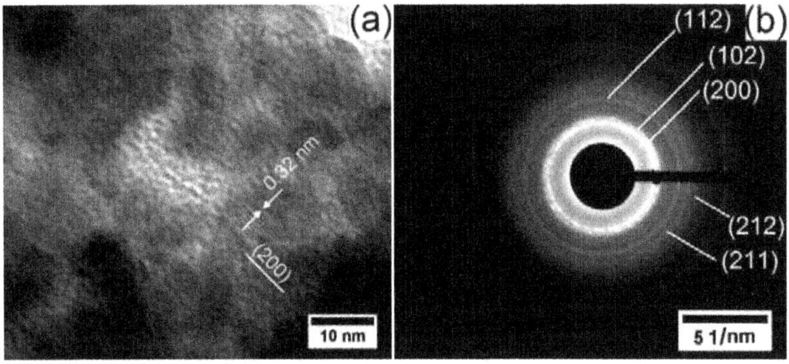

Figure 4. a). HRTEM of Core-4 shell $La_{75\%}Eu_{25\%}PO_4$ nanoparticles and **b)** SAED Core-4 shells $La_{40\%}Ce_{45\%}Tb_{15\%}PO_4$ nanoparticles

4. Conclusions

$LnPO_4$ core shell nanostructures doped with Ce-Tb and Eu, exhibiting green and red luminescence respectively, had been synthesized by an aqueous route. The obtained nanoparticles were evaluated by UV-Vis, TEM, HRTEM, XRD and SAED in order to study the growth, morphology and crystallinity. The time evolution of the absorbance peak at 273 nm for the $La_{40\%}Ce_{45\%}Tb_{15\%}PO_4$ nanoparticles and 230 nm for the $La_{75\%}Eu_{25\%}PO_4$ showed an increasing absorbance. This behavior evidenced the evolution of the amount of $LnPO_4$ in the aqueous suspension with time. In addition nanoparticles with a mean size of 4.3 ± 0.7 nm and, 4.0 ± 0.6 nm were obtained for the core structures while a particle size of 5.9 ± 0.8 nm and 5.8 ± 0.8 nm were found for the core-4 shells nanoparticles. The results were in close agreement with particle size found in XRD by the Scherrer Equation. Further evaluation of the crystalline structure with HRTEM and SAED lead to a d_{200} ~0.32 nm. The nanoparticles exhibited a rhabdophane crystalline structure, a hydrated form of hexagonal $LnPO_4$.These red and green emitters can be used not only in applications such as luminescent devices but also in the medical field, where core shells structures are often needed to encapsulate drugs while their light emission allows to track the drug carriers.

5. Acknowledgements

Research financed with the support of the mechanical and nuclear engineering department from Virginia Commonwealth University (VCU) and the NRC-HQ-84-14-FOA-002 faculty development program in radiation detection and health physics. Also the authors will like to acknowledge to the material's characterization facility NCC (Nanomaterials Core Characterization Center) at VCU.

6. References

[1] Shigeo Shionoya and William M. Yen. *Phosphor Handbook*, (Boca Raton, FL: CRC Press, 1998), 231-233

[2] M. Kawase, T. Suzuki and K. Miura, "Growth mechanism of lanthanum phosphate particles by continuous precipitation," *Chemical Engineering Science*, 62 (2007), 4875–4879.

[3] K. Byrappa, C. Chandrashekar, B. Basavalingu, K. LokanathaRai, S. Ananda and M. Yoshimura, "Growth, morphology and mechanism of rare earth vanadate crystals under mild hydrothermal conditions," *Journal of Crystal Growth*, 306 (1) (2007), 94-101.

[4] X. Xu, L. Hu, N. Gao, S. Liu, S. Wageh, A. A. Al-Ghamdi, A. Alshahrie and X. Fang, "Controlled growth from ZnS nanoparticles to ZnS–CdS nanoparticle hybrids with enhanced photoactivity," *Advanced Functional Materials*, 25 (3) (2014), 445-454

[5] V. Buissette, M. Moreau, T. Gacoin, J.-P. Boilot, J.-Y. Chane-Ching and T. Le Mercier, "Colloidal Synthesis of Luminescent Rhabdophane $LaPO_4$:Ln^{3+}•xH2O (Ln = Ce, Tb, Eu; x ≈ 0.7) Nanocrystals," *Chemistry of materials*, 16 (19) (2004), 3767-3773.

[6] Z. Li, X. Hou, L. Yu, Z. Zhang and P. Zhang, "Preparation of lanthanum trifluoride nanoparticles surface-capped by tributyl phosphate and evaluation of their tribological properties as lubricant additive in liquid paraffin," *Applied Surface Science*, 292 (2014), 971–977.

[7] M. Anbia, M. K. Rofouei and S. W. Husain, "Synthesis of Mesoporous Lanthanum Phosphate and Its Use as a Novel Sorbent," *Chinese journal of chemistry*, 24 (8) (2006), 1026-1030.

[8] K. Carroll, Core-Shell Nanoparticles: Synthesis, Design, and Characterization, (PhD thesis Virginia Commonwealth University, 2010), 2-3

[9] Pekka Hänninen and Harri Härmä, *Lanthanide luminescence: Photophysical, analytical and biological aspects*, (Berlin: Springer Science & Business Media, 2011), 118-119.

[10] V. Buissette, D. Giaume, T. Gacoin and J.-P. Boilot, "Aqueous routes to lanthanide-doped oxide nanophosphors," *Journal of Materials Chemistry*, 16 (6) (2006), 529-539.

[11] Igor S. Kulaev, Vladimir Vagabov and Tatiana Kulakovskaya, *The biochemistry of inorganic polyphosphates* (Chichester: John Wiley & Sons 2005), 11-12.

[12] J. Rojas, J. Woodward, N. Chen, A. Rondinone, C. Castano and S. Mirzadeh, "Synthesis and characterization of lanthanum phosphate nanoparticles as carriers for ^{223}Ra and ^{225}Ra for targeted alpha therapy," *Nuclear Medicine and Biology*, 42, (7) (2015), 614-620.

[13] M. Cayless, Lamps and lighting (New York, NY: Routledge, 2012), 149-151.

[14] S. K. Gupta, P. S. Ghosh, M. Sahu, K. Bhattacharyya, R. Tewari and V. Natarajan, "Intense red emitting monoclinic $LaPO_4$:Eu^{3+} nanoparticles: host–dopant energy transfer dynamics and photoluminescence properties," *RSC Advances*, 5, (72) (2015), 58832–58842.

[15] N. T. Thanh, N. Maclean and S. Mahiddine, "Mechanisms of Nucleation and Growth of Nanoparticles in Solution," *Chemical Reviews*, 114 (15) (2014) 7610–7630.

[16] M. Califano, "Re-examination of the Size-Dependent Absorption Properties of CdSe Quantum Dots," *The Journal of Physical Chemistry C*, 113 (45) (2009), 19468–19474.

[17] V. Pankratov, A. Popov, S. Chernov, A. Zharkouskaya and C. Feldmann, "Mechanism for energy transfer processes between Ce^{3+} and Tb^{3+} in $LaPO_4$: Ce, Tb nanocrystals by time-resolved luminescence spectroscopy," *Basic Solid State Physics*, 247, (9) (2010), 2252-2257.

[18] S. Syrotyuk, Y. Chornodolskyy, V. Vidtovskyy, A. Voloshinovskii and A. Gektin, "Band Structure of $LaPO_4$," *Functional Materials*, 20, (3) (2013), 373-377.

[19] C. T. Petit, R. Lan, P. I. Cowin and S. Tao, "Structure and conductivity of strontium-doped cerium orthovanadates $Ce_{1-x}Sr_xVO_4$ ($0 \leq x \leq 0.175$)," *Journal of Solid State Chemistry*, 183, (6), (2010), 1231–1238.

[20] **A. M. Ibrahim, Z. Lenčéš, P. Šajgalík, L. Benco** and M. Marsman, "Electronic structure and energy level schemes of RE^{3+}:$LaSi_3N_5$ and RE^{2+}:$LaSi_3N_{5-x}O_x$ phosphors (RE= Ce, Pr, Nd, Pm, Sm, Eu) from first principles," *Journal of Luminescence*, (164) (2015), 131–137.

[21] V. Pankratov, A. Popov, A. Kotlov and C. Feldmann, "Luminescence of nano and macrosized $LaPO_4$: Ce, Tb excited by synchrotron radiation," *Optical Materials*, (33), (2011), 1102-1105.

[22] M. Ramaiah, "Luminescence studies of europium doped with yttrium oxide nano phosphor," (Atlanta, GA: Scholarly Editions, 2013), 238-240.

[23] Z.-W. Chiu, Y.-J. Hsiao and L.-W. Ji, "Photoluminescence and Optoelectronics Characteristics of Eu-doped $InBO_3$ Nanocrystals," *International Journal of Electrochemical Science*, (10), 2391-2399.

[24] M. Runowski, K. Dabrowska, T. Grzyb, P. Miernikiewicz and S. Lis, "Core/shell-type nanorods of Tb^{3+} doped $LaPO_4$, modified with amine groups, revealing reduced cytotoxicity," *Journal of nanoparticle research*, 15 (11) (2013), 1-16

[25] Zhu, X. Liu, X. Liu, Q. Li, J. Li, S. Zhang, J. Meng and X. Cao, "Facile sonochemical synthesis of $CePO_4$:Tb/$LaPO_4$ core/shell nanorods with highly improved photoluminescent properties," *Journal of Nanotechnology*, 17 (16) (2006), 4217-4222.

INFLUENCE OF SYNTHESIS PARAMETERS ON MORPHOLOGY, CRYSTALLINE STRUCTURE AND COLLOIDAL STABILITY OF CORE AND CORE-SHELL $LaPO_4$ NANOPARTICLES

Miguel Toro [a], Jessika Rojas [a]
[a] Virginia Commonwealth University, Mechanical and Nuclear Engineer Department
Richmond, Virginia, United States

ABSTRACT

Lanthanum phosphate ($LaPO_4$) nanoparticles (NPs) with core and core-shell structure have shown potential as carriers of radionuclides, such as ^{225}Ac and ^{223}Ra, for targeted radiotherapy and imaging. However, their ability to retain both the parent and daughter isotopes may be influenced by the synthesis parameters. In this work, $LaPO_4$ NPs were prepared in aqueous solution using lanthanum nitrate and sodium tripolyphosphate precursors, at moderate temperature and purified by dialysis. Various synthesis procedures modifying the synthesis time, synthesis temperature, and purification procedure were designed and their influence on the size, shape, and stability of the NPs was evaluated through statistical analysis. The crystalline structure, morphology, and colloidal stability of the NPs were characterized with X-Ray diffraction (XRD), transmission electron microscopy (TEM), and zeta potential (ZP). The particle size was found to increase with both the addition of shells and the increment of either the synthesis temperature or time. Nanoparticles as small as 3.4 nm were obtained at 80 °C for 2 h while NPs synthesized at 90 °C for 3 h with the addition of 4 shells of $LaPO_4$ were 6.5 nm in diameter. The synthesis parameters did not show any effect on the crystalline structure nor on the colloidal stability. Similarly, the purification procedures did not have influence on the particle size, furthermore their effect on the colloidal stability was not statistically different.

INTRODUCTION

Materials at the nanoscale level have proven to exhibit remarkable properties not observed at the bulk level. This fact has brought the opportunity to utilize nanomaterials in applications where their specific properties lead to the development of new technologies or the enhancement of current ones. Metallic nanoparticles with various geometries, for instance, have shown numerous properties not obtained in the bulk [1,2]. Thus, they have been investigated and implemented in various areas such as energy, electronics, catalysis, and medicine, among others [3-6]. Nanomaterials have been studied and used in various medical fields such as radioimmunotherapy, target radiotherapy, and diagnostic imaging. For example, super paramagnetic iron-oxide NPs have shown potential as contrast agents for MRI and gold NPs have demonstrated unique size-dependent optical and photothermal properties that make them candidates as contrast agents [7]. Moreover, in the nuclear medicine field, inorganic NPs such as quantum dots, metallic, oxides, and phosphates have generated interest for theranostics as carriers for radioisotopes [8]. Radioactive NPs have been previously synthesized containing the radioactive material on the surface through a linker or as part of the structure as a dopant [9,10].

Lanthanide phosphate NPs have been studied for target alpha therapy (TAT) as carriers for radioisotopes that decay by the emission of multiple alpha particles such as ^{223}Ra, ^{225}Ac, and ^{225}Ra [10-13]. These NPs have shown resistance to radiation damage, size-shape dependent properties. In addition, their surface may be modified with bioconjugates to target specific organs [10,11,14]. Among the different lanthanides phosphates, lanthanum and gadolinium phosphate NPs have been tested and exhibited promising results retaining the radionuclides during in vitro and in vivo experiments [10–13]. Understanding the effect of the synthesis parameters on the NPs size, crystalline structure, and colloidal stability is crucial for the future application of this type of NPs in medicine. For instance, the size and shape of the NPs may influence the enhance permeability and retention effect (EPR) that is critical for the accumulation of the NPs in the tumor tissue [15]. Moreover, the type of crystalline structure of the NPs might influence the radiation resistance of radioactive NPs used in theranostics [16]. Furthermore, the colloidal stability is critical since NPs tend to aggregate increasing the average size and losing some of their characteristics at the nanoscale level [17].

This study aims to extend the knowledge of the influence of synthesis parameters such as synthesis temperature, synthesis time, and purification procedure on the characteristics of $LaPO_4$ NPs with core and core-shells structures. Lanthanum phosphate NPs were synthesized using an aqueous route at moderate temperature by adapting a procedure developed by Buissette et al [18]. The synthesis temperature, synthesis time, and purification procedure were varied in two different levels to obtain a factorial experiment 2^3. The levels for synthesis temperature and time were defined by tracking the growth kinetics of lanthanide phosphate ($LnPO_4$) NPs with Ultra Violet – Visible spectroscopy (UV-Vis) and the purification procedures were defined based on a preliminary qualitative study in which the stability and sedimentation of NPs was followed during 8 weeks. The influence of these parameters on the NPs was characterized using X-Ray diffraction (XRD), transmission electron microscopy (TEM), and zeta potential (ZP).

EXPERIMENTAL PROCEDURE

Materials

Lanthanum chloride heptahydrate ($LaCl_3$-$7H_2O$), terbium chloride hexahydrate ($TbCl_3$-$6H_2O$, 99.9% purity), cerium chloride heptahydrate ($CeCl_3$-$7H_2O$, 99.9% purity), and sodium tripolyphosphate (Na-TPP) were obtained from Sigma Aldrich. Lanthanum nitrate hexahydrate ($La(NO_3)_3$-$6H_2O$, 99.99% purity, REacton) was from Alfa Aesar. All the chemicals were analytical chemical grade and used without further purification. Deionized water (18 MΩ) was obtained from a MilliPore MilliQ water purification system. A 10 kDa molecular weight cutoff regenerated cellulose dialysis membrane from ®Spectra/Por was used for the purification of the colloidal suspension; the membrane was washed before being used to remove preservatives.

Sample preparation

Core, core + 2 shells, and core + 4 shells $LaPO_4$ NPs were synthesized following a procedure proposed by Buissette et al [18]. For the synthesis of core NPs, 0.1 M $La(NO_3)_3$-$6H_2O$ solution and 0.1 M Na-TPP, in a 1:1 volume ratio, were mixed and vortexed until a clear solution was obtained. Then, the solution was heated at the respective temperature for certain amount of time at continuous stirring. After synthesis, the turbid solution of core NPs was transferred to a dialysis membrane and dialyzed overnight against DI water. The purified solution was split into two equal parts for the later characterization of the core NPs and synthesis of the core-shells NPs.

Two purification procedures to remove unreacted species from both core and core + shells were implemented and evaluated. In the first purification procedure, dialysis was carried out only after the addition of the selected number of shells, i.e. no dialysis purification within the addition of each $LaPO_4$ shell. In this context, dialyzed core suspension was mixed with shells solution prepared with 0.1 M $La(NO_3)_3$-$6H_2O$ and 0.1 M Na-TPP in a volume ratio of 1:2 and vortexed for ~2 mins. Subsequently, the solution was heated at the selected temperature and time. Once the synthesis time had elapsed for the first shell, a second shell was added. This was done by adding shells solution to the ongoing core + 1 shell reaction at the selected temperature. The process continued for additional few more hours. For characterization purposes a portion of the core + 2 shells NPs was dialyzed overnight, while the rest was used for the synthesis of core + 3 shells NPs. Core + 3 shells and core + 4 shells $LaPO_4$ solutions were prepared following the same proportions and steps described above, i.e. adding shells solution to the ongoing core-shells reaction at the respective temperature and heated for few more hours. Finally, the core + 4 shells solution was dialyzed overnight after the synthesis. For the second purification procedure, similar solutions (chemicals and proportions) and synthesis steps were used for the core-shell structures. The procedure consisted on dialyzing the suspensions for a period of 10 h after the addition of each shell. Therefore, a dialyzed solution was used for the synthesis of each core-shell structure. These additional dialysis steps were implemented to reduce the concentration of unreacted species before the synthesis of each shell. Samples of core, core + 2 shells, and core + 4 shells NPs prepared by the two different purification procedures and combinations of times and temperatures were collected for characterization purposes. The ratio of La content in the core to that of the shells was 1:1.

Design of experiments (DOE)

In order to evaluate the influence of the synthesis parameters, time and temperature, on the characteristics of $LaPO_4$ NPs, two levels of temperature and time were selected. For this purpose a screening experiment was carried out using UV-Vis spectroscopy to determine the lower and upper parameters where crystalline lanthanide phosphate was obtained. $LnPO_4$ core NPs (40% $LaCl_3$-$7H_2O$, 45% $CeCl_3$-$7H_2O$, and 15% $TbCl_3$-$6H_2O$ in volume) were synthesized at temperatures of 75, 80, 85, and 90 °C. The upper limit parameters of 90 °C and 3 h were selected based on the procedure proposed by Buissette et al [19]. The growth kinetics was studied following the evolution of the UV-Vis absorption bands of cerium-trypolyphosphate free complexes and crystalline lanthanide phosphate on the absorption spectrum of the $LnPO_4$ NPs [19]. The spectra were collected for aliquots taken every half an hour throughout the synthesis of the core NPs. Finally, the two purification procedures were implemented and their effect on the NPs colloidal stability was evaluated.

Materials characterization

A PANalytical X'Pert Pro MPD X-ray diffractometer with a copper anode (Cu Kα, λ = 1.5401 Å) was used to investigate the crystalline structure of the NPs. The samples were dried in an oven under vacuum (~30 mm Hg) for 2 days at a constant temperature of 35 °C. Afterwards, the samples were ground until a fine white powder was obtained. The powder was deposited on a zero background plate and the measurements were carried out using the sample stage Reflection-Transmission Spinner PW3064/60 and rotatory speed of 0.5 rev/s. The X-ray tube was operated at a voltage of 45 kV and a current of 40 mA. Crystalline structure and crystallite size analysis were performed on X'Pert HighScore Plus software.

The morphology and particle size were studied by TEM with a Zeiss Libra 120 Plus operating at 120 kV. For TEM imaging, a solution of the purified and filtered NPs was diluted in acetone and sonicated for 20 min. The samples were prepared by immersing a formvar-carbon coated copper grid into the diluted suspension and allowed to dry at room temperature. Particle size analysis was carried out using the Image J 4.18v software. A Malvern Zen 3600 Zetasizer (Zetasizer nano) was used as a tool to evaluate the colloidal stability of the colloidal suspensions through zeta potential measurements. The samples were filtered with a 0.2 μm pore size filter and measured without further preparation steps.

RESULTS AND DISCUSSION

Synthesis of $LaPO_4$ NPs

The synthesis of $LaPO_4$ NPs in aqueous media is based on a precipitation reaction mechanism. Initially, a homogeneous solution containing La precursor and the complexing agent sodium tripolyphosphate is prepared in order to obtain La-TPP polymeric species. The complexing agent serves two main functions during the synthesis, it restricts the growth of the NPs and forms a stabilizing shell to avoid NPs aggregation. The subsequent thermal decomposition of the solution at moderate temperature promotes the hydrolysis of the tripolyphosphate to form pyrophosphate and orthophosphate ions. The interaction of the orthophosphate (PO_4^{3-}) and La^{3+} ions contribute to the formation of crystallites, while a fraction of the polyphosphate groups help to restrict the growth of the crystallites and provides stability to the NPs in solution [18]. Moreover, an increase in synthesis time results in an increment in the concentration of orthophosphate ions and a change in the solution appearance into a white suspension. Additionally, it promotes the aggregation of NPs since the orthophosphate ions increase at the expense of polyphosphate groups reducing the stabilizing shell around the NPs [18]. The addition of shells to the core NPs was proposed to replicate the core-shell structures studied elsewhere [10,11] and evaluate the effect of the synthesis parameters on these core-shells structures.

Growth kinetics of $LnPO_4$ NPs

The growth kinetics of $LnPO_4$ was carried out to follow the evolution of the UV-Vis absorption bands corresponding to cerium-tripolyphosphate free complexes (248 and 302 nm) and the one associated with the formation of crystalline lanthanide phosphate phase (272 nm). The UV-Vis spectra obtained for the core $LnPO_4$ NPs synthesized at various temperatures of 75, 80, 85, and 90 °C, and times of 1.5, 2, and 3 h are shown in Figure 1 a). The absorption spectrum showed at 0 h correspond to the mixture of the lanthanide salt and the surfactant before the synthesis. For this mixture, the absorption bands of cerium-tripolyphosphate free complexes located at 248 and 302 nm are observed, while the band at 272 nm is not present. Once the solution is heated at temperatures greater than 80 °C the vanishing of the cerium-tripolyphosphate bands and the growth of the band at 272 nm can be observed after 1.5 h of synthesis. For temperatures below 80 °C, the time required for the formation of the crystalline phase is not of interest for the study since the transition from the free complexes to the crystalline phase started after 3 h of synthesis. As it was stated before, variations in synthesis temperatures and times for the DOE might promote the synthesis of NPs with different sizes, i.e. NPs synthesized using the lower and upper parameters might have different sizes. As a consequence, the lower temperature and time selected for the experiment were 80 °C and 2 h respectively. In Figure 1 b) the absorption spectra of 80 and 90 °C for synthesis times of 0, 1, 2, and 3 hours are shown. This plot shows in detail the formation of the crystalline phase between the selected parameters. The combination of synthesis parameters used for the DOE are shown on Table I.

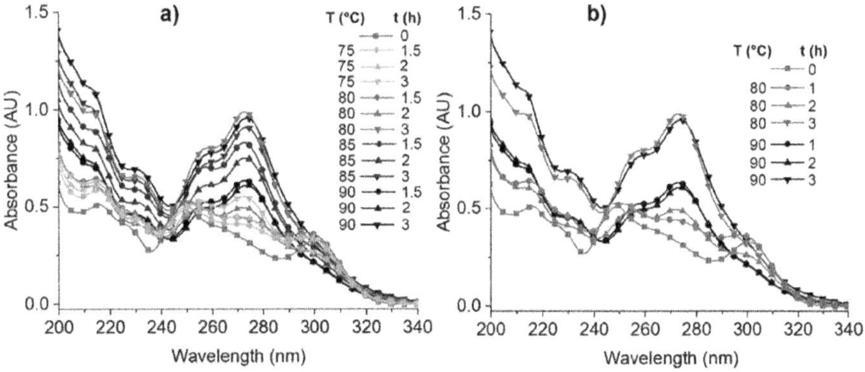

Figure 1. Absorption spectra for a) 75, 80, 85, and 90 °C synthesis temperatures and b) 80 and 90 °C at 3 synthesis times.

Table I. Combination of parameters defined by DOE.

Sample	Temperature (°C)	Time (h)	Purification procedure	
A	90	2	1	90-2-1
B			2	90-2-2
C	90	3	1	90-3-1
D			2	90-3-2
E	80	2	1	80-2-1
F			2	80-2-2
G	80	3	1	80-3-1
H			2	80-3-2

Crystalline structure and crystallite size

Core, core + 2 shells, and core + 4 shells lanthanum phosphate NPs for the different combinations of synthesis parameters were analyzed using XRD to evaluate the crystalline structure and crystallite size of the NPs. The XRD spectra for the core and core-shells NPs are shown in Figure 2 to 4 for the different parameters of temperatures and times. For the purification procedures there is no difference on the spectra, and hence on the crystalline structure and crystallite size. The broadening of the peaks, characteristic of nanocrystalline materials, is observed in all the spectra for the different samples. The spectra for the core NPs synthesized at the four combinations of parameters are shown in Figure 2. Comparing the spectra for the different parameters, it can be observed that for the lower temperature and time (80 °C and 2 h), the peaks are broader and less intense compare to the others. Moreover, the increase of temperature and time leads to sharper and well-defined peaks. A similar behavior can be observed on Figure 3 and 4 for core + 2 and core + 4 shells respectively.

The analysis of crystalline structure was done on the X'Pert HighScore Plus software using the ICDD database version number 2.0902. A crystalline structure of lanthanum phosphate with a hexagonal crystal system and a space group $P6_222$ (ICDD database Ref. code: 00-004-0635) was obtained for the core and core-shells structures synthesized at the different parameters combinations.

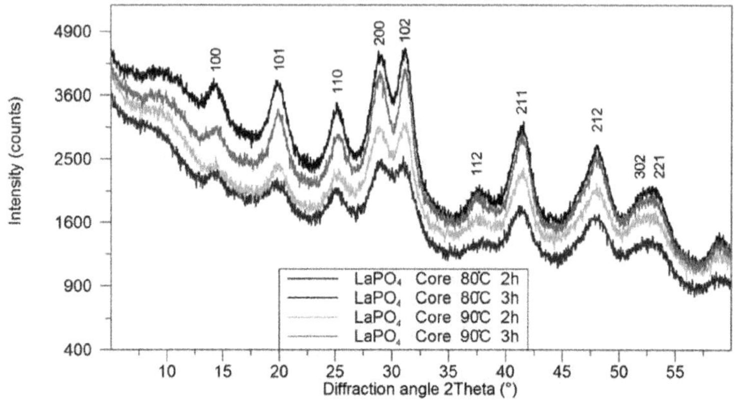

Figure 2. XRD spectra for $LaPO_4$ core NPs at the 4 combinations of synthesis temperature and time.

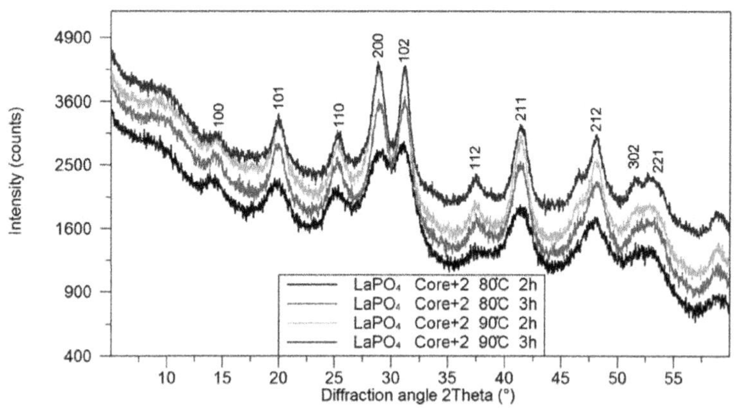

Figure 3. XRD spectra for $LaPO_4$ core + 2 shells NPs at the 4 combinations of synthesis temperature and time.

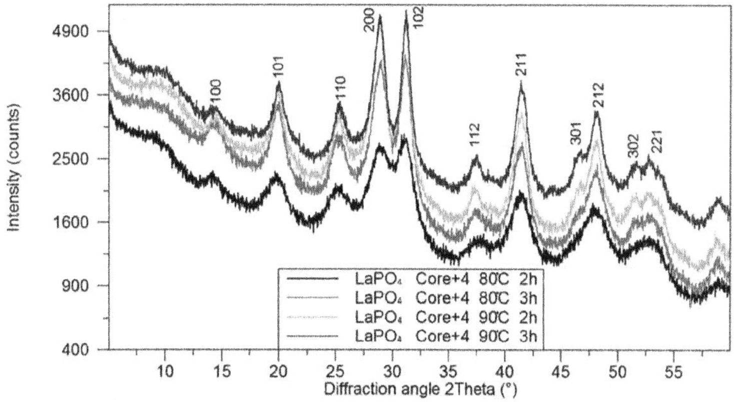

Figure 4. XRD spectra for $LaPO_4$ core + 4 shells NPs at the 4 combinations of synthesis temperature and time.

The XRD spectra were used to calculate the crystallite size using Scherrer equation [20]. For the crystallite size analysis, the peak located at ~ 41° was used. The results obtained for the different core and core-shells NPs at different synthesis parameters are shown on Figure 5. Comparing the values obtained for the core crystallite size for the parameters 80 °C - 3 h, 90 °C - 2 h, and 90 °C - 3 h there is not a statistical difference between them. Furthermore, an increase in the size of the NPs with the addition of shells for all the parameters was evidenced. Moreover, the increment of synthesis temperature or time has a direct influence with the growth of bigger NPs for the core and core-shells since the greatest values were obtained for 90 °C - 3 h.

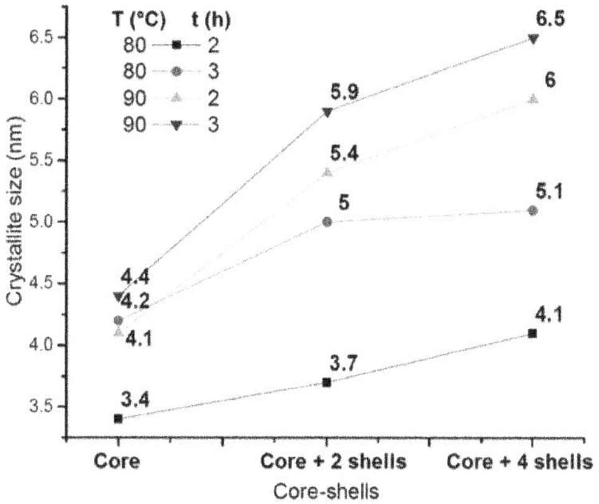

Figure 5. Crystallite size for core, core + 2 shells, and core + 4 shells NPs for different synthesis parameters.

Size, morphology, and distribution

TEM micrographs along with the particle size distributions of core and core + 4 shells synthesized at different synthesis temperatures and times are shown on Figures 6 to 8. The sample preparation on the formvar-carbon coated copper grid promotes the aggregation of NPs due to the surface dewetting and the "coffee-ring effect" as can be observed on Figures 6 and 7 [21]. The "coffee-ring effect" consists in the segregation of the NPs in the outer ring of the grid, increasing the density of NPs in this area and hence promoting aggregation. Although this makes the particle size measurement difficult, the contrast arising from electron diffraction in the crystalline NPs allows to distinguish and measure the size of individual NPs. In Figure 8 the larger presence of unreacted species from the phosphate precursor, built up through the addition of 4 shells to the core $LaPO_4$, becomes noticeable. The core and core-shells NPs can be approximated to a spherical morphology with an average diameter of 3.7 nm (± 0.6 nm), 3.5 nm (± 0.6 nm), and 4.9 nm (± 1 nm) for core 90 °C – 2 h, core + 4 shells 80 °C – 2 h, and core + 4 shells 90 °C – 3 h respectively. These results are slightly different to the ones obtained by XRD using Scherrer equation since XRD provides a statistical representation of the particle size due to the amount of sample that contributes to the diffraction, while TEM is a local measurement of the average NP diameter.

Figure 6. TEM micrograph and particle size distribution for core $LaPO_4$ synthesized at 90 °C for 2 hours.

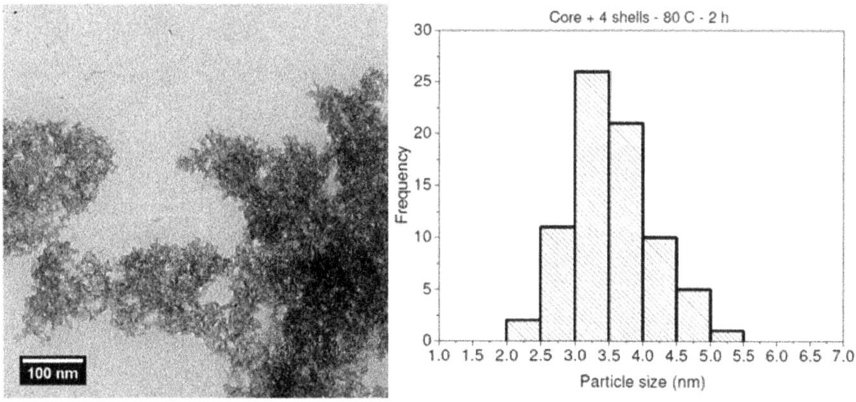

Figure 7. TEM micrograph and particle size distribution for core + 4 shells $LaPO_4$ synthesized at 80 °C for 2 hours.

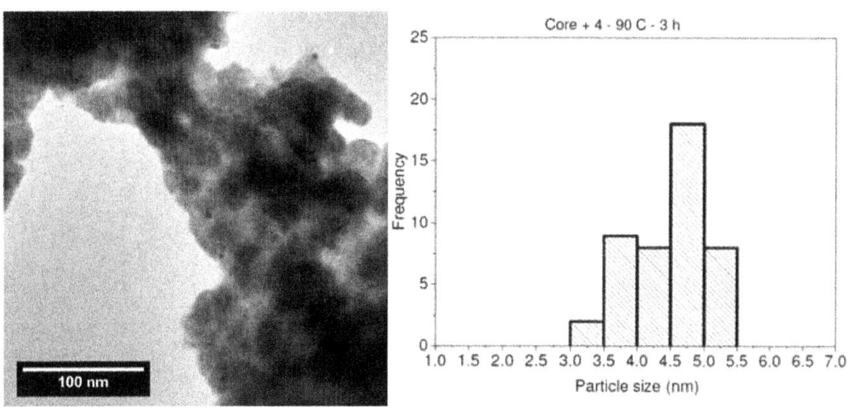

Figure 8. TEM micrograph and particle size distribution for core + 4 shells $LaPO_4$ synthesized at 90 °C for 3 hours.

Stability colloidal suspension

The colloidal behavior of the NPs in water was studied to determine the stability of all the different combination parameters using Zeta Potential. The pH of the various samples was measured prior to the zeta potential measurement giving a value of 6. Zeta potentials (ζ) were measured by laser Doppler velocimetry (Malvern Zen 3600 Zetasizer) using Henry's equation:

$$U_E = \frac{2\varepsilon\zeta f(ka)}{3\eta} \quad (1)$$

where ε is the dielectric constant of water, ζ is the zeta potential, *f*(ka) is the Henry's function, η is the medium viscosity, and U_E is the electrophoretic mobility. In Table II a summary of the average zeta potential and standard deviation for the different parameters is presented. According to Larsson et al., a suspension is considered stable when zeta potentials are greater than 30 mV or smaller than -30 mV [22]. Nonetheless, this assumption may not be accurate since NPs can be either monodisperse or in an aggregated state without significantly changing the magnitude of zeta potential [18]. Moreover, the NPs can loss their mechanical stability by showing sedimentation due to the difference in densities without the formation of aggregates. Besides, the magnitudes obtained are related with the conditions used to synthesize or prepare the NPs for the measurements, which restrict the possibility to compare results with the literature. For example, F. Meiser et al. studied the zeta potential of functionalize and bioconjugate lanthanide phosphate NPs as a function of pH obtaining a maximum value of ~ -40 mV [14]. Also, M. McLaughlin et al. reported the zeta potential of lanthanide phosphate NPs without surface modification, functionalized NPs, and bioconjugate NPs with a magnitude of -63.2, -56.4, and -27.9 mV respectively [11]. Therefore, two statistical tests were done to evaluate if the purification procedures affect the colloidal stability due to the influence of the concentration of unreacted species [23]. The ANOVA test and t-test revealed that the difference between the samples variance and means are not statistically different respectively.

Table II. Zeta potentials for different synthesis parameters and purification process for core, core + 2 shells, and core + 4 shells in a colloidal suspension with pH = 6.

		ZP (mV)	
		Purification process 1	Purification process 2
Core	80 °C - 2 h	-28.6 (±1.6)	
	80 °C - 3 h	-11.8 (±4.6)	
	90 °C - 2 h	-29.4 (±0.5)	
	90 °C - 3 h	-33.5 (±1.3)	
Core + 2 shells	80 °C - 2 h	-33.7 (±1.5)	-35.9 (±3.2)
	80 °C - 3 h	-34.4 (±1.5)	-35.9 (±1.5)
	90 °C - 2 h	-34.1 (±1.0)	-32.3 (±0.4)
	90 °C - 3 h	-34.4 (±0.7)	-32.4 (±1.2)
Core + 4 shells	80 °C - 2 h	-35.7 (±1.4)	-32.6 (±2.0)
	80 °C - 3 h	-32.2 (±1.3)	-32.6 (±0.8)
	90 °C - 2 h	-34.0 (±1.0)	-34.4 (±0.4)
	90 °C - 3 h	-34.1 (±1.4)	-33.0 (±1.2)

CONCLUSIONS

In this study, $LaPO_4$ core and core-shells NPs were synthesized with the aim to evaluate the influence of synthesis and purification parameters on the size, morphology, crystalline structure, and colloidal stability. The synthesis parameters, temperature and time, have a direct effect on the particle size, which increase with the increment of the temperature and/or time according to the hydrolysis reaction of tripolyphosphate groups to form orthophosphate ions.

Furthermore, there is no effect of these parameters on the crystalline structure and colloidal stability for the proposed combinations since a hexagonal crystalline structure of lanthanum phosphate was obtained for all the core and core-shell structures and the zeta potential is not statistically different among one another based on an ANOVA test. Finally, the proposed purification procedures did not show to have any influence in the morphology of the nanoparticles. Moreover, the difference on the zeta potential among the samples is not statistically significant. Therefore, the proposed purification procedures do not evidenced any effect in the colloidal stability of the NPs. However, for bioconjugation and in vivo studies the reduction of unreacted species may play an important role.

ACKNOWLEDGEMENTS
This work is being financed by the Virginia Commonwealth University with the support of the Mechanical and Nuclear Engineering department and the NRC-HQ-84-14-FOA-002, Faculty Development Program in Radiation Detection and Health Physics at VCU.

REFERENCES

1. L. Wang and Y. Yamauchi, Metallic nanocages: synthesis of bimetallic Pt-Pd hollow nanoparticles with dendritic shells by selective chemical etching, *J. Am. Chem. Soc.* **135,** 16762–5 (2013).

2. M. Mahmoud, D. O'Neil, and M. El-Sayed, Hollow and solid metallic nanoparticles in sensing and in nanocatalysis. *Chem. Mater.* **26,** 44–58 (2014).

3. A. Mesbahi, A review on gold nanoparticles radiosensitization effect in radiation therapy of cancer. *Reports Pract. Oncol. Radiother.* **15,** 176–180 (2010).

4. G. Cao, Nanostructures and Nanomaterials - Synthesis, Properties and Applications. *Imperial College Press* (2004).

5. H. Xu, L. Zeng, S. Xing, G. Shi, Y. Xian, and L. Jin, Microwave-radiated synthesis of gold nanoparticles/carbon nanotubes composites and its application to voltammetric detection of trace mercury(II). *Electrochem. commun.* **10,** 1839–1843 (2008).

6. K. Okuyama, W. Lenggoro, and T. Iwaki, Nanoparticle Preparation and Its Application - A Nanotechnology Particle Project in Japan. *2004 Int. Conf. MEMS, NANO Smart Syst.* 1–4 (2004).

7. H.-C. Huang, S. Barua, G. Sharma, S. K. Dey, and K. Rege, Inorganic nanoparticles for cancer imaging and therapy. *J. Control. Release* **155,** 344–357 (2011).

8. S. S. Kelkar and T. M. Reineke, Theranostics: combining imaging and therapy. *Bioconjug.*

Chem. **22,** 1879–903 (2011).

9. M. F. McLaughlin, J. Woodward, R. A. Boll, A. J. Rondinone, S. Mirzadeh, and J. D. Robertson, Gold-coated lanthanide phosphate nanoparticles for an 225 Ac in vivo, *Radiochim,* **101**, 595-600 (2013).

10. Woodward, J. *et al.* LaPO4 nanoparticles doped with actinium-225 that partially sequester daughter radionuclides, *Bioconjug. Chem.* **22,** 766–776 (2011).

11. M. F. McLaughlin, J. Woodward, R. A. Boll, J. S. Wall, A. J. Rondinone, S. J. Kennel, S. Mirzadeh, and J. D. Robertson, Gold Coated Lanthanide Phosphate Nanoparticles for Targeted Alpha Generator Radiotherapy, *PLoS One* **8,** 2–9 (2013).

12. J. V. Rojas, J. D. Woodward, N. Chen, A. J. Rondinone, C. H. Castano, and S. Mirzadeh, Synthesis and characterization of lanthanum phosphate nanoparticles as carriers for 223Ra and 225Ra for targeted alpha therapy, *Nucl. Med. Biol.* **42,** 614–620 (2015).

13. M. F. McLaughlin, D. Robertson, P. H. Pevsner, J. S. Wall, S. Mirzadeh, and S. J. Kennel, LnPO4 nanoparticles doped with Ac-225 and sequestered daughters for targeted alpha therapy, *Cancer Biother. Radiopharm.* **29,** 34–41 (2014).

14. F. Meiser, C. Cortez, and F. Caruso, Biofunctionalization of fluorescent rare-earth-doped lanthanum phosphate colloidal nanoparticles, *Angew. Chemie - Int. Ed.* **43,** 5954–5957 (2004).

15. C. Heneweer, J. P. Holland, V. Divilov, S. Carlin, and J. S. Lewis, Magnitude of enhanced permeability and retention effect in tumors with different phenotypes: 89Zr-albumin as a model system, *J. Nucl. Med.* **52,** 625–633 (2011).

16. A. E. Grechanovsky, N. N. Eremin, and V. S. Urusov, Radiation Resistance of LaPO4 (Monazite Structure) and YbPO4 (Zircon Structure) from Data of Computer Simulation, *Physics of the solid state,* **55,** 1929–1935 (2013).

17. F. Gambinossi, S. E. Mylon, and J. K. Ferri, Aggregation kinetics and colloidal stability of functionalized nanoparticles, *Adv. Colloid Interface Sci,* (2014).

18. V. Buissette, D. Giaume, T. Gacoin, and J.-P. Boilot, Aqueous routes to lanthanide-doped oxide nanophosphors, *J. Mater. Chem,* **16,** 529 (2006).

19. V. Buissette, M. Moreau, T. Gacoin, J.-P. Boilot, J.-Y. Chane-Ching, and T. Le Mercier, Colloidal synthesis of luminescent rhabdophane LaPO4:Ln 3+·xH2O (Ln = Ce, Tb, Eu; x

≈ 0.7) nanocrystals, *Chem. Mater,* **16,** 3767–3773 (2004).

20. B. D. Cullity and R. S. Stock, *Elements of X-Ray Diffraction.* (Addison-Wesley Publishing Company, 1956).

21. B. Michen, C. Geers, D. Vanhecke, C. Endes, B. Rothen-Rutishauser, S. Balog, and A. Petri-Fink, Avoiding drying-artifacts in transmission electron microscopy: Characterizing the size and colloidal state of nanoparticles, *Sci. Rep.,* **5,** 9793 (2015).

22. M. Larsson, J. Duffy, and A. Hill,n Suspension stability : Why particle size , zeta potential and rheology are important. *Www.Malvern.Com* **20,** 1–35 (2011).

23. Instruments, B. & Technical, C. Colloidal Stability in Aqueous Suspensions : Two Variables. **1,** 1–4 (1896).

ZINC OXIDE NANOPARTICLES FOR SPACE SATELLITE SOLAR PANEL PROTECTION LAYER

Phillip Clift*, Jordan Wladyka, Tyler Payton, and Dale Henneke**
New Mexico Institute of Mining and Technology
Socorro, New Mexico, USA

ABSTRACT

Space solar panels have a protective layer (i.e., a flexible, cover-glass material), which protects the panel from solar radiation. Unfortunately, current formulations of this layer are inadequate due to an accumulation of electrical charge over time, which causes the protective layer to decay and reduces its effectiveness. This work describes a synthesis method for zinc oxide (ZnO) nanoparticles, their characterization, and incorporation into a novel composite material that could serves as a replacement for the current cover glass technology. The ZnO nanoparticles provide a conduit within the composite for the electrical charge to dissipate. It was found that composites having as little as 2wt% ZnO nanoparticle provided greater than 10,000× improvement in the electrical conductivity of the protective layer while reducing the optical transmission of usable light by approximately 10%.

INTRODUCTION

State of the art solar panels that are used in space applications are rolled or folded into a pre-launch configuration, which is unfurled once the satellite reaches orbit. These solar panels have a protective layer that is composed of a composite material containing a silicone-based matrix with glass bead filler; this composite is called pseudomorphic glass (PMG). Figure 1 depicts a cross-section of a typical solar panel using a PMG protective layer and substrate.

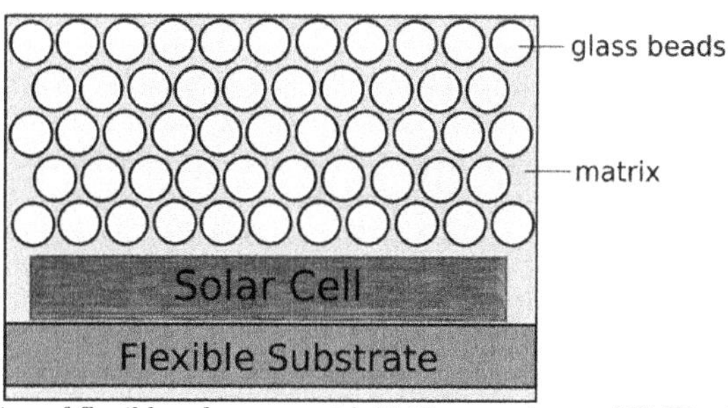

Figure 1. Conceptual design of flexible solar array with PMG coverglass and PMG substrate for hybrid configuration.

PMG Formulations

The PMG provides protection against space radiation (i.e., cosmic radiation), it is flexible, will not fracture when rolled, and it is transparent to light that is converted to electricity by the solar cell (i.e., EM radiation on the order of 200 to 2500 nm). The layer must be optically transparent over this range of wavelengths or the solar array will not function. The traditional PMG layer accumulates electrical charge, which leads to degradation overtime. To mitigate the accumulation of electrical charge, the layer should be electrically conductive but must also retain

1 Author e-mail address: pclift@nmt.edu
2 Corresponding author. E-mail: henneke@nmt.edu

its mechanical flexibility and optical transparency.

Wilt and Levin[1,2] improved the electrical conductivity of PMG by incorporating aluminum-zinc oxide (AZO) and indium-tin oxide (ITO) into the PMG layer. Multiple samples were made by loading a variety of weight fractions of AZO and ITO into the composite; the loading varied from 0.001 to 0.5wt%. At 0.1wt% AZO loading, the electrical conductivity was found to be around 9×10^{-13} S/cm at 500V, while the optical transparency compared to unloaded PMG was found to be 85% at a wavelength of 500nm. For a loading of 0.1wt% ITO, the electrical conductivity was found 2.8×10^{-13} S/cm at 500V, while the optical transparency at 500nm was reduced to 90%. It was shown that electrical conductivity can be improved by incorporating electrically conductive materials into the PMG layer, but there is a loss of optical transparency even when using materials that are traditionally considered to be transparent.

Zinc oxide could be used as an alternative to AZO and ITO composite formulations due to its optical transparency over the relevant wavelength range and its electrical conductivity. ZnO has a band gap around 3.37 eV at 300K [2] and is commonly used as a transparent thin film electrode layer for amorphous Si solar cells and varistors[3,4]. Nanoparticles of ZnO can be made with a variety of morphologies that would effectively dissipate accumulated charge in a PMG layer.

Zinc Oxide Agglomerates

Swati *et al.*,[5] produced ZnO agglomerates (ZnO-A) and characterized their chemical and physical properties with both organic or inorganic precursors. ZnO-A were produced using a variety of organic and inorganic precursors (e.g., zinc acetate, zinc nitrate, and zinc chloride). The precursors were mixed together and placed into a pre-heated furnace at ~900°C in a pure oxygen environment until the reaction was complete. After synthesis, the ZnO-A powder was studied using differential scanning calorimetry (DSC), X-ray diffraction (XRD), scanning electron microscopy (SEM), and photoluminescence emission (PLE).

From the DSC data, it was determined that the optimal temperature for this sol-gel reaction was ~900°C, due to the overall weight loss reaching a maximum of 92.4% from thermal decomposition. This temperature represents the highest yield for ZnO-A that can be formed using this method. XRD was used to verify the crystallite size, as well as, impurities in the ZnO-A. It was evident from the XRD data that the use of organic precursors yielded the smallest crystallite size (~13.45nm) when compared to other inorganic precursors. PLE was used to determine the optical properties of the ZNO-A. The emission spectra intensities increased and shifted from 2.28eV to 2.82eV, from green to blue emission of the visible light spectra, as crystallite size was reduced. SEM provided visual evidence of the surface morphologies for the ZnO-A synthesized from various precursors (i.e., zinc acetate, zinc chloride, zinc nitrate). The ZnO-A, with zinc acetate and zinc nitrate precursors, possessed similar morphologies with primary and secondary agglomerates attached to one another. Whereas, the zinc chloride was found to have primary and secondary particles, in the form of platelets, stacked on one another.

Zinc Oxide Tetrapods

Wang *et al.*,[6] synthesized and characterized ZnO tetrapods (ZnO-T) to better understand the formation mechanism of ZnO nanocrystals. ZnO-T have a morphology that is a polyhedral caltrop with a zinc blende core crystal structure and wurtzite crystals protruding from the [111] faces. The wurtzite legs are separated at an angle of 120° with respect to each of their positions. Wang *et al.* made the ZnO-T by placing zinc powder in quartz boat and heating it in a tube furnace to ~850°C in an oxygen / argon environment. The gas was set to flow through the quartz tube at rate of 10 sccm for O_2 and 100 sccm for Ar. As the zinc vaporized, it reacted with oxygen in the tube furnace forming ZnO-T, which was collected onto a silicon substrate that was placed above the zinc powder. The collected material was examined by XRD and TEM.

XRD data verified the composition of the ZnO-T showing with three major peaks at 32°, 34°, and 36° (2θ) representing the (100), (002), and (101) planes for the wurtzite structure, respectively. Selected area diffraction (SAD) was done using TEM, which showed evidence of a wurtzite crystal structure for the ZnO-T legs with a growth direction of the legs in the {0001} direction with a lattice parameter of 0.259 nm. There was evidence that the oxygen content played a major role in the growth of ZnO-T. Higher concentrations of oxygen produced longer the legs on the ZnO-T, while the growth rates that were observed corresponded to kinetically controlled growth.

EXPERIMENTAL PROCEDURE

Zinc oxide agglomerates were synthesized using a sol-gel process adapted from Swati *et al.* [7]. The reagents that were used, both of which were obtained from Alfa Aesar, are: ACS grade zinc acetate and ACS grade citric acid. An aqueous solution, having a molar ratio of 1:1:10 zinc acetate dehydrate to citric acid to water, was mixed in a beaker. 40mL of this solution was poured into a quartz boat, placed into a tube furnace, and heated to 900°C for 30 minutes. The material that was removed from the boat was used as the ZnO-A material for composite testing.

Zinc oxide tetrapods were synthesized using an adaption of the method described by Wang *et al.* [6]. The experimental setup (see Figure 2) permitted control of the reaction temperature, ambient gas flowrate, and reaction pressure. ACS grade Zinc Powder (99.9%) having particles ranging in size from 1 to 5 μm was obtained from Atlantic Equipment Engineers Inc. The Zn powder was placed into a quartz boat and heated to ~900°C at a pressure of 1 atm with a gas mixture having $5\%_{vol}$ oxygen / $95\%_{vol}$ argon flowing at a rate of 100 sccm.

After synthesis both of the ZnO products were characterized to determine particle size and morphology. SEM was used to determine the morphology of the particles as well as relative particle size. XRD was performed to confirm the crystal structure of both nanocrystalline materials and to confirm the purity of the final product after synthesis.

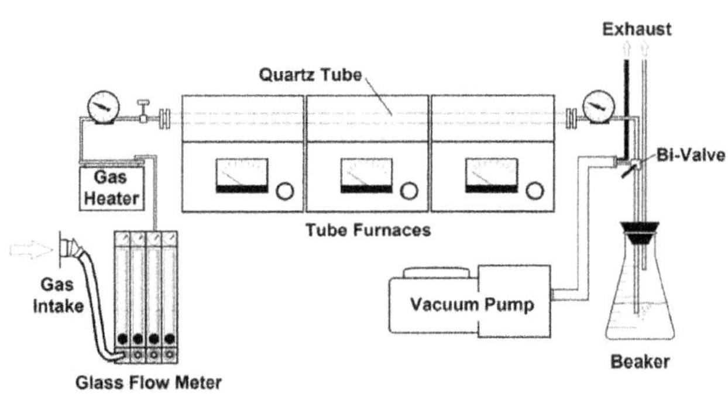

Figure 2. Schematic of tube furnace and synthesis apparatus.

The synthesized ZnO-A and ZnO-T materials incorporated into the PMG composite in various weight percentages. The PMG composite was made using a two-part space grade silicone resin from Dow-Corning (DC93-500) mixed with fused silica beads ranging in size from 1-20μm. The synthesized zinc oxide materials were mixed with the resin and beads, then tape cast onto a glass substrate. The casts were made 0.127mm thick and the composite material was left to cure for 24 hours. The cast composite material was then analyzed using UV/VIS spectrometry to determine it transmissivity and with a conventional four-point resistivity probe tester to find its conductivity.

RESULTS AND DISCUSSION

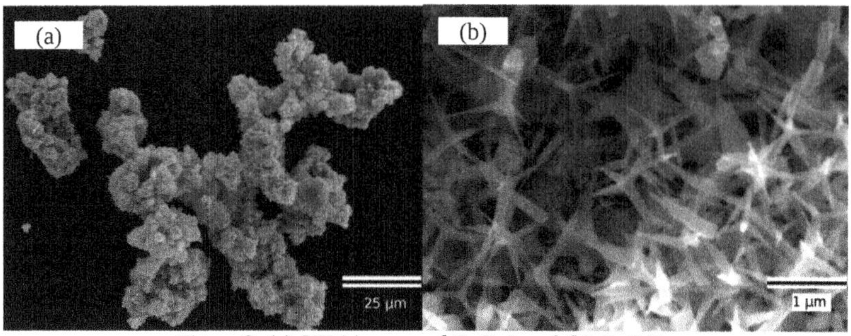

Figure 3. (a) SEM images of ZnO-A and (b) ZnO-T.

The morphologies of the zinc oxide agglomerates (ZnO-A) and tetrapods (ZnO-T) were analyzed by SEM; Figure 3 shows images of the resulting materials. Figure 3a depicts the ZnO-A material collected after synthesis, while Figure 3b shows the ZnO-T collected on a silicon wafer placed ~5 cm downstream from the quartz boat. In figure 3a, the resulting ZnO-A material was found to be similar to the material made by Swati *et al.* with the primary and secondary spherical particles agglomerated in the same fashion. The synthesized ZnO-T material (shown in figure 3b) is also quite similar to what was made by Wang *et al.* and it possesses the same polyhedron caltrop geometry. These ZnO materials were then mixed by milling with fused silica beads and reanalyzed by SEM, as shown in Figure 4.

Figures 4. (a) SEM images of 2wt% ZnO-A with fused silica beads, and (b) 2wt% ZnO-T with fused silica beads.

The ZnO-A and fused silica bead blended at 2wt% can be seen in figure 4a; the overall size of the 1-20 µm beads can be seen in contrast with the ZnO-A material. Figure 4b shows the same weight percent blend with ZnO-T and glass beads. The ZnO-T is clearly visible and seems to adhere to the larger beads, but not the smaller beads.

The ZnO materials and fused silica were mixed with the PMG composite and tape cast to form 5 mil thick sheets. Four tape cast composites were made, two with ZnO-A and two with

ZnO-T. Each pair of tape casts, with respective ZnO nanoparticles, was incorporated with either 2wt% or 6wt% of the ZnO materials. After curing the samples, they were sectioned into smaller specimens and analyzed using UV/Vis spectrometry and four-point probe conductivity testing.

Figure 5 depicts the transmissivity of the tape-cast composite materials. The transmittance percentage for the ZnO-A 2wt% and 6wt% at 500 nm are: 86.5% and 81.6%; respectively. For the ZnO-T, the transmittance for the ZnO-T at 2wt% and 6wt% at 500 nm are: 84.9% and 77.4%; respectively. The slight deviations in the UV/Vis spectra between the 2wt% and the 6wt%, of the ZnO-A and ZnO-T, may be attributed to the agglomeration of ZnO materials during the tape casting process.

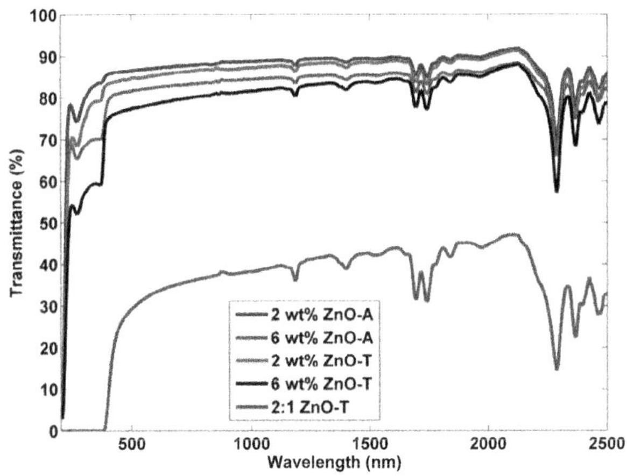

Figure 5. UV/Vis spectra of composite with various loadings of ZnO nanocrystals.

Table I. Resistance, conductivity, and transmissivity at 500 nm for ZnO-A and ZnO-T PMG composites at different loadings.

Sample	Thickness [mm]	Resistance [ohm/sq]	Conductivity [S/cm]	Transmissivity [%] (at 500nm)
2wt% ZnO-A	0.12	1.14×10^{8}	7.30×10^{-7}	86.5
6wt% ZnO-A	0.12	7.44×10^{7}	1.12×10^{-6}	81.6
2wt% ZnO-T	0.11	1.50×10^{8}	6.06×10^{-7}	84.9
6wt% ZnO-T	0.12	6.12×10^{7}	1.36×10^{-6}	77.4

The conductivity of the composite material as determined by four-point probe measurement (without a bias voltage applied to the sample), can be seen in Table I. The conductivity of ZnO-A composite was found to increase by 21.1% as the ZnO loading increased from 2wt% to 6wt%. For the ZnO-T composites, an increase of 38.3% for the same change in ZnO loading was observed. When comparing the two forms of ZnO (ZnO-A to ZnO-T) at 2wt%, a 9.3% increase in the conductivity was found. At 6wt%, the increase in conductivity was found to be 9.7% for the tetrapods when compared to the agglomerates.

Figure 6 compares the conductivity (S/cm) of various composite materials as a function of conductive filler loading within the composite (wt%). The data from Table I is plotted in Figure 6 along with data from Wilt *et al.* The four data points corresponding to the ZnO-A and ZnO-T composites appear in the top-center and top-right corner of the plot. It is clear that these samples have orders of magnitude improvement in the conductivity while having a similar loss in the optical transmissivity.

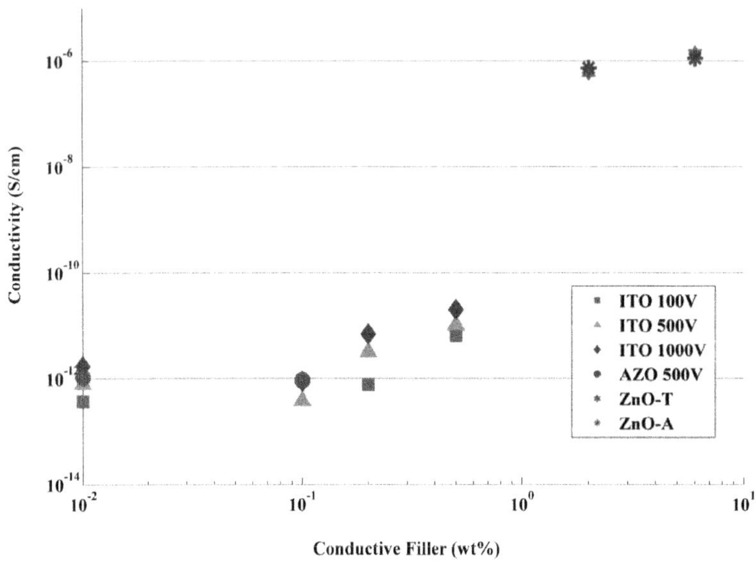

Figure 6. Conductivity plot of ZnO composite materials. ITO and AZO composite data is from Wilt *et al.* [1], these composites were tested with and without an induced voltage applied to the sample. The ZnO-A and ZnO-T composites were only tested without an applied voltage.

It is possible that a percolation network within the composite may have permitted a complete circuit from one ZnO site to another, which would increase the conductivity of the composite while maintaining much of its optical transparency. This is especially true since the optical transparency of bulk ZnO is substantially less than that of nanometer sized ZnO particles in small dispersed quantities. However, it is still possible to produce 1D thin films of ZnO and have the material possess full transparency [15].

CONCLUSION

ZnO-A and ZnO-T composites had higher conductivities values (~µS/cm) compared to the ITO and AZO composites made by Wilt *et al.* (~pS/cm) [1]. The optical transmissivity of the ZnO composites was similar to that of ITO and AZO, but at a much higher material loading. ZnO-A and ZnO-T composites are clearly a better solution than either ITO or AZO for replacing current coverglass materials.

These results indicate ZnO nanomaterials outperform both AZO and ITO, but an optimal loading amount still needs to be determined, as well as, a method to uniformly disperse the ZnO materials within the PMG. This data shows significant gains in transmittance and conductance when compared to ITO and AZO at the same level of loading.

The data may be indicative of a percolation conduction where the n-type ZnO is behaving as the conduction network through the polymer composite matrix. It should also be noted with the intrinsic defects of ZnO, the material will be dominated by oxygen vacancies. However, since ZnO-T was synthesized in 5% oxygen and the remainder was Zn and Argon vapor. It is possible that the ZnO-T may have extra Zn atoms at oxygen vacancy sites. This may increase the concentration of available electrons, but reduces hole concentration and mobility. In other literature, it has been shown that ZnO used as a conductive network in a polymer matrix does increase conductivity for thin film applications, as described by Oosterhout *et al.*[13]. Hong *et al.*[14] describe the conduction of a charge, via a percolation network of micron sized to nano-sized particles, is mainly attributed to the tunneling effect.

More research and testing is in progress to verify the data and observations made thus far.

ACKNOWLEDGEMENTS

We would like to extend a special thanks to Dave Wilt, of the Kirtland Air Force Research Lab, for supplying the some of the raw materials; and for allowing us to use the characterization equipment for this research. Without his help none of this would have been possible.

REFERENCES

[1] Z. S. Levin, D. Wilt, R. Hoffman, and D. Ferguson, Conductive Solar Cell Cover Glass Replacement For Electric Propulsion Environments, *Phys, Sim, and Photon. Eng. Photo. Devices.,* **8981**, 1-5 (2014).

[2] D. Wilt, A. Howard, N. Snyder, T. Sahlstrom, N. Heersema, L. Nathan, T. Ohshima, S. Sato, and M. Imaizumi. PseudoMorphic Glass to enable high efficiency space photovoltaic devices. *PVSC,* **37th IEEE**, 001949-54 (2011).

[3] B. Rech and H. Wagner, Potential of amorphous silicon for solar cells, *Appl. Phys. A,* **69**, 155-167 (1999).

[4] K. Ellmer, A. Klein, and B. Rech, Transparent Conductive Zinc Oxide: basics and Applications in Thin Film Solar Cells, Springer Series, **104**, 70 (2008).

[5] G. Swati, S. Mishra, D. Yadav, R.K. Sharma, D. Dwivedi, N. Vijayan, J.S. Tawale, V. Shanker, and D. Haranath, High yield synthesis and characterization of aqueous zinc oxide nano-crystals using various precursors, *J. Alloys and Cmpnds*, **571**, 1-5 (2013).

[6] F.Z. Wang, Z.Z. Ye, D.W. Ma, L.P. Zhu, and F. Zhuge, Novel morphologies of ZnO nanotetrapods, *Mater. Lett.,* **59**, 560 – 563 (2005).

[7] M. Kitano, T. Hamabe, and S. Maeda, Growth of large tetrapod-like ZnO Crystals I: Experimental considerations on kinetics of growth. *J. Crystal Growth*, **102**, 965-973 (1990).

[8] M. Kitano, T. Hamabe, S. Maeda, and T. Okabe, Growth of large tetrapod-like ZnO crystals II: Morphological considerations on growth mechanism. *J. Crystal Growth*, **108**, 277-284 (1991).

[9] M.C. Newton, S. Ferth, T. Matsuura, and P. A. Warburton, Synthesis and characterization of zinc oxide tetrapod nanocrystals. *J. Phys.*, **26**, 251-255 (2006).

[10] N.A. Samat, R. M. Nor, Sol–gel synthesis of zinc oxide nanoparticles using Citrus aurantifolia extracts. *Ceramics Intl.*, **39**, 545-548 (2013).

[11] Q. Wan, K. Yu, T.H. Wang, and C.L. Lin, Low-field electron emission from tetrapod-like ZnO nanostructures synthesized by rapid evaporation. *Appl. Phys. Lett.*, **83-11**, 2253-2255 (2003).

[12] Z. Zhang, H. Yuan, Large-scale synthesis and optical behaviors of ZnO tetrapods. *Appl. Phys. Lett.,* **90**, 1-3 (2007).

[13] S.D. Oosterhout, L. Koster, S.S. van Bavel, J. Loos, O. Stenzel, R. Thiedmann, V. Schmidt, B.

Campo, T.J. Cleij, L. Lutzen, and D. Vanderzande, Controlling the morphology and efficiency of hybrid ZnO: polythiophene solar cells via side chain functionalization. *Advanced Energy Materials*, **1(1)**, 90-96 (2011).
[14] J.I. Hong, L.S. Schadler, R.W. Siegel, and E. Mårtensson, Rescaled electrical properties of ZnO/low density polyethylene nanocomposites. *Applied physics letters*, **82(12)**, 1956-1958 (2003).
[15] E.M. Fortunato, P.M. Barquinha, A.C.M.B.G. Pimentel, A.M. Gonçalves, A.J. Marques, L.M. Pereira, and R.F. Martins, Fully Transparent ZnO Thin-Film Transistor Produced at Room Temperature. *Advanced Materials*, **17(5)**, 590-594, (2005).

Dielectronic Materials and Electronic Devices

SYNTHESIS AND CHARACTERIZATION OF $BaTiO_3$-BASED CERAMICS DOPED IN B SITE BY $BaTi_{1-x}Nb_xO_3$

F. R. Barrientos-Hernández[1], M. Ortiz-Domínguez[2], M. Pérez-Labra[1], E. O. Ávila-Dávila[3], J. P. Hernández-Lara[1], L. A. Cruz-Gutiérrez[1]

[1] Universidad Autónoma del Estado de Hidalgo, AACTyM, Carretera Pachuca–Tulancingo km. 4.5, C.P. 42184, Mineral de la Reforma, Hidalgo, México.
[2] Universidad Autónoma del Estado de Hidalgo, Campus Sahagún, Carretera Cd. Sahagún-Otumba s/n, Hidalgo, México.
[3] Instituto Tecnológico de Pachuca, Departamento de Ingeniería Industrial, Carretera México-Pachuca Km. 87.5, C.P. 42080, Pachuca, Hidalgo, México.

ABSTRACT

In this work, the substitution of Nb^{5+} ions into the B site of $BaTiO_3$ ceramics by mechanism $BaTi_{1-x}Nb_xO_3$ (x= 0.001, 0.003, 0.005, 0.04, 0.08, 0.20, and 0.25) have been investigated regarding their microstructural and dielectric properties. The ceramics powders were successfully prepared by a conventional solid-state synthesis in air atmosphere. The main objective was to study the effect of the addition of Nb^{5+} on the structure and electrical properties of barium titanate. The phases formed, the crystal structure and the microstructure of the samples were characterized using X-ray diffraction (XRD), raman spectroscopy and scanning electron microscopy (SEM). The modified $BaTiO_3$ dielectric ceramics prepared by conventional solid-state route in air atmosphere exhibit single perovskite structure and improved electrical properties; the results shows that the addition of a small amount of Nb^{5+} gradually changes the crystal structure from a typical tetragonal-pseudocubic structure. The results of permittivity measurements shows the difficulty in obtaining, electrically homogeneous samples by solid-state method. A secondary phase ($Ba_8Ti_3Nb_4O_{24}$) start to appear as the doping content of $Nb^{5+} \geq 0.25$, indicating that the solubility limit of Nb^{5+} in $BaTiO_3$ reached. Pellets derived of the samples with x= 0.005 and sintered at 1500 °C exhibit optimum properties of dielectric constant decreasing to 90°C curie point. The effects of dielectric and ferroelectric properties of samples are also investigated.

INTRODUCTION

In recent decades, there has been great interest in the electroceramic materials because of its dielectric, ferroelectric and piezoelectric properties; as these are vital in the communication industries, energy, electronics, optics and automation, among others. Within the electroceramic materials, there are ferroelectric ceramic, they use properties that result from ferroelectricity as electro-optical, dielectric, piezoelectric and pyroelectric. The ferroelectricity is a phenomenon which was discovered in 1921 by Valasek[1]; this phenomenon also has been called Seignette electricity or Rochelle salt (potassium sodium tartrate) this salt was the first material in which the ferroelectric properties such as spontaneous polarization below the Curie point, the presence of ferroelectric domains which cause polarization and hysteresis curves were discovered. Ferroelectricity defined as the reversibility of the spontaneous polarization under application of an applied electric field[2,3,4]; the barium titanate ($BaTiO_3$), is one of the most widely used ferroelectric ceramic material with perosvkite-type structure (ABO_3). The $BaTiO_3$ presents several phase transitions, above its Curie point (approximately 130 °C) the unit cell is cubic. Below the Curie point the structure is slightly distorted to the tetragonal form with a dipole moment along the c direction. Other transformations occur at temperatures close to 0 °C and -80 °C: below 0 °C the unit cell is orthorhombic with the polar axis parallel to a face diagonal and below -80 °C it is rhombohedral with the polar axis along a body diagonal[5]. The conventional method for synthesizing $BaTiO_3$ based on the solid-state method between TiO_2 and $BaCO_3$ at high temperatures (>1400°C)[6], the reaction is as follows:

$$BaCO_3 + TiO_2 \xrightarrow{\Delta} BaTiO_3 + CO_2 \uparrow \tag{1}$$

It has been shown that adding elements donors or acceptors in the $BaTiO_3$ perovskite-type structure considerably improves the properties of this material[7-13]. In recent decades several methods have been used for the synthesis of ferroelectric $BaTiO_3$ ceramics, such as solid-state synthesis[14], sol-gel[15,16], spark plasma sintering[17] hydrothermal[18,19]. The addition of pentavalent cation Nb^{5+} which replaces A or B site in perovskite $BaTiO_3$ structure, according to Hume-Rothery rules; modifies the microstructure and electrical properties of $BaTiO_3$ ceramics. In previous studies some authors studied the incorporation of foreigners ions into the perovskite structure of barium titanate[20], for example, V. Paunovic[21] et al. reported the microstructure and dielectric properties of Dy/Mn doped $BaTiO_3$ ceramics; the samples were prepared by the conventional solid state reaction and sintered at 1290 °C and 1350 °C in air atmosphere for 2h. They found the highest dielectric constant of 5600 at room temperature and greatest change of Tc (εr=9130) were measured in 0.1 at% Dy/$BaTiO_3$, sintered at 1350 °C. On the other hand, G. Schileo[22] et al. investigated the relation between structure/property in $BaTiO_3$ ceramics with doping simultaneous (Bi-Gd); the ceramics were prepared by solid state reaction method. They found a maximum of solubility of Bi/Gd in $BaTiO_3$ at $x=0.1$, also a change of symmetry from tetragonal to pseudocubic at $x=0.08$; with $x \geq 0.06$ demonstrated the presence of the secondary phase $Bi_{1.55}Gd_{0.45}O_3$. In the previous work[14], we have reported the synthesis of the phase $Ba_8Ti_3Nb_4O_{24}$ using a series of $Ba(Ti_{1-5x}Nb_{4x})O_3$ solid solutions; we found when the doped level is increased to $x \geq 0.02$, a peak displacement is showed, this is due to the phase transformation of tetragonal to cubic in to the unit cell. Also, with $x=0.125$ the crystal structure transforms to the hexagonal phase of $Ba_8Ti_3Nb_4O_{24}$. Thus, in this paper we have fabricated Nb^{5+} doped barium titanate ceramic powders by solid-state reaction method, In addition, the effect of niobium on microstructure, morphology and electrical properties of barium titanate were investigated.

EXPERIMENTAL PROCEDURES

The effect of doping (Nb^{5+}) on the B site of the perovskite structure of $BaTiO_3$ was investigated; the ceramic raw materials were high purity carbonates and oxides ($BaCO_3$ Sigma-Aldrich CAS No. 513-77-9 99.99%, TiO_2 Sigma-Aldrich CAS No. 1317-80-2 99.99 % and Nb_2O_5 Sigma-Aldrich CAS No. 1313-96-8 99.99% powders. The different compositions were synthesizes by solid-state reaction method according to the nominal formula $BaTi_{1-x}Nb_xO_3$ ($0 \leq x \leq 0.25$). The ceramic powders were dried in a type muffle oven for 24 hours at 200 °C, except Nb_2O_5 dried at 900 °C for 24 hours in a high temperature oven in order to completely remove moisture. Subsequent to drying, the powders are weighed in different stoichiometry corresponding, to according to the mechanisms used to obtain 5 g of powder quantities; the precursors powders were milled in a polyethylene terephthalate (PET) containers during 5 h using yttria-stabilised zirconia milling media (5 mm balls diameter) as the milling media in analytical grade acetone as control medium. The mixtures were calcined in an platinum crucible at 900 °C in air for 12 h to remove the carbonate phase and initiate reaction; reground in an agate mortar and pestle, sintering and reground again, finally fired at 1500°C/12 h and then cooled down to room temperature. The sintered powders obtained for each composition were then mixed using an organic binder (5% solution of polyvinyl alcohol -PVA- in water), and pressed into green cylindrical disks of 10 mm diameter and about 5 mm height, at a pressure of about 200 MPa for 2 min in a stainless steel cylindrical die. The final sintering conditions of pellets were of 1500 °C during 12 h with a heating rate of 10 °C/min followed by furnace cooling at room temperature. To study the evolution of phases of the ceramic powders, and to identify the secondary phases, was used a diffractometer Inel Equinox 2000 with Cu K_α (λ-1.5406 Å) radiation, which was used as the first indication of phase purity. The microstructure of sintered pellets was observed using a scanning electron microscopy (SEM JEOL 6300). In order to measure the electrical properties, platinum electrodes were collocated on two opposite pellets faces with Ag/Pt paste (ESL Electro-science) and drying in a muffle at 200°C for 30 min. Capacitance was measured over a temperature range from room to 200°C by using an LCRmeter (GW Instek LCR 819), with a frequency of 1 kHz in tube furnace with air. Electrical Hysteresis curves were recorded using an automatic P-E loop tracer model RT-66A of Radiant Technologies. Raman studies were carried out using a spectrophotometer Perkin Elmer Spectrum Gx; in a range of 0-1200 wavelength, because in this range the vibrational modes characteristic of barium titanate occur.

RESULTS AND DISCUSSION

Figure 1 shows the distribution of particles sizes of the precursors ceramics powders; all ceramics exhibit a distribution of particle sizes. In particular, Nb_2O_5 and TiO_2 exhibit a tri-modal distribution of sizes, whereas $BaCO_3$ shows a bi-modal distribution. The d_{50} values were 2.7, 7.4 and 8.1 respectively.

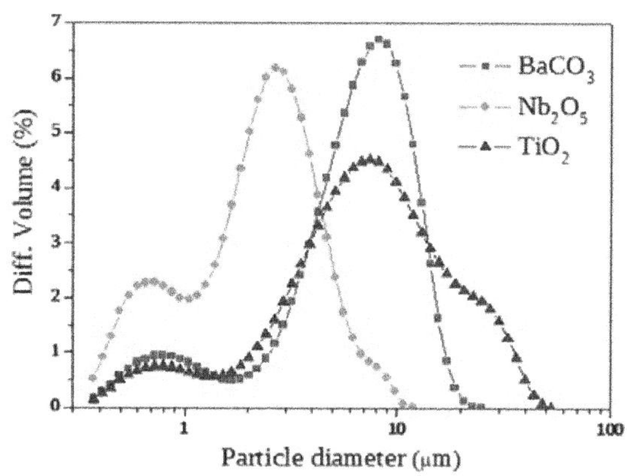

Figure 1. Distribution of particle sizes of raw materials.

Figure 2 show the SEM images of raw materials, $BaCO_3$ and TiO_2 are acicular morphology with sizes > 5 μm long, while for Nb_2O_5 has rounded particles with sizes ≥ 2 microns, this images are consistent with the result of particle size. X-ray diffraction patterns of the mechanism $BaTi_{1-x}Nb_xO_3$ ceramics are show in the Figure 3; XRD patterns for $0.001 \leq x \leq 0.04$ shows the presence of a double peak at 45° indicating the tetragonal phase. With the change of x=0.04 to x=0.08 exists the distortion of the tetragonal to cubic cell. Therefore, x=0.001, 0.003, 0.005 and 0.04 are tetragonal (JCPDS 050626) at room temperature, x= 0.08 and 0.20 are cubic (JCPDS 310174). While compositions with $x \geq 0.25$ are the hexagonal phase[23,24,25] $Ba_8Ti_3Nb_4O_{24}$ (JCPDS 841068). When Nb^{5+} is induced to $BaTiO_3$ lattice, it prefers to substitute at the Ti^{4+} site with oxygen vacancy compensation, due to their similar ionic radii in octahedral coordination ($r(Nb^{5+})$ = 0.64 Å, $r(Ti^{4+})$ = 0.61 Å). As a result, the lattice expansion across the solid solution is evident as the Nb^{5+} content increases.

Figure 2. SEM images of raw materials a) $BaCO_3$, b) TiO_2 c) Nb_2O_5.

This expand explains in the inset in Figure 3, this shows the XRD patterns of $BaTiO_3$ doped Nb^{5+} in the 44-46° 2 theta range. For the samples with $x \leq 0.04$, splitting of the (2 0 0)/(0 0 2) diffraction peaks was observed, indicating that the sample was stabilized in the tetragonal phase or pseudo cubic phase with slight tetragonal distortion at room temperature. For 0.08 and 0.2, only a single diffraction peak was observed, and the effect of Nb^{5+} substitution on the $BaTiO_3$ structure shifts the (2 0 0) reflection, which is characteristic of the cubic lattice of $BaTiO_3$. Figure 4 shows that the cell volume increases with the addition of Nb^{5+} ions in the perovskite structure, (Ti^{4+} by Nb^{5+} ions).

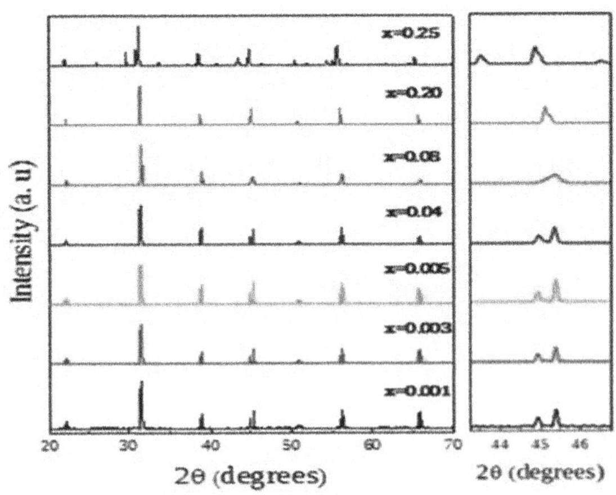

Figure 3. XRD patterns of $BaTi_{1-x}Nb_xO_3$ ($0 \leq x \leq 0.25$).

Like an alternative method, Raman Spectroscopy is capable of measuring the lattice vibrational spectra to confirm the phase transformation of $BaTiO_3$ doped Nb^{5+}. $BaTiO_3$ is a typical ferroelectric with perovskite structure. Above 120 °C, it is cubic and belongs to space group *Pm3m*(O^1_h). Between 5 and 120 °C, it is tetragonal with *P4mm*(C^1_{4v}) symmetry. In the cubic

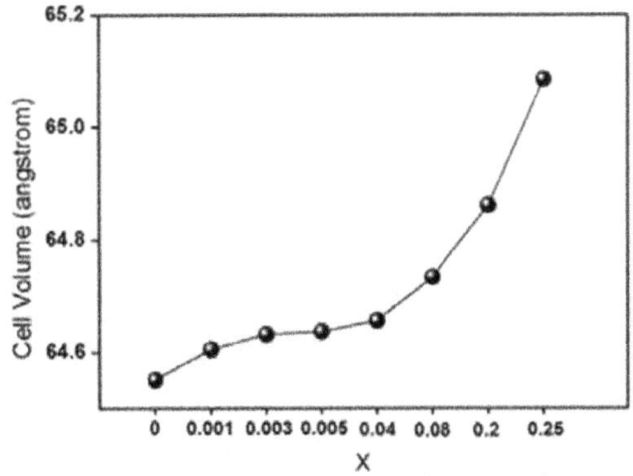

Figure 4. Unit cell volume as a function of x.

(paraelectric) phase, the 12 optic modes transform as the $3F_{1u}(IR) + 1F_{2u}$ irreducible representations, and there are no first-order Raman active modes. In the tetragonal (ferroelectric) phase, each F_{1u} mode splits into a A_1 mode and an E mode, while the F_{2u} mode splits into B_1 and E, resulting in $3A_1(IR,R) + B_1(R) + 4E(IR,R)$ modes. There are further splittings (LO, TO splitting) of the vibrational modes due to long-range electrostatic forces associated with lattice ionicity. As a consequence, each A_1 or E mode splits into a pair of modes, i.e., $A_1 \rightarrow A_1(TO) + A_1(LO)$ and $E \rightarrow E(TO) + E(LO)$, where TO and LO refer, respectively, to the transverse and longitudinal optic modes[26]. The Raman spectra, for $BaTiO_3$ ceramic powders samples obtained by the solid state synthesis are presented in Figure 5. This plots present the Raman peaks situated at 205 cm^{-1} (E(TO + LO), A1(LO)), 265 cm^{-1} (A1(TO)), 304 cm^{-1} (B1, E(TO + LO)), 513 cm^{-1} (A1(TO), E(TO)) and 717 cm^{-1} (A1(LO), E(LO)).

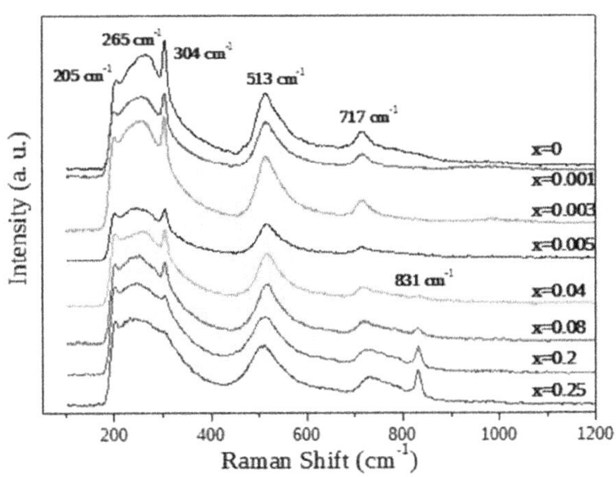

Fig. 5. Raman spectra of samples with $0 \leq x \leq 0.25$.

When $BaTiO_3$ passes through the tetragonal-cubic phase transition point, all the active Raman modes, in the tetragonal phase (P4mm), will be inactive in the perfect cubic phase (Pm3m), owing to forbidding of Raman selection rules. However, the two broad A1 (TO) bands at 265 cm^{-1} and 513 cm^{-1} persist into the cubic phase above T_C, which is attributed to intrinsic disorder in the cubic phase, and they become broad and weak with increasing Nb^{5+} containing. The band at 304 cm^{-1} disappear above T_C, which is considered the signature of t–c phase transition. These results agree well with the XRD results described above. From Figure 5 it is clearly seen that the addition of Nb^{5+} ($x \geq 0.04$) a new Raman band[27,28] appeared at ≈ 831 cm^{-1}; the difference of cation radii between Ba^{2+}, Ti^{4+} and Nb^{5+} can lead to the changes of the volume of unit cell in perovskite, which causes the internal deformation/distortion of TiO_6 octahedron. This deformation can be responsible for the Raman band at ≈ 831 cm^{-1}. The Raman band positions and assignments for all samples are summarized in Table I.

Table I. Raman band positions and corresponding band assignments of ($0 \leq x \leq 0.25$) samples.

Sample code	Band positions (cm^{-1}) (E(TO + LO), A1(LO)),	(A1(TO))	(B1, E(TO + LO	(A1(TO), E(TO))	(A1(LO), E(LO))	
X=0	205	265	304	513	717	----
X=0.001	200	258	303	515	717	----
X=0.003	202	256	303	514	717	----
X=0.005	201	246	304	516	713	----
X=0.04	203	259	304	517	714	834
X=0.08	202	253	304	517	719	831
X=0.20	203	248	304	515	722	833
X=0.25	203	----	----	511	731	833

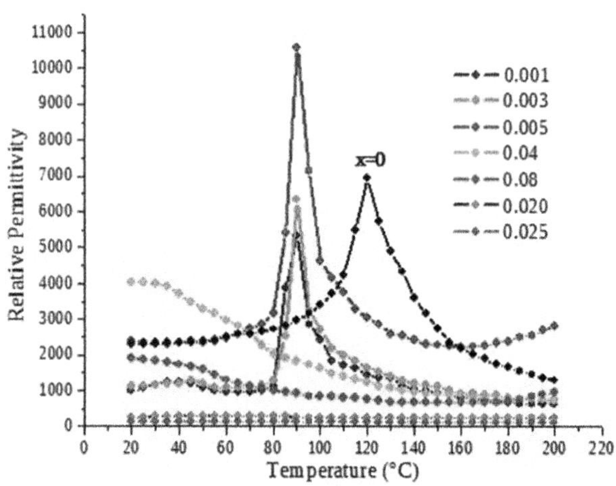

Fig. 6. Permittivity against temperature plots of samples, at 1 kHz.

Figure 6 summarizes the dielectric constants of $BaTiO_3$ ceramics with different Nb_2O_5 concentrations, measured as a function of temperature, in the range between 25 and 200°C at 1 kHz. The tetragonal phase (ferroelectric) to cubic phase (paraelectric) transition at Curie temperature (Tc) was Tc = 120 °C and a permittivity of 6944 for base material ($BaTiO_3$ pure). The dielectric constant of $BaTiO_3$ ceramics was increased with an increase of Nb^{5+} ions concentration; $BaTiO_3$ ceramics with Nb^{5+} doping concentration of 0.001, 0.003, and 0.005 showed 5348, 6337, and 10560 in dielectric constant at the Tc, (the phase transition temperature of cubic-paraelectric to tetragonal-ferroelectric), respectively. With x=0.20 and 0.25 the paraelectric behavior is clear and according to Raman, DRX and Hysteresis plots. The ferroelectric hysteresis loops (P versus E) for the prepared $BaTi_{1-x}Nb_xO_3$ samples for $0 \leq x \leq 0.25$ are displayed in Figures 7 and 8; it was observed that the remnant polarization (Pr) decreased from ≈5.0 to ≈1.0 (μC/cm²) for x=0 and x=0.005 respectively.

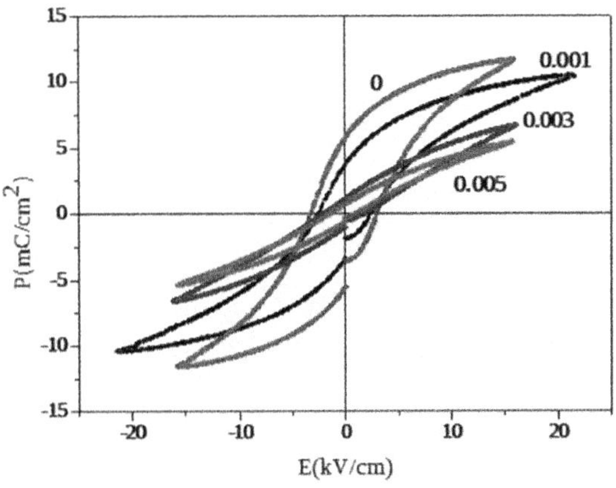

Fig. 7. Hysteresis loops of $BaTi_{1-x}Nb_xO_3$ in a range of 0-0.005.

Therefore, substituting Ti^{4+} ions with Nb^{5+} ions decreases the ferroelectric behavior of the $BaTiO_3$. Well saturated loops and comparable remnant polarization were obtained from the doped ceramics by solid state route. As the Nb^{5+} content increases, however, the loop becomes narrower and more slanted with decreased remnant polarization Pr and coercive field Ec (Figure 8), indicating the change of ferroelectric to dielectric properties. Similar results have been reported by Chen et al.[29] The less saturated hysteresis loops and the weaker ferroelectricity may be attributed to the formation of non-polar regions induced by the substitution of Nb^{5+} for B sites in $BaTiO_3$ ceramics. With increasing the dopant quantity the transition is most notable. Figure 9 illustrates the morphology of thermally etched surfaces of doped $BaTiO_3$ ceramics sintered at 1500 °C for 5 h with different doping content ($0 \leq x \leq 0.25$). The sample (9a) show the microstructure of $BaTiO_3$ with low dopant (x=0.001). The conventional solid-state method, relatively dense and homogenous grains with an average grain size of 3 μm could be obtained in Figure 9b. With doping content increasing, slight grain growth occurs and regular grains of ≈ 10 μm can be obtained for x = 0.08. However, the grains tend to be bigger and irregular when the doping content achieves 0.25.

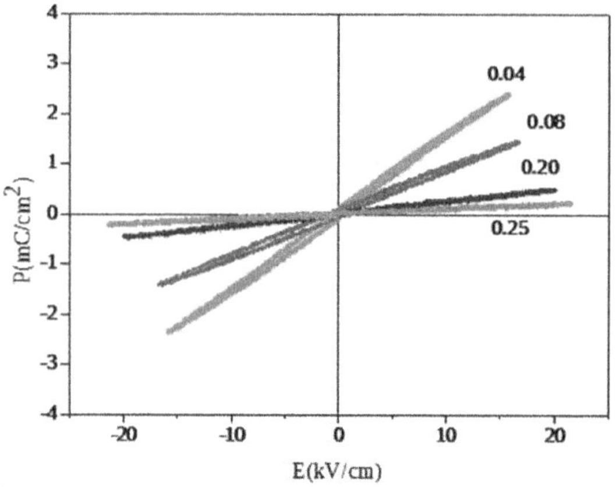

Fig. 8. Hysteresis loops of $BaTi_{1-x}Nb_xO_3$ in a range of 0.04-0.25.

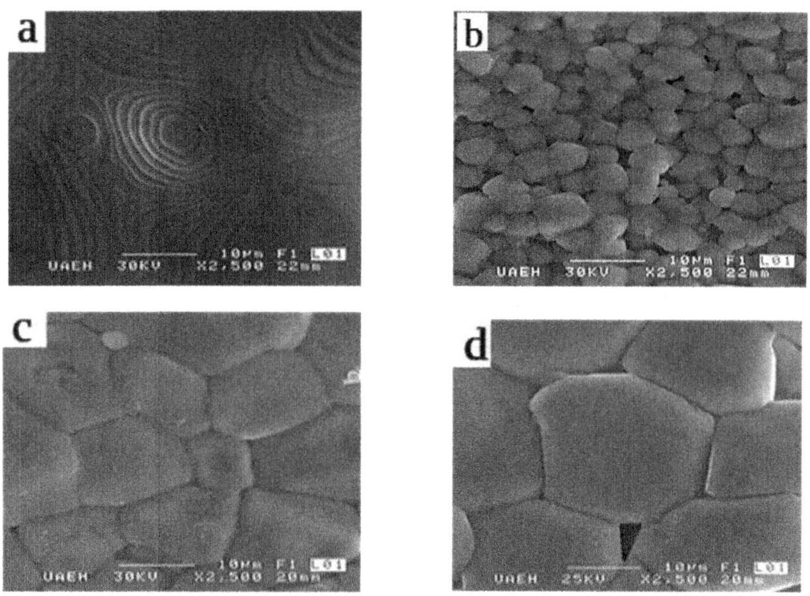

Fig. 9. SEM images of $BaTi_{1-x}Nb_xO_3$, a) 0.001, b) 0.005, c) 0.08, d) 0.25

CONCLUSIONS

The $BaTiO_3$ doped Nb^{5+} with the mechanism $BaTi_{1-x}Nb_xO_3$; (x= 0.001, 0.003, 0.005, 0.04, 0.08, 0.20 and 0.25) solid solutions were synthesized using a solid state route. The synthesized ceramic investigated by XDR, SEM, Raman and Electrical measurements; it was confirmed that increasing of Nb^{5+} content in the perovskite structure results in transformation of the symmetry, from tetragonal to cubic plus $Ba_8Ti_4Nb_3O_{24}$. XRD results have shown that compositions with $x \geq 0.08$ are cubic after heating at 1350 °C, but they transform to the tetragonal phase during cooling; with $x \leq 0.04$ the sample is tetragonal. Electrical measurements showed that Niobium content drops both the Curie temperature and permittivity values. In compositions with $x \geq 0.25$, an extra phase appeared which was identified as a hexagonal phase with formula $Ba_8Nb_4Ti_3O_{24}$. The samples with x = 0.20 and 0.25 shows an insulating behavior.

Acknowledgements

The authors gratefully acknowledge to PRODEP Mexico, for the financial support for this work.

REFERENCES

[1]J. Valasek, Piezoelectric and Allied Phenomena in Rochelle Salt, *Phys. Rev.*, **17**, 475–81 (1921).

[2]L. E. Cross and R. E. Newnham, History of Ferroelectrics; pp. 289–305 in Ceramics and Civilization, Vol. III, *High-Technology Ceramics—Past, Present, and Future*. American Ceramic Society, Westerville, OH, 1987.

[3]G. Busch, Early History of Ferroelectricity, *Ferroelectrics*, **74**, 267–84 (1987).

[4]W. Kanzig, History of Ferroelectricity 1938–1955, *Ferroelectrics*, **74**, 285–91 (1987).

[5]A. von Hippel, Ferroelectricity, Domain Structure, and Phase Transitions of Barium Titanate, *Rev. Mod. Phys.*, **22** [3] 221–37 (1950).

[6]A. J. Moulson and J. M. Herbert, Electroceramics: Materials, Properties, Applications, John Wiley & Sons (2003).

[7]L. Padilla-Campos, D.E. Diaz-Droguett, R. Lavín, and S. Fuentes, Synthesis and structural analysis of Co-doped $BaTiO_3$, *J. Mol. Struct.* **1099,** 502-509 (2015).

[8]J.M. De Paoli, R.E. Carbonio, and R.D. Sánchez, Oxygen vacancy doping effect on the electrical and magnetic behavior of $Ba_{5-x}La_xNb_{4-x}Ti_xO_{15}$, *J. Phys. Chem. Solids.*, **68,** 124-130 (2007).

[9]Cailan Tian, Zhenxing Yuen, Yuanyuan Zhou, and Longtu Li, Crystal structures and microwave dielectric properties of Zn, W co-substituted $BaTiO_3$ perovskite ceramics, *Journal of Solid State Chem.*, **197,** 242-247 (2013).

[10]F.R. Barrientos Hernández, A. Arenas Flores, and E. Cardoso Legorreta, Synthesis and Characterization of $BaTiO_3$ Doped in the B Site by the Barium Titanate Route, *Integr. Ferroelectr.*, **126,** 1-6 (2011).

[11]K. Samuvel, and K. Ramachandran, High calcination of ferroelectric $BaTiO_3$ doped Fe nanoceramics prepared by a solid-state sintering method, *Spectrochim. Acta, Part A*, **146,** 113-118 (2015).

[12]M. Saleem, I.S. Kim, J.S. Song, S.J. Jeong, M.S. Kim, and S. Yoon, Synthesis, sintering and dielectric properties of a $BaTiO_3$–Ni composite, *Ceram. Int.*, **40,** 7329-7335 (2014).

[13]Meng Zhang, Ji Zheng, Lu Liang, Fei Jiang, and Ying Wang, Preparation and microwave absorption properties of Ag-doped $BaTiO_3$ nanocomposites, *J. Magn. Magn. Mater.*, **368,** 198-201 (2014).

[14]F.R. Barrientos Hernández, I.A. Lira Hernández, C. Gómez Yáñez, A. Arenas Flores, R. Cabrera Sierra, and M. Pérez Labra, Structural evolution of $Ba_8Ti_3Nb_4O_{24}$ from $BaTiO_3$ using a series of $Ba(Ti_{1-5x}Nb_{4x})O_3$ solid solutions, *J. Alloys Compd.*, **583,** 587-592 (2014).

[15]Adelina Carmen Ianculescu, Catalina Andreea Vasilescu, Maria Crisan, Malina Raileanu, Bogdan Stefan Vasile, Mihai Calugaru, Dorel Crisan, Nicolae Dragan, Lavinia Curecheriu, and Liliana Mitoseriu, Formation mechanism and characteristics of lanthanum-doped $BaTiO_3$ powders and ceramics prepared by the sol–gel process, *Mater.Charact.,* **106,** 195-207 (2015).
[16]Inas K. Battisha, Ali B. Abou Hamad, and Ragab M. Mahani, Structure and dielectric studies of nano-composite Fe_2O_3:$BaTiO_3$ prepared by sol–gel method, *Physica B,* **404,** 2274-2279 (2009).
[17]Songhak Yoon, Jürgen Dornseiffer, Yan Xiong, Daniel Grüner, Zhijian Shen, Shoichi Iwaya, Christian Pithan, and Rainer Waser, Spark plasma sintering of nanocrystalline $BaTiO_3$-powders: Consolidation behavior and dielectric characteristics, *J. Eur. Ceram. Soc.,* **31,** 1723-1731 (2011).
[18]Boram Min, S.M. Moon, and N.H. Cho, Structural and dielectric features of Nb-doped nano-sized $BaTiO_3$ powders prepared by hydro-thermal synthesis methods, *Curr. Appl. Phys.,* **11,** S193-S196 (2011).
[19]Marjeta Maček Kržmancn, Dejan Klement, Boštjan Jančar, and DaniloSuvorov, Hydrothermal conditions for the formation of tetragonal $BaTiO_3$ particles from potassium titanate and barium salt, *Ceram. Int.,* **41,** 15128-15137 (2015).
[20]M.T. Buscaglia, V. Buscaglia, M. Viviani, P. Nanni, and M. Hanuskova, Infuence of foreign ions on the crystal structure of $BaTiO_3$, *J. Eur. Ceram. Soc.,* **20,** 1997-2007 (2000).
[21]V. Paunovic, V.V. Mitic, Z. Prijic, and Lj. Zivkovica, Microstructure and dielectric properties of Dy/Mn doped $BaTiO_3$ ceramics, *Ceram. Int.,* **40,** 4277-4288 (2014).
[22]Giorgio Schileo, Antonio Feteira, Klaus Reichmann, Ming Li, and Derek C. Sinclair, Structure–property relationships in $(1-x)BaTiO_3$–$xBiGdO_3$ ceramics, *J. Eur. Ceram. Soc.,* **35,** 2479-2488 (2015).
[23] Liang Fang, Congxue Su, Zhenhai Wei, Huanfu Zhou, and Hui Zhang, Phase structure, band gap and microwave dielectric properties of $Ba_8Ti_3Nb_{4x}Sb_xO_{24}$ ceramics, *Ceram. Int.,* **39,** 579-583 (2013).
[24] Nicolas Teneze, Philippe Boullay, Vaclav Petricek, Gilles Trolliard, and Danièle Mercurio, Structural study of the cation ordering in the ternary oxide $Ba_8Ti_3Nb_4O_{24}$, *solid state sci.,* **4,** 1129-1136 (2002).
[25] G. Trolliard, N. Teneze, Ph. Boullay, and D. Mercurio, TEM study of cation-deficient-perovskite related $A_nB_{n-1}O_{3n}$ compounds: the twin-shift option, *J. Solid State Chem.,* **177,** 1188-1196 (2004).
[26]M.S. Chen, Z.X. Zhen, S.H. Tang, W.S. Shi, D.F. Cui, and Z.H. Chen, Stress effect on Raman spectra of Ce-doped $BaTiO_3$ films, *J. Phys. Condens. Matter.,* **12,** 7013-7023 (2000).
[27]Da-Yong Lu, Xiu-Yun Sun, and Masayuki Toda, A novel high-k 'Y5V' barium titanate ceramics co-doped with lanthanum and cerium, *J. Phys. Chem. Solids,* **68,** 650-664 (2007).
[28] Ru Z. Hou, Paula Ferreira, and Paula M. Vilarinho, Nanoporous $BaTiO_3$ Crystallites, *Chem. Mater.,* **21,** 3536-3541 (2009).
[29] L. Chen, B.C. Luo, N.Y. Chan, J.Y. Dai, M. Hoffman, S. Li, and D.Y. Wang, Enhancement of photovoltaic properties with Nb modified (Bi, Na) TiO_3–$BaTiO_3$ ferroelectric ceramics, *J. Alloys Compd.,* **587,** 339-343 (2014).

INFLUENCES OF THE CERAMIC MATRIX IN THE PROPERTIES OF FERROELECTRIC COMPOSITES BASED ON PVDF POLYMERS

Danilo Umbelino Figueiredo, Evaristo Alexandre Falcão, Eriton Rodrigo Botero
Federal University of Grande Dourados, Faculty of Science and Technology, Dourados-Itahum road km 11, Dourados-MS, Brazil

José Antonio Eiras, Fabio Luis Zabotto and Ducinei Garcia
Federal University of Sao Carlos, Department of Physics, Group of Ferroic Materials, Washington Luiz road, km 235, Sao Carlos- SP, Brazil

ABSTRACT

Ferroelectric composites of Polyvynilidene Fluoride (PVDF) polymers and Lanthanum modified Lead Titanate Zirconte ceramics were synthesized based the solution method. The ratio of PVDF/PLZT was adopted as 25/75 of mass, with the ceramic powders having the concentration of lanthanum varying according the ratio La/Zr/Ti equal to 5/65/35, 9/65/35 and 11/65/35. The composites were characterized considering the structural and electrical factors. From the structural point of view it was found that the polymer composites have concentration of the polar phase (β phase) higher than 70%. It is also noted that synthesis route favors the formation of an amorphous phase, but it was not noticed a significant influence of the ceramics composition on the structural properties of composites. In the electrical point of view, it was clear that the variance of lanthanum concentration in the ceramics powders significantly modifies the values of dielectric constant and polarization of the composites.

INTRODUCTION

Polyvynilidene Fluoride polymers, (-H2C-CF2), or PVDF, is a very known ferroelectric polymer, and have been studied since the 60's due to their biggest piezo and pyroelectric coefficient values, flexibility, excellent processability, chemical stability and mechanical resistance[1]. This material has a crystalline polymorphism, resulting from the different conformations of their chains that crystalize at four phases, called α, β, γ, and δ[2]. The most common is α phase, usually obtained from the crystallization of the melted polymer, or from crystallization in solution [3, 4]. For the last case, in general, yields a semicrystalline polymer, where the degree of crystallinity is around 40%. But, when well crystallized this value can reach 90%. In α phase the chains assume a conformation TG + TG- where the electric dipoles are paralleled aligned in the polymer chain, resulting in null electric dipole moment. Differently, the β phase is polar, since the all-trans conformation of their chains results in dipole moment perpendicular to the polymer chain. In this case the ferroelectricity can be reach by an application of an electric field performing a rotation of the polymeric chain[6]. Due to these prominent ferroelectric properties, the β phase is the most desirable for technological applications in sensors and actuators[5, 6].

When crystallized in β phase PVDF polymers present the greatest piezo and pyroelectric coefficients among the ferroelectric polymers[4]. However, when compared to other classes of ferroelectric materials, such as ceramics and single-crystals, the low value of those coefficients restricts the PVDF applications. Even with higher piezo and pyroelectric coefficients, conventional ferroelectric materials have the limitation concerning about the mechanical flexibility for some technological applications[7]. In this way, the development of composite materials involving ferroelectric ceramics and polymers has become an alternative to bypass these limitations. In this way, it is interesting the combination of properties of conventional ferroelectric with the flexibility of ferroelectric polymers, looking forward to the development of a flexible material with great ferroelectric properties.

Lantanhum modified Lead Titanate Zirconate $(Pb_{1-x}La_x)(Zr_{1-y}Ti_y)_{1-1/4x}O_3e$, or PLZT, is a ferroelectric system formed from a solid solution of Lead Titanate ($PbTiO_3$) and Lead Zirconate ($PbZrO_3$) with addition of Lanthanum. This material has an ABO_3 perovskite structure type [11], with ions of Zr $^{4+}$ and Ti $^{4+}$ occupying the B sites, which and are involved by one oxygen octahedron, while the sites A are filled by Pb^{2+} and La^{3+}. This difference between the valence of these ions of the A sites generates structural vacancies in both A and/or B sites of the material, which gives to PLZT unique properties in the electronic and structural point of view[8-11]. In the beginning of the 70's, the transparency was achieved on PLZT ceramics, and this system got technological interest due to its excellent electrical and optical properties[10, 11]. Of the several compositions for the PLZT system, those with the molar ratio Zr/Ti of 65/35 and the molar lanthanum concentration between 4 and 12, called PLZT x/65/35 with 4 < x > 12, are located in a tri-critical region (junction point of three crystalline symmetries) of the phase diagram, which gives special properties compared to other compositions, which makes them very interesting applications for electronic or optoelectronic devices[8].

In this work were prepared ferroelectric-based polymer composite of PVDF and PLZT ceramics (with 25% PVDF and 75% PLZT) through crystallization method in solution. In order to verify the influence of ceramic matrix on the electrical properties of composites PVDF base, were used three different stoichiometric ratios of PLZT.

EXPERIMENTAL

Materials

The composites were synthesized by dissolving the PVDF powder (Solef 11010/1001) in dimethylformamide (DMF) Vetec (UV / HPLC) for the ratio of 80 mg of polymer per mL of solvent, under mechanical stirring at room temperature. After a complete dissolution of polymer, an aliquot of the solution was transferred to a petri dish (pure sample) while the remaining solution was kept under stirring for further addition of ceramic powder[12] of PLZT (5/65/35; 9/65/35 and 11/65/35), at the mass ratio of 0.25 PVDF/0.75PLZT. After the addition, the system was kept under stirring for 5 minutes at room temperature for a well dispersion of the ceramic particles. The solutions were transferred to Petri dishes and dried off for 6 hours at 90°C to eliminate the solvent and begin the polymer crystallization.

Characterizations

The XRD patterns of the samples were obtained on a Rigaku diffractometer with CuK_α radiation (1.5406 Å) in 2θ range between 10 and 40, at room temperature. The structural characterization of the phases of the polymer was performed by FT-IR using a spectrophotometer Model FT/IR- 4100 , Jasco Corporation, at room temperature, with resolution of 2 cm^{-1}, 200 scans per measure and spectral region of 1000 to 400 cm^{-1}. The quantification of the content of β phase compared to α phase of the polymer (F_β(%))was carried out by indexation of the absorption peaks and using Equation 1, proposed by Salimi and Yousefi from the Beer Lambert Law[13, 14].

$$F_{\beta(\%)} = \frac{Abs_\beta}{1{,}26Abs_\alpha + Abs\beta} \times 100 \qquad \text{(Equation 1)}$$

where Absα and Absβ are absorbance of α phase (at 763 cm $^{-1}$) and β phase (at 840 cm^{-1}), respectively. The absorbance value of each phase was determined by the Gaussian fitting of the infra-red absorbance spectrum[15], using peak analyzer tool of the OriginPro software version 8.5.0. This calculus is just an approximation, since the absorbance of other crystalline phases (λ and δ) was negligible. For the electrical characterizations Au electrodes were deposited by sputtering on the surfaces of samples. The real and imaginary parts of the electrical permittivity, as a function of frequency (10 kHz to 1 MHz), at room temperature, were obtained in an impedance analyzer (4194A Gain Phase Analyzer HP). The polarization curves as a function of electric field bipolar, hysteresis loop were obtained with a bipolar electric field 10 kV/cm of 10 Hz, at room temperature, based on a Sawer-Tower circuit[16].

RESULTS

XRD

The XRD patterns of PVDF/PLZT composites, in the ratio 25% PVDF/75% PLZT, crystallized at 90 ° C, for different compositions of PLZT are shown in Figure 1. The diffraction peaks at 2θ values of 21.8 ° (100), 31.0 ° (110) and 38.1 ° (111) are typical of PLZT perovskite structure with pseudo-cubic symmetry[10] can be easily identified in all diffraction patterns. For these three compositions, no significant changes in the diffraction peaks, resulting from stoichiometry difference of PLZT, can be observed. Looking to the structure of polymeric matrix, the peaks at 18.4 ° (020) and 19.9 ° (021) are characteristics of the α phase and at 20.3 ° (110) / (200) are characteristic of the β phase[17]. These peaks only can be observed with magnification of the image (inset in Figure 1). This difficulty in identifying the crystalline phases of PVDF is partly due to the proximity between the 2θ values of those three main reflections relative to α and β phases, but especially to low crystallinity of the polymer matrix. An amorphous halo overlying the crystalline peaks can be observed, indicating that the crystallization process adopted is not efficient. However, influences of the ceramic phase on the polymer phase, resulting in the appearance of new diffraction peaks, or enlargement of existing diffraction peaks are not observed in these profiles, indicating that the integrity of both phases was kept for synthesis conditions employed. It is remarkable that new methods of crystallization are already being implemented to this system, but the characterizations have not been carried out.

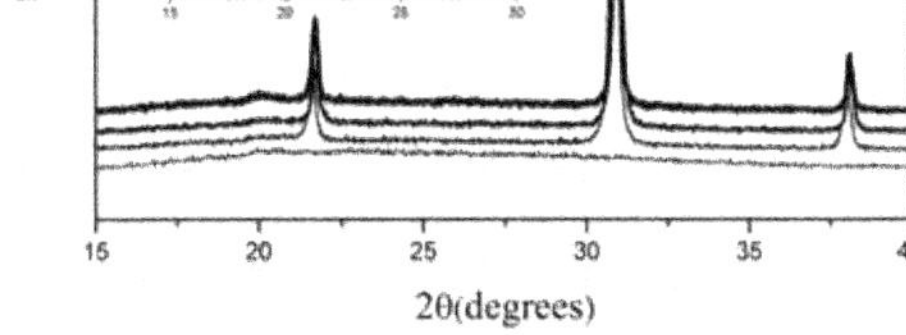

Figure 1: XRD patterns of PVDF/PLZT composites crystallized at 90° C, for 25% PVDF/75% PLZT. The inset is a zoom of the PVDF pattern identifying the diffraction peaks of α and β phases.

Infrared Spectroscopy

The transmission spectra in the infrared range of the PLZT/PVDF composites in the ratio 25% PVDF/75% PLZT, crystallized at 90 ° C, for different compositions of PLZT are shown in Figure 2. It is noted that the intensity of the bands related to the β phase of PVDF at 511 and 840 cm^{-1} are more noticeable than the bands typical of the nonpolar α phase at 976, 855, 796, 764, 614 and 532 cm^{-1}, showing that the adopted route of synthesis favors formation of the ferroelectric phase in all composites[3, 18]. The values of relative concentration of β phase was obtained according to Equation 1 and identifies around 70% of the β phase of PVDF are present in the composites. Traces of PLZT can be identified in the region of 595 cm^{-1}, characteristic of the strain of the Ti bonds[19]. No significant differences in the FTIR spectra concerning to changes of the stoichiometry of PLZT could be identified. In Figure 2 can also be identified absorptions relating to amorphous phase of the PVDF (peaks between 488 and 880 cm^{-1}), more intense than the peaks related to the crystalline phases of the PVDF, which corroborates with the previous discussions on XRD measurements. No changes concerning the typical absorptions peaks of PVDF as PLZT could be found, indicating that mixed phases were not formed between the compositions.

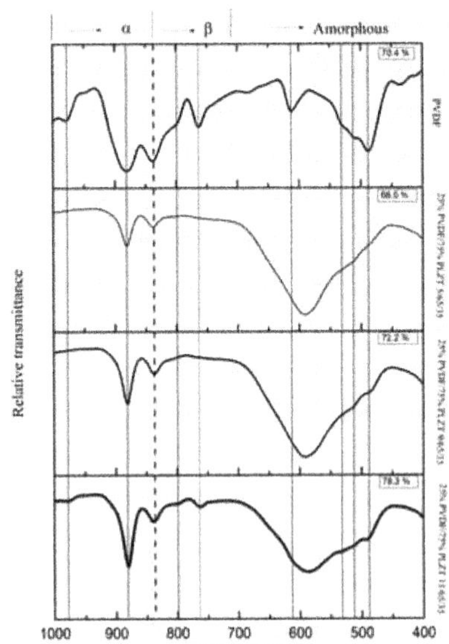

Figure 2: FTIR spectra of PVDF/PLZT composites crystallized at 90° C, for 25% PVDF/75% PLZT.

Dielectric Spectroscopy

Figure 3 shows the real and imaginary part of dielectric permittivity (ε 'and ε ") of the samples. All compositions have a typical behavior of dielectrics, as proposed by Debey[20]. The spectra show similar behavior of the relaxation of PLZT bodies, where the values of both the real and imaginary parts of permittivity decrease as the lanthanum is added to the system, which usually achieves the relaxor behavior of the material[21]. All compositions have permittivity values higher than the PVDF, which shows the great influence that ceramic matrix exerts on the composites.

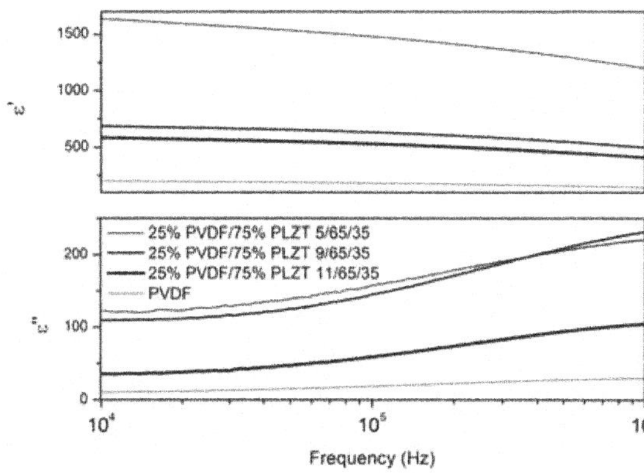

Figure 3: Real and Imaginary parts of the electrical of PVDF/PLZT composites crystallized at 90° C, for 25% PVDF/75% PLZT.

Electric polarization

The electric polarization curves as a function of a bipolar electric field, known as ferroelectric hysteresis, are presented in Figure 4. All measurements were performed with maximum field of 10kV/cm and 10 Hz. The maximum values for the polarization was reached by the pure PVDF films at $14\mu C/cm^2$, for the other samples the values are not too high. However the values of the coercive field (Ec) and the remaining polarization (Pr) had their values changed according to the stoichiometric of PLZT added to the polymers. The both Ec and Pr decreases as a function of the Lanthanum content, similar to PLZT ceramics[9]. The shape of the hysteresis loop also changes according to the PLZT added, moving from a typical ferroelectric hysteresis loop for the PLZT 5/65/35 doping composition to a close to slim loop curve for higher dopant concentrations.

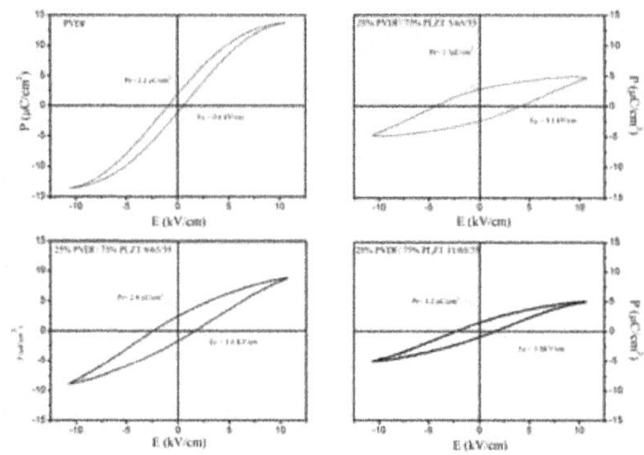

Figura 4: Hysteresis loop of the electrical of PVDF/PLZT composites crystallized at 90° C, for 25% PVDF/75% PLZT.

CONCLUSIONS

The properties of 25PVDF/75PLZT composites for the stoichiometric ceramic matrix La/Zr/Ti equal to 5/65/35, 9/65/35 and 11/65 / 35 were analyzed taking into account the structural and electrical factors. For the synthesis route employed was possible to obtain films with β phase concentration, knows as a polar ferroelectric, on the order of 70%, but substantial amount of amorphous phase of PVDF are present in the samples, which was confirmed by the XRD profiles. The structural characterizations also show there was no visible influence of each structural phase into the other. In the electrical point of view, all characterizations proved that the compounds follow the typical properties of the ceramic system employed, but with higher coefficients than PVDF and lower than the PLZT ceramics.

ACKNOWLEDGEMENTS

The authors are grateful for the financial support of CAPES (071/2013-183082), FUNDECT (23/200.528/2014), and GFM-UFSCar for the technical support.

REFERENCES

1- A. J. Lovinger, Ferroelectric Polymers. *Science*, **220**, **4602**, 1115-21 (1983).
2- H. S. Nalwa, Ferroelectric Polymers: Chemistry, Physics and Applications. 1. ed. New York: Marcel Dekker, Inc. (1995).
3- R. Gregorio, D. D. Borges, Effect of crystallization rate on the formation of the polymorphs of solution cast poly(vinylidene fluoride), *Polymer*, **49**, **18**, 4009-16 (2008).
4- V. Sencadas, R. Gregorio, L. Méndez, α to β Phase Transformation and Microestructural Changes of PVDF Films Induced by Uniaxial Stretch, *J. Macromol. Sci. B Phys.*, **48**, **3**, 514-25 (2009).
5- T. Furukawa, M. Date, E. Fukuda, Hysteresis Phenomena in Polyvinylide Fluoride Under High Electric Field, *J. App. Phys.*, **51**, **2**, 1135-42 (1980).

6- E. Fukada, History and recent progress in piezoelectric polymers, *IEEE Ultrasonics, Ferroelectrics, and Frequency Control Society*, **47**, **6**, 1277-90 (2000).
7- Y. Xu, Ferroelectric materials and their applications. Amsterdan: North-Holand (1991).
8- G. H. Haertling, Ferroelectric Ceramics: History and Technology, *J. Am. Ceram. Soc.*, **82**, **4**, 797-818 (1999).
9- G. H. Haertling, Electronic Ceramics: Properties, Devices, and Applications. New York: Marcel Dekker, INC (1988).
10- G. H. Haertling, C. E. Land, Hot-Pressed (Pb,La)(Zr,Ti)O3 ferroelectric ceramics for electro-optic applications, *J. Am. Ceram. Soc.*, **54**, **1**, 1-11 (1971).
11- G. H. Haertling, C. E. Land, Recent Improvements in the optical and electro-optical properties of PLZT ceramics, *Ferr.*, **3**, p. 269-80 (1972).
12- G. R. Salmazzo, Síntese e caracterização estrutural de compósitos a base de poli(fluoretode vinilideno)/ $(Pb_{0,91}La_{0,09})(Zr_{0,65}Ti_{0,35})_{0,98}O_3$, *Ceram.*, **60**, 83-7 (2014).
13- A. Salimi, A. A. Yousefi, Analysis Method: FTIR studies of β-phase crystal formation in stretched PVDF films, *Pol. Test.*, **22**, 699-704 (2003).
14- A. Salimi, A. A. Yousefi, Conformational Changes and Phase Transformation Mechanisms in PVDF Solution-Cast Films, *J. Pol. Scien.: Part B: Pol. Phys.*, **42**, 3487–95 (2004).
15- B. Stuart, Infrared Spectroscopy: Fundamentals and Applications. [S.l.]: John Wiley & Sons (2004).
16- C. B. Sawyer, C. H. Tower, *Phys. Rev.* **35,** 269 (1930).
17- D. R. Dillon, K. K. Tenneti, C. Y. Li, F. K. Ko, I. Sics, B. S. Hsiao, On the structure and morphology of polyvinylidene fluoride–nanoclay nanocomposites, *Pol.*, **47,** 1678–88 (2006).
18- W. M.Prest Jr, D. J. Luca, The Morphology and Thermal Response of High-Temperature-Crystallized Poly (vinylidene fluoride), *J. Appl. Phys*, **46**, 4136-43 (1975).
19- Y. Zhang, X. Zeng, X. Y. He, A. L. Ding, Q. R. Yin, The microstructure and properties of Dysprosin ions doped PLZT ceramics, *Int. Ferr.: And Internation Journal*, **78**, **1**, (2006)
20- A. R. Von Hippel, Dielectric Materials and Applications, New York, Jhon Wiley and Sons (1954).
21- K. Binder, A. P. Young, Spin Glasses: Experimental facts, theorical concepts, and open questions, *Rev. Mod. Phys.*, **58**, **4**, 801-76 (1986).

PIEZOELECTRIC RESPONSE OF SN AND MN MODIFIED LEAD TITANATE PIEZOELECTRIC CERAMICS

Deepam Maurya[1*], Hyun-Cheol Song[1], Min-Gyu Kang[1], Yongke Yan[1], Robert Bodnar[2], Ilan Levine[3], Edward Behnke[3], Haley Borsodi[3], Juan I. Collar[4], and Shashank Priya[1*]

[1]Bio-inspired Materials and Devices Laboratory (BMDL), Center for Energy Harvesting Materials and Systems (CEHMS), Virginia Tech, Blacksburg, VA 24061, USA
[2]Department of Geosciences, Virginia Tech, Blacksburg, VA 24061, USA
[3]Department of Physics and Astronomy, Indiana University South Bend, South Bend, Indiana 46634, USA
[4]Enrico Fermi Institute, KICP and Department of Physics, University of Chicago, Chicago, Illinois 60637, USA

ABSTRACT

In this paper, we report the electromechanical properties of Sn and Mn modified $PbTiO_3$ (PT) piezoelectric ceramics. XRD analysis on sintered ceramics confirmed the formation of a tetragonal perovskite phase. The co-doping of Sn and Mn was found to increase the Curie temperature (T_c) and tetragonality favoring the long range ordering. The modified PT sample, containing 20 at% Sn and 5 at% Mn, was found to exhibit high planar coupling constant k_p ~ 42% with enhanced Curie temperature T_c ~ 470°C. The modified piezoelectric PT specimen with high T_c can be a good candidate for sensing applications in high temperature environments.

INTRODUCTION

Piezoelectric materials are widely used as sensors, actuators, energy harvesters, etc.[1,2] The high temperature stability of piezoelectric sensors becomes critical while deployed in harsh environments such as jet engines, rocket exhausts, nuclear Fast Breeder Reactors (FBRs), geothermal exploration and production and deep bore-drilling.[3,4,5] Another important application of piezoelectric materials being developed is related to sensing of the acoustic emissions in bubble chambers for dark matter detection.[6,7] The piezoelectric ceramics based on $PbTiO_3$ exhibit a very small aging rate as compared to other piezoelectric materials due to their high Curie temperature (T_c) ~ 490°C.[8] This makes the $PbTiO_3$ (PT) based piezoelectric materials exceptionally stable for long periods of time over a wide temperature regime. Moreover, a high Curie temperature of 490°C and a dielectric constant of less than 200 makes PT-based ceramics attractive for high temperature and high frequency transducer applications.[8] However, it is difficult to achieve a dense ceramic body of PT due to its high tetragonality (6%).[9] In this study, we describe the development of Sn and Mn doped PT based ceramics that exhibit good sinterability and high temperature stability. In our formulation, substitution of Sn was made on the A-site and Mn on the B-site of PT. Interestingly, increasing the content of Sn doping in the presence of Mn was found to increase the Curie temperature.

Substitution of Sn on the A-site was inspired by predictions made for the compound $SnTiO_3$ (SNO). Theoretical calculations predicted SNO as a promising Pb-free material with higher functional response than that of $PbTiO_3$.[10,11] However synthesis of $SnTiO_3$, which requires stabilization of Sn(II), has been extremely challenging due to high temperature (>1000°C) processing of these ceramics. At such high temperatures, Sn^{4+} is highly favorable over Sn^{2+}

rendering difficulty in the synthesis of $SnTiO_3$.[12] This also leads to the loss in lone pair electrons associated with Sn^{2+}. The presence of lone pair of electrons has been related to the high functionality of the Pb-based perovskite type piezoelectric materials.[13] Here, our goal was to substitute Sn on the A site of $PbTiO_3$ and investigate the functional response and local structural changes. Unfortunately, $Pb_{1-x}Sn_xTiO_3$ ceramics exhibited high leakage current and were difficult to pole. Mn doping has been found to be successful in reducing the leakage current in perovskite type piezoelectric materials.[14] Thus, we decided to attempt Mn substitution on the B-site of Sn doped PT.

EXPERIMENTAL

Sn and Mn doped PT was synthesized using a conventional solid state reaction method. Stoichiometric amounts of the raw materials were ball milled in a polyethylene bottle using ethanol medium for 24h. The dried powders were calcined at 800°C (2h) and ball milled again for 48h. After sieving, calcined powder was compacted to form cylindrical disks and sintered at 1050 and 1150°C (2h). A LEO Zeiss 1550 (Zeiss, Munich, Germany) scanning electron microscope was used to investigate the surface morphology of the sintered samples. In order to confirm the crystallographic phase for various modified PT ceramic samples, XRD-spectra were collected at room temperature (RT) using a Philips Xpert Pro X-ray diffractometer (Almelo, The Netherlands).[15] For electrical measurements, silver paste was applied on flat surfaces of sintered and polished ceramic disks (diameter ~10.0 mm, thickness ~1.0 mm) and fired at 650°C for 30min. Poling was done by applying an *E*-field of 5kV/cm on the doped PT specimen dipped in silicone oil bath heated at 140°C. The longitudinal piezoelectric constant (d_{33}) of Sn and Mn doped PT ceramics was measured using the Berlincourt method. We used a computer controlled HP 4194A impedance analyzer coupled with a high temperature oven for measuring the dielectric response as a function of temperature at various frequencies with oscillation level of 500 mV. Polarization–electric field (*P-E*) hysteresis measurements were conducted using the modified Sawyer-Tower bridge Precision II (Radiant Technologies). We conducted Raman measurements in the Virginia Tech Vibrational Spectroscopy Laboratory using a Jobin-Yvon HR800 Raman microprobe. This Raman spectrometer was equipped with a 514 nm laser, which was focused onto the sample through a microscope lens. The X-ray photoelectron spectroscopy (XPS) measurements were performed using a PHI Quantera SXM Scanning Photoelectron Spectrometer Microprobe.

RESULTS AND DISCUSSION

Figure 1 (a) and (b) show the XRD spectra recorded on the calcined and sintered samples of $Pb_{1-x}Sn_xTiO_3$ ceramics, respectively. The peaks of the secondary phase are marked in the XRD spectra. The sharp Bragg reflections in the XRD-spectra of the sintered samples suggest a higher degree of crystallinity. The wide splitting in (200) Bragg reflections suggests a high tetragonality of the specimen.[16] We synthesized $Pb_{1-x}Sn_xTi_{1-y}Mn_yO_3$ (x=0.05, 0.10, 0.15, 0.20; y = 0.05) ceramics under similar synthesis conditions. In these compositions, Mn was substituted on the Ti-site of various Sn doped PT ceramics. Fig. 1(c) and (d) show the XRD-spectra recorded at RT for the $Pb_{1-x}Sn_xTi_{1-y}Mn_yO_3$ ceramics calcined at 800°C (2h) and sintered at 1150°C (2h), respectively. The XRD-spectra recorded for the calcined samples of $Pb_{1-x}Sn_xTi_{1-y}Mn_yO_3$ ceramics were found to indicate a higher degree of crystallinity than that of calcined $Pb_{1-x}Sn_xTiO_3$ samples. The peaks representing an impurity phase are marked in the spectra. The sintered samples were found to exhibit a smaller amount of the impurity phase as confirmed by the lower peak intensity, suggesting that Mn doping assists in stabilizing the perovskite phase. Figure 2 shows variation in tetragonality for various compositions of $Pb_{1-x}Sn_xTi_{1-y}Mn_yO_3$ ceramics. Note that the tetragonality decreases significantly for the $Pb_{1-x}Sn_xTi_{1-y}Mn_yO_3$ composition with x =0.05 and y =0.05 (compared to pure $PbTiO_3$), and the tetragonality then increases slightly with increasing Sn.

Figure 3 (a) and (b) show SEM micrographs of $Pb_{1-x}Sn_xTiO_3$ (x=0.0) ceramic samples after sintering at 1050 and 1150°C for 2h. Due to the high tetragonality (~6%) and resulting high spontaneous strain, this specimen was not sintered. On the other hand, the samples with increased Sn content on the A-site were found to have higher porosity at lower sintering temperature. Fig. 3 (c), (e), and (g) show SEM micrographs of the samples sintered at 1050°C for 2h. From these micrographs, one can clearly see the porosity. In order to reduce the porosity, we increased the sintering temperature to 1150°C (2h), which resulted in lower porosity (i.e. higher density) as observed in SEM micrographs (Fig. 3 (d), (f), (h)). The increased sintering temperature not only reduced the porosity but it also resulted in grain coarsening, with grain size ~1.5 µm. The increase in grain size with temperature can be explained by the phenomenological kinetic grain growth equation[17]:

$$\log D = \frac{1}{p}\log t + \frac{1}{p}[logK_o - 0.434\frac{E_a}{RT}] \quad (1)$$

where D is the grain size at time t, p is kinetic growth exponent, K_o a constant, E_a apparent activation energy, R the gas constant, and T is the absolute temperature.[17] However, the dielectric measurements suggested a higher leakage current in $Pb_{1-x}Sn_xTiO_3$ ceramics sintered at 1150°C, as shown in Fig. 4. All of the $Pb_{1-x}Sn_xTiO_3$ compositions were found to exhibit high loss tangent factor > 22%, indicating high dielectric losses. The increased leakage current is expected to be due to higher porosity and oxygen vacancies in the system. The PT ceramic samples sintered in air may have Ti^{3+} and oxygen vacancies (using Kröger–Vink notation): $2Ti_{Ti}^{x} + O_O^{x} \leftrightarrow 2Ti_{Ti}^{'} + V_O^{\cdot\cdot} + \frac{1}{2}O_2$.[18] The charge compensation in Mn doped PT samples can occur as follows:[19] $Mn^{4+} + 2Ti^{3+} \rightarrow Mn^{2+} + 2Ti^{4+}$ reducing their leakage current.[18] Therefore, in order to reduce the leakage current, we substituted 5 at% Mn on the Ti site of $Pb_{1-x}Sn_xTi_{1-y}Mn_yO_3$ (x =0.05, 0.10, 0.15, 0.20) ceramics. Here, we note that the composition with x =0.05 and y = 0.05 did not sinter well and is thus excluded from further analysis. Figure 5(a), (b) and (c) show SEM micrographs of the $Pb_{1-x}Sn_xTi_{1-y}Mn_yO_3$ (x =0.10, 0.15, 0.20, y = 0.05) ceramics sintered at 1150°C (2h). These ceramic samples were found to have >95% relative density. The high relative density in Mn doped PT samples can be attributed to the liquid phase sintering.[20] The presence of a liquid phase facilitates mass transport, resulting in densification and grain coarsening. The grain size of Mn doped ceramic samples was found to be 2.5 µm, 1.7 µm, and 1.2 µm for x =0.10, 0.15, and 0.20, respectively. Thus, the grain size was found to decrease with increasing substitution of Sn on the A-site of Mn doped PT. This may be attributed to the inhibited grain boundary motion due to the point defects created by the elemental substitution. Since other specimens did not exhibit high relative density, therefore, we mainly focused on $Pb_{1-x}Sn_xTi_{1-y}Mn_yO_3$ (x =0.10, 0.15, 0.20, y = 0.05) ceramics in this work.

Figure 6 show XPS spectra recorded in the $3d$ region of Sn for $Pb_{1-x}Sn_xTi_{1-y}Mn_yO_3$ (x =0.10, 0.15, 0.20, y = 0.05) ceramics. One can clearly observe two peaks corresponding to $Sn3d_{5/2}$ and $Sn3d_{3/2}$ at 486.2 eV at 494.7 eV, respectively. The $Sn3d_{5/2}$ and $Sn3d_{3/2}$ peaks were separated by 8.5 eV, which was similar to a standard specimen of Sn.[21] These peaks are the signature of Sn^{4+} indicating Sn^{4+} substitution on the Ti^{4+} site during high temperature processing.[22] As such, the starting composition $Pb_{1-x}Sn_xTi_{1-y}Mn_yO_3$ will have Pb non-stoichiometry in the system, as Sn is not substituting on the Pb-site. In order to investigate the effect of Sn and Mn substitution on local structure[23], we performed Raman scattering measurements, as shown in Figure 7. PT has a tetragonal structure with space group C_{4v}^1. In tetragonal PT, the transformation of the optical lattice modes occurs as 3A1+4E+B1 species. The three T1u modes of high temperature cubic phase give rise to the three A1+E modes.[24] However, one T2u mode gives rise to one E mode and one B1 mode (these are silent modes as they are not Raman and infrared active). Furthermore, in a ferroelectric phase, the long range ferroelectric phase further splits all A1+E modes into longitudinal (LO) and transverse (TO) components. The differences in the Raman scattering results for various samples can be interpreted as:

1. Below 400 cm^{-1}, there is no significant change. This may indicate that Sn substitution has little effect on the local structure of the Pb site. Since Sn^{4+} is more stable in the high temperature regime, the substitution of Sn^{4+} occurred mainly on the T^{4+} site, which is consistent with the XPS results.
2. Above 400 cm^{-1} significant differences were observed for different samples. This frequency regime is sensitive to the octahedral dynamics in ABO_3-type ferroelectric perovskites. With increasing Sn content, A1(TO3) and E (LO3) modes become sharper with the disappearance of a broad mode in between them. This broad mode could be related to the impurity phase. However, a band at 760 cm^{-1} increases with increasing Sn content, which can be attributed to the Sn-O bonds.[25] Concomitantly, a band at 700 cm^{-1} E(LO3) is reduced in intensity. This may be the signature of the perturbation in octahedral dynamics due to Sn doping in PT.

Some of the modes (E(TO2), A1(TO3), E(LO3)) were found to shift towards higher frequency (hardening of modes) with increasing Sn, as shown in Table 1. With increasing Sn content, hardening of modes along with the increased sharpness of the modes (especially E(TO3) and A1(TO3)) indicates stabilization of the ferroelectric phase. In order to investigate the Curie temperature (T_c) of these compositions, relative permittivity versus temperature measurements were performed. Figure 8 shows the relative permittivity as a function of temperature for $Pb_{1-x}Sn_xTi_{1-y}Mn_yO_3$ (x =0.10, 0.15, 0.20, y = 0.05) ceramics at various frequencies. From these plots, one can clearly observe the sharp peak related to the T_c. Interestingly, the T_c was found to shift towards higher temperature with the increasing Sn content. This is unusual compared to the conventional Sn doped (B-site) ferroelectric perovskite systems, where, T_c shifts towards room temperature with increasing Sn content.[16] The Curie temperature for x=0.1, 0.15, 0.20 was found to be 440°C, 458°C, and 470°C, respectively. The increased T_c is in line with the increasing trend of tetragonality observed with increasing Sn content, as shown in Figure 2. High Curie temperature (> 440°C) demonstrates high temperature stability of the piezoelectric ceramics. The increase in Curie temperature with increasing Sn content in Mn-substituted PT ceramics is quite intriguing. In a previous work, Sn^{2+}-doped $BaTiO_3$ (sintered in a reducing atmosphere) was found to exhibit a shift of T_c towards higher temperature. [26] However, in the present case the samples were sintered in air, therefore, the presence of the reduced Sn species, Sn^{2+}, is highly unlikely, as confirmed by Raman and XPS results. This suggests that the shift of T_c towards high temperature is probably a synergic effect of the Mn and Sn doping in PT ceramics. However, the Pb non-stoichiometry due to the substitution of Sn^{4+} on Ti^{4+} site may also affect the T_c.

Next, we measured piezoelectric properties of these ceramics as shown in Table 2. All of the samples were found to have a small loss tangent factor (<5%) with the dielectric constant in the range of 137-146. The longitudinal piezoelectric constant (d_{33}) was found to be 27, 24, and 21 pC/N for x = 0.10, 0.15 and 0.20, respectively. The $Pb_{1-x}Sn_xTi_{1-y}Mn_yO_3$ ceramics with x = 0.10, 0.15, and 0.20 were found to have planar coupling constants (k_p) of ~ 24, 29, and 42%, respectively. Fig. 9(a), (b), (c) show polarization hysteresis loops recorded using a bipolar signal at 1 Hz for $Pb_{1-x}Sn_xTi_{0.05}Mn_{0.95}O_3$ ceramic samples with x = 0.10, 0.15, 0.20. The hysteresis loops were not opening at room temperature, presumably, due to the high coercive field and large activation energy required to switch domains. With increasing temperature, the hysteresis loops started to open, probably due to the reduced activation energy required for domain switching.

The higher activation energy required to switch domains can be understood by considering the higher T_c of the lead titanate based ceramics. In the case of systems with T_c significantly distant from RT, the domain structure is homogeneous as there is very limited interaction with defects. The increase in Curie temperature rapidly decreases the domain wall energy $\sim(\Delta T)^{3/2}$ (ΔT is the distance from Curie point) in comparison with the temperature dependence of volume energy $\sim E_d P \sim (\Delta T)^{1/2}$[Ref.[27]]. These two energies compete with each other. Therefore, the domain walls in materials having high T_c will be thinner, as compared to the domain walls in their lower Curie temperature counterparts. Moreover, the velocity of the lateral motion of domain walls depends on various parameters[27]:

$$\left.\begin{aligned} v &= v_{\infty}\exp\left(-w\exp\left[-\frac{3\pi^2\delta}{4a}\right]\right) \\ \text{with},\ w &= 2(\gamma a^2)(\varepsilon)^{-\frac{1}{2}}T^{-1}(8\pi^4\gamma)^{\frac{3}{4}}(\delta/a)^{9/4}E^{-1} \end{aligned}\right\} \qquad (2)$$

where, γ is the surface energy of the domain wall, δ domain wall width, a is the size of elementary cell, ε is the permittivity, T is the temperature and E is the applied field.[27] Equation (2) shows that the velocity of the domain wall in a given regime very strongly depends on its width (a functional dependence of exponent in exponent).[27] Therefore, even a relatively small decrease in the width of the walls could result in a considerable decrease in the velocity of the domain wall, v, resulting in a lower piezoelectric response and inhibiting domain switching, as observed in our PT systems. Therefore, the temperature of the poling bath was increased to effectively align domains in preferred orientation during poling process. We note that the samples poled at room temperature did not exhibit piezoelectric response. In order to investigate domain morphology, we performed piezoresponse force microscopy (PFM). Fig. 10 (a), (d), and (g) show the PFM amplitude image for $Pb_{1-x}Sn_xTi_{0.95}Mn_{0.05}O_3$ (x =0.10, 0.15, and 0.20) ceramics. Fig. 10 (b), (e), and (h) show the PFM phase images indicating domain morphology. The magnified view of the ferroelectric domain is depicted in Figure 10 (c), (f) and (i). It can be seen from these results that the domain size was relatively reduced with increasing Sn contents and Mn doping in various samples. The reduced domain size appears to be the result of the finite interaction between domains and doping induced point defects. The presence of point defects and increased T_c are expected to increase the activation energy for the domain wall movement. According to equation (2), at higher temperature, domain walls have higher surface energy and larger domain wall width (smaller domain size),[28] which increases their mobility.[29,30] Therefore, for different PT samples, different measurement temperatures were required for achieving polarization switching during *PE* hysteresis loop measurements.

CONCLUSIONS

In summary, we substituted Sn on the A-site and Mn on the B-site of $PbTiO_3$ piezoelectric ceramics with starting composition $Pb_{1-x}Sn_xTi_{1-y}Mn_yO_3$ (x =0.10, 0.15, 0.20, y = 0.05). XPS results confirmed the substitution of Sn on the B-site as Sn^{4+}. Raman scattering results suggested that Sn and Mn co-doping favored the ferroelectric phase shifting towards higher temperature. The ceramic sample containing 20 at% Sn and 5 at% Mn was found to exhibit a planar coupling constant ~ 42% with high T_c ~ 470°C. These results further suggest a pathway to design new piezoelectric materials with high Curie temperatures.

ACKNOWLEDGEMENTS
The authors gratefully acknowledge financial support from the National Science Foundation through grant number PHY-1242637. Thanks to Charles Farley for helping with the Raman scattering measurements. Thanks to ICTAS Nanoscale Characterization and Fabrication Laboratory, Virginia Tech, for helping with the microstructural characterization.

REFERENCES:

1 Deepam Maurya, Yuan Zhou, Yongke Yan, and Shashank Priya, Journal of Materials Chemistry C **1** (11), 2102 (2013).
2 Deepam Maurya, Yuan Zhou, Yaojin Wang, Yongke Yan, Jiefang Li, Dwight Viehland, and Shashank Priya, Sci. Rep. **5** (2015).
3 Bernard Jaffe, *Piezoelectric Ceramics*. (Academic Press, 1971).
4 Udayan De, Kriti Ranjan Sahu, and Abhijit De, Solid State Phenomena **232**, 235 (2015).
5 Vishwas Bedekar, Josiah Oliver, Shujun Zhang, and Shashank Priya, Japanese Journal of Applied Physics **48** (9R), 091406 (2009).
6 Coupp Collaboration, E. Behnke, J. Behnke, S. J. Brice, D. Broemmelsiek, J. I. Collar, A. Conner, P. S. Cooper, M. Crisler, C. E. Dahl, D. Fustin, E. Grace, J. Hall, M. Hu, I. Levine, W. H. Lippincott, T. Moan, T. Nania, E. Ramberg, A. E. Robinson, A. Sonnenschein, M. Szydagis, and E. Vázquez-Jáuregui, Physical Review D **86** (5), 052001 (2012).
7 Pico Collaboration, C. Amole, M. Ardid, D. M Asner, D. Baxter, E. Behnke, P. Bhattacharjee, H. Borsodi, M. Bou-Cabo, S. J Brice, D. Broemmelsiek, K. Clark, J. I Collar, P. S Cooper, M. Crisler, C. E Dahl, S. Daley, M. Das, F. Debris, N. Dhungana, J. Farine, I. Felis, R. Filgas, M. Fines-Neuschild, F. Girard, G. Giroux, M. Hai, J. Hall, O. Harris, C. M Jackson, M. Jin, C. B Krauss, M. Lafrenière, M. Laurin, I. Lawson, I. Levine, W. H Lippincott, E. Mann, J. P Martin, D. Maurya, P. Mitra, R. Neilson, A. J Noble, A. Plante, R. B Podviianiuk, S. Priya, A. E Robinson, M. Ruschman, O. Scallon, S. Seth, A. Sonnenschein, N. Starinski, I. Štekl, E. Vázquez-Jáuregui, J. Wells, U. Wichoski, V. Zacek, and J. Zhang, Physical Review Letters **114** (23), 231302 (2015).
8 Ichiro Ueda and Seiji Ikegami, Japanese Journal of Applied Physics **7** (3), 236 (1968).
9 V. R. Palkar, S. C. Purandare, and R. Pinto, Materials Letters **43** (5–6), 329 (2000).
10 William D. Parker, James M. Rondinelli, and S. M. Nakhmanson, Physical Review B **84** (24), 245126 (2011).
11 S. F. Matar, I. Baraille, and M. A. Subramanian, Chemical Physics **355** (1), 43 (2009).
12 Thomas Fix, S. Lata Sahonta, Vincent Garcia, Judith L. MacManus-Driscoll, and Mark G. Blamire, Crystal Growth & Design **11** (5), 1422 (2011).
13 John B Goodenough and Jianshi Zhou, Science and Technology of Advanced Materials **16** (3), 036003 (2015).
14 Y. D. Hou, M. K. Zhu, J. L. Tang, X. M. Song, C. S. Tian, and H. Yan, Materials Chemistry and Physics **99** (1), 66 (2006).
15 Deepam Maurya, Yuan Zhou, Bo Chen, Min-Gyu Kang, Peter Nguyen, Mantu K. Hudait, and Shashank Priya, ACS Applied Materials & Interfaces (2015).
16 Deepam Maurya, Cheol-Woo Ahn, Shujun Zhang, and Shashank Priya, Journal of the American Ceramic Society **93** (5), 1225 (2010).

17 Tetsuya Senda and Richard C. Bradt, Journal of the American Ceramic Society **73** (1), 106 (1990).

18 Alan Templeton, Xiaoru Wang, Stuart J. Penn, Stephen J. Webb, Lesley F. Cohen, and Neil McN Alford, Journal of the American Ceramic Society **83** (1), 95 (2000).

19 Yao Shuai, Shengqiang Zhou, Danilo Bürger, Helfried Reuther, Ilona Skorupa, Varun John, Manfred Helm, and Heidemarie Schmidt, Journal of Applied Physics **109** (8), 084105 (2011).

20 Deepam Maurya, M. Murayama, and Shashank Priya, Journal of the American Ceramic Society **94** (9), 2857 (2011).

21 Michael A. Stranick and Anthony Moskwa, Surface Science Spectra **2** (1), 50 (1993).

22 Congkang Xu, Guoding Xu, Yingkai Liu, Xiaolin Zhao, and Guanghou Wang, Scripta Materialia **46** (11), 789 (2002).

23 Deepam Maurya, Ashok Kumar, Valeri Petkov, James E. Mahaney, Ram S. Katiyar, and Shashank Priya, RSC Advances **4** (3), 1283 (2014).

24 Mingrong Shen, G. G. Siu, Z. K. Xu, and Wenwu Cao, Applied Physics Letters **86** (25), 252903 (2005).

25 Baithy Mallesham, Putla Sudarsanam, Gangadhara Raju, and Benjaram M. Reddy, Green Chemistry **15** (2), 478 (2013).

26 Shoichiro Suzuki, Toshikazu Takeda, Akira Ando, and Hiroshi Takagi, Applied Physics Letters **96** (13), 132903 (2010).

27 A. S. Sidorkin, *Domain Structure in Ferroelectrics and Related Materials*. (Cambridge International Science Publishing, 2006), p. 250.

28 Deepam Maurya, M. Murayama, A. Pramanick, W. T. Reynolds, Ke An, and Shashank Priya, Journal of Applied Physics **113** (11), 114101 (2013).

29 Deepam Maurya, Natthapong Wongdamnern, Rattikorn Yimnirun, and Shashank Priya, Journal of Applied Physics **108** (12) (2010).

30 Deepam Maurya, Abhijit Pramanick, Mikhail Feygenson, Joerg C. Neuefeind, Robert J. Bodnar, and Shashank Priya, Journal of Materials Chemistry C **2** (39), 8423 (2014).

Table 1. List of some of the Raman modes that were shifted towards higher frequency for various compositions of $Pb_{1-x}Sn_xTi_{1-y}Mn_yO_3$ (x =0.10, 0.15, 0.20, y = 0.05) ceramics.

Modes / Sample	x =0.10	x =0.15	x =0.20
E(TO2)	203 cm^{-1}	205 cm^{-1}	207 cm^{-1}
A1(TO3)	609 cm^{-1}	614 cm^{-1}	614 cm^{-1}
E(LO3)	693 cm^{-1}	695 cm^{-1}	697 cm^{-1}

Table 2. Piezoelectric properties of various compositions of $Pb_{1-x}Sn_xTi_{1-y}Mn_yO_3$ (x =0.10, 0.15, 0.20, y = 0.05) ceramics.

Sample / Properties	Tanδ (%)	ε_{33}^{T}	k_p (%)	d_{33} (pC/N)	g_{33} $(V.m/N10^3)$
0.10	4.0	141	24	27	21
0.15	2.4	137	29	24	19
0.20	5	146	42	21	16

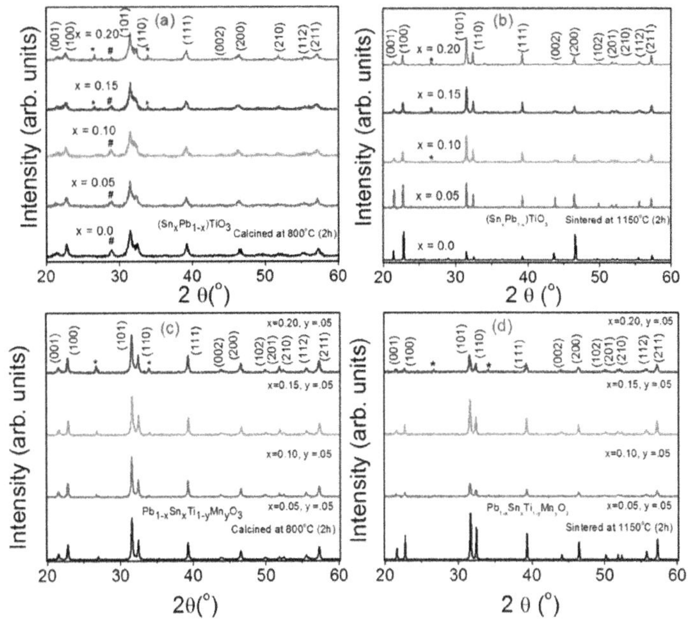

Figure-1 The XRD spectra recorded at room temperature (RT) for $Pb_{1-x}Sn_xTiO_3$ ceramics (a) Calcined at 800°C (2h) and (b) sintered at 1150°C (2h). The XRD spectra recorded at RT for $Pb_{1-x}Sn_xTi_{1-y}Mn_yO_3$ (x =0.10, 0.15, 0.20, y = 0.05) ceramics (a) Calcined at 800°C (2h) and (b) sintered at 1150°C (2h). The impurity peaks are marked on the spectra with # or *. The large separation between (002) and (200) Bragg reflections suggest a large tetragonality, and therefore, a large spontaneous strain in the system.

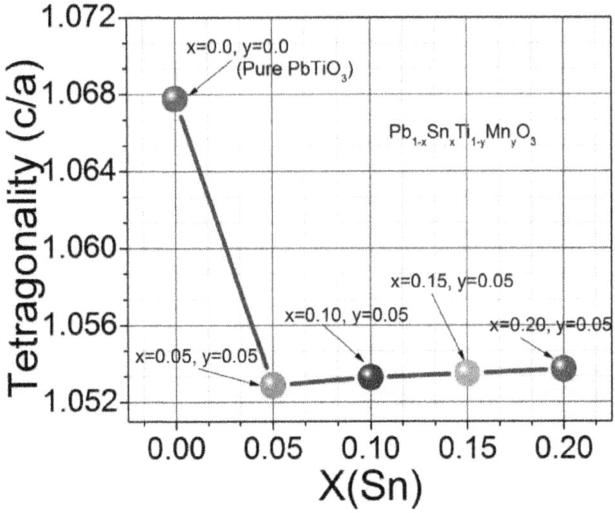

Figure-2 Tetragonality versus Sn content in piezoelectric $Pb_{1-x}Sn_xTi_{1-y}Mn_yO_3$ ceramics. The tetragonality decreased drastically for the lightly doped sample. However, increased doping concentration resulted in a gradual increase in the tetragonality.

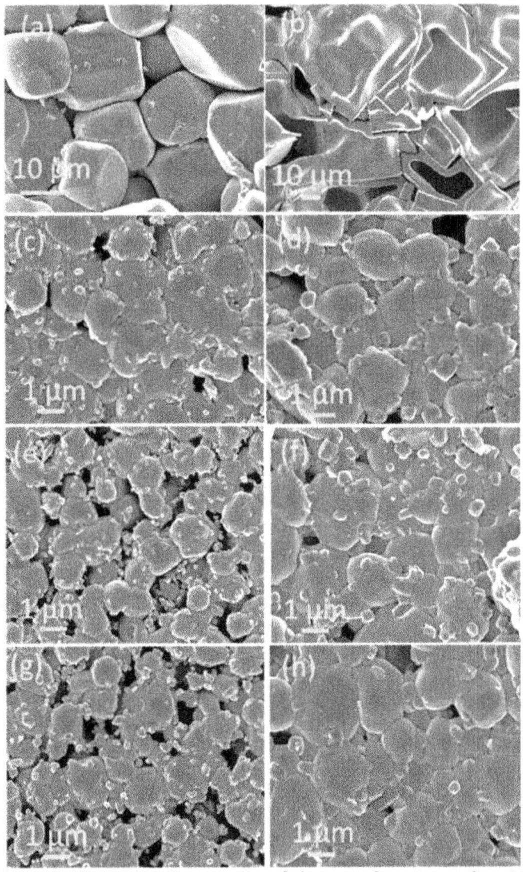

Figure-3 (a), (c), (e), and (g) SEM micrographs of the samples sintered at 1050°C (2h) for $Pb_{1-x}Sn_xTiO_3$ with x = 0.00, 0.10, 0.15, and 0.20, respectively. (b), (d), (f) and (h) SEM micrographs of the samples sintered at 1150°C for $Pb_{1-x}Sn_xTiO_3$ with x = 0.00, 0.10, 0.15, and 0.20, respectively. The undoped PT (x = 0.00) was not sintered due to its high spontaneous strain as a result of large tetragonality. However, doped samples with reduced tetragonality exhibit evidence of densification.

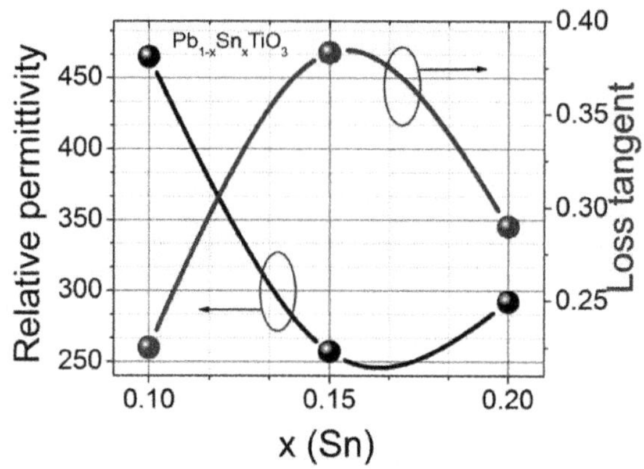

Figure-4 Relative permittivity and loss tangent at 1 *k*Hz as a function of x (Sn) for $Pb_{1-x}SnTiO_3$ ceramics at RT. The Sn modified PT samples exhibited higher loss tangent factor and reduced relative permittivity.

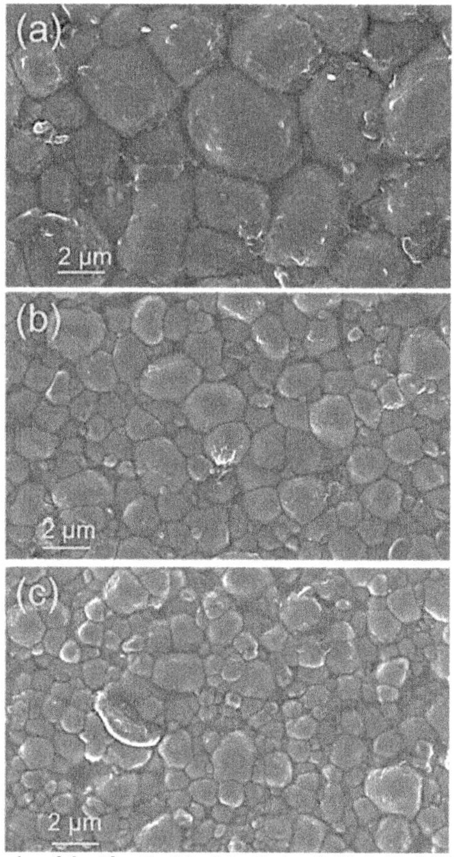

Figure-5 SEM micrographs of the $Pb_{1-x}Sn_xTi_{1-y}Mn_yO_3$ (x =0.10, 0.15, 0.20, y =0.05) ceramics sintered at 1150°C (2h). Mn and Sn co-doping resulted in a high density PT ceramic body.

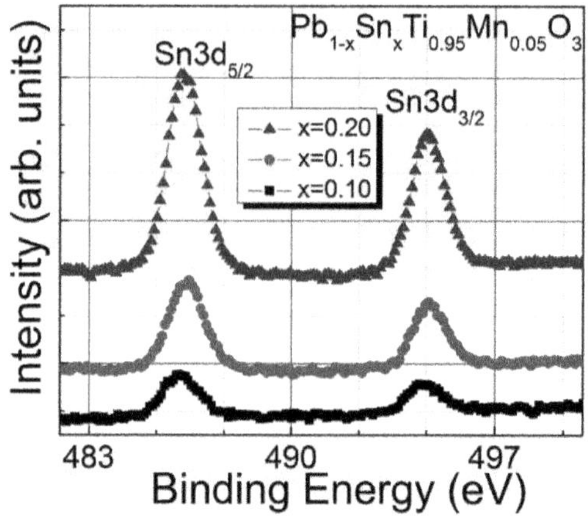

Figure- 6 XPS-spectra recorded at RT for $Pb_{1-x}Sn_xTi_{1-y}Mn_yO_3$ (x = 0.10, 0.15, 0.20, y = 0.05) ceramics. These results suggest the presence of the Sn^{4+} in the modified PT systems. The increased intensity for $Pb_{1-x}Sn_xTi_{1-y}Mn_yO_3$ (x = 0.20, y = 0.05) ceramics can be attributed to the increased concentration of Sn doping.

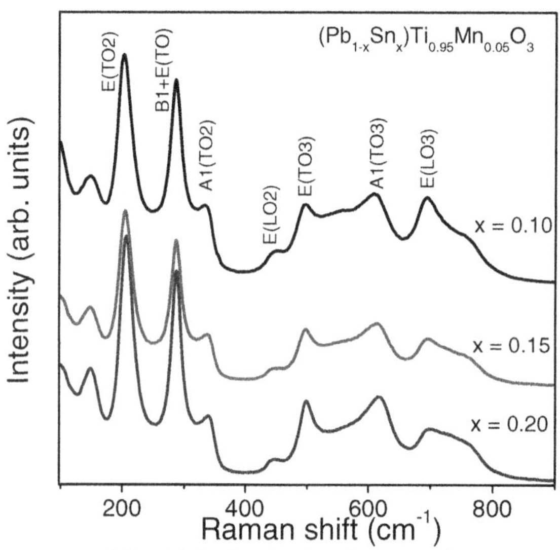

Figure-7 Raman spectra recorded at RT for the $Pb_{1-x}Sn_xTi_{1-y}Mn_yO_3$ (x = 0.10, 0.15, 0.20, y = 0.05) ceramics. Note significant changes in the high frequency region. Sn substitution mainly perturbed the modes related to the octahedral dynamics.

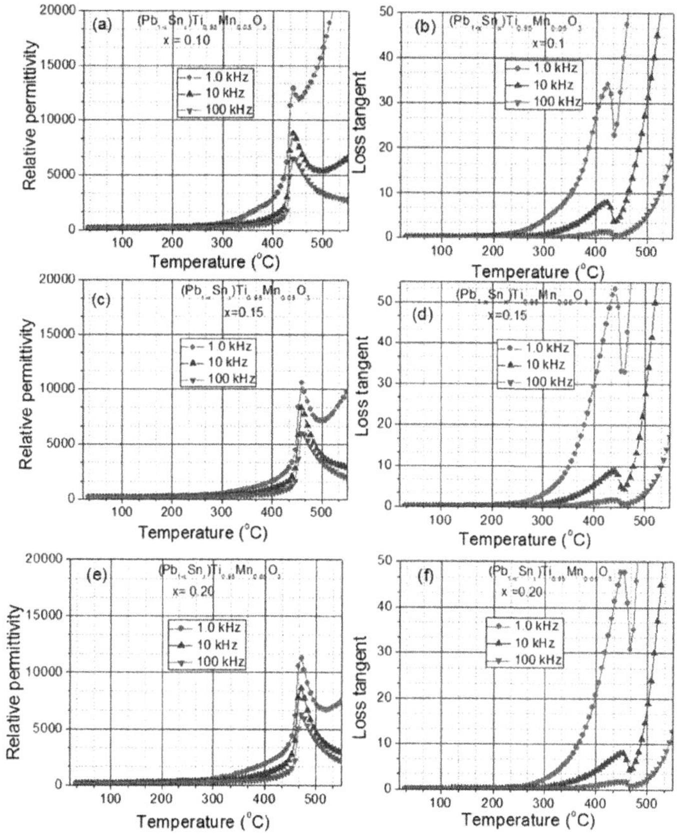

Figure-8 Temperature dependence of the dielectric response at various frequencies for $Pb_{1-x}Sn_xTi_{1-y}Mn_yO_3$ (0.10, 0.15, 0.20, y = 0.05) with (a)-(b) x = 0.10, (c)-(d) x = 0.15, (e)-(f) x = 0.20. Note the shift of the Curie temperature towards higher temperature with increasing Sn content.

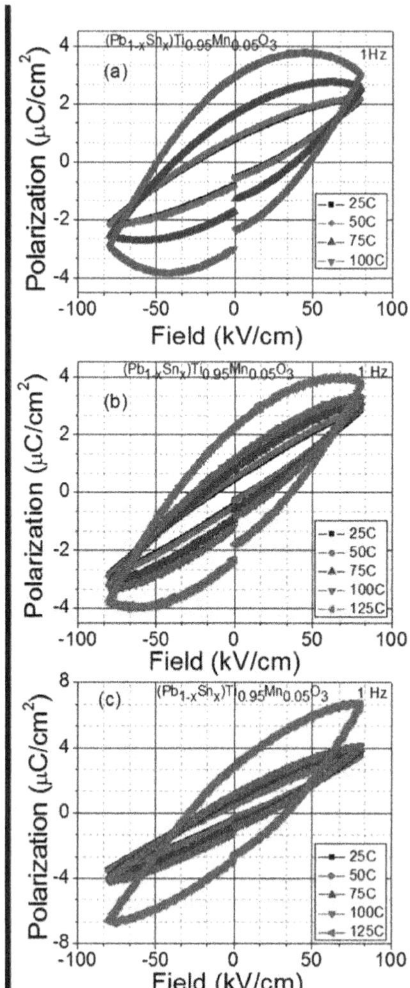

Figure-9 P-E loop of $Pb_{1-x}Sn_xTi_{1-y}Mn_yO_3$ (0.10, 0.15, 0.20, and y = 0.05) at 1Hz for (a) x = 0.10, (b) x = 0.15, (c) x = 0.20. The P-E hysteresis loops for different systems were found to open at different temperatures indicating dissimilar activation energies.

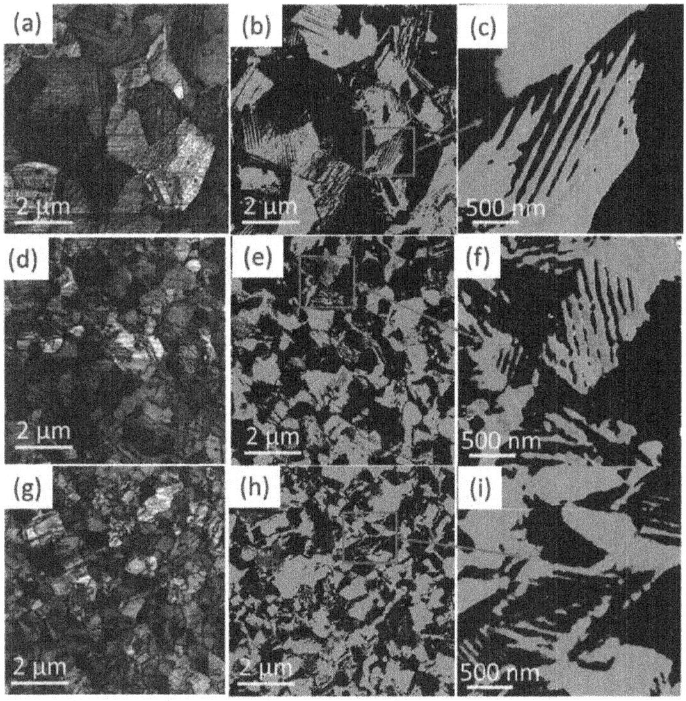

Figure-10 PFM amplitude and phase images for $Pb_{1-x}Sn_xTi_{1-y}Mn_yO_3$ (0.10, 0.15, 0.20, y = 0.05) with (a)-(b) x = 0.10, (d)-(e) x = 0.15, (g)-(h) x = 0.20. The magnified view of the corresponding domain morphology is presented in (c), (f) and (i). The heavily doped samples exhibited smaller domain size due to finite interaction between domains and doping induced defects.

COMPARISON OF GRAIN SIZE EFFECTS ON MICROSTRUCTURE AND DIELECTRIC PROPERTIES OF $Y_{2/3}Cu3Ti_{4-x}Fe_xO_{12}$ (X = 0.00, 0.05 AND 0.15) CERAMICS SYNTHESIZED BY GLYCINE ASSISTED SEMI WET ROUTE

S. Sharma[a], M.M. Singh[a], Narsingh B. Singh[b], K.D. Mandal*[a]

[a]Department of Chemistry, Indian Institute of Technology (Banaras Hindu University), Varanasi, 221005, (U.P.), India.

[b]Department of Chemistry and Bio-chemistry, University of Maryland, Baltimore County, USA

ABSTRACT

The Precursor powder of $Y_{2/3}Cu_3Ti_{4-x}Fe_xO_{12}$ ceramics (x = 0.00, 0.05, 0.15) were fabricated by semi-wet route using metal nitrate solution and solid TiO_2 powder at a relatively lower sintering temperature 950 °C. The effects of Fe^{3+} doping at different concentration level upon the microstructure and dielectric properties of yttrium copper titanate ceramics (YCTO) have been studied over the temperature and frequency range 35-950 °C and 2Hz-5MHz, respectively. The heterovalent doping of Fe^{3+} on Ti^{4+}-site in the YCTO ceramics resulted into an appreciable decrease in the grain size and dielectric constant but a significant lowering of dielectric loss was observed. The loss tangent ($\tan\delta$) of $Y_{2/3}Cu_3Ti_{4-x}Fe_xO_{12}$ (x = 0.15) ceramic drops to 0.58 at a frequency of 1KHz, which is reduced by 19% in comparision to pure CCTO ceramic. The origin of dielectric dispersion phenomena in iron doped YCTO ceramic has been discussed.

1. INTRODUCTION

With recent advancements in miniaturization and integration in device manufacturing processes, it is intriguing to develop some new type of capacitors with high dielectric constant and stable temperature characteristics. Considerable research efforts have been distinctly manifested to develop new materials exhibiting high dielectric constant for applications in microelectronics industries leading to a fast growing field of science, both in the field of novel materials and their applications in future packing technologies [1-3]. Traditional $BaTiO_3$ (BT) and $PbZr_xTi_{1-x}O_3$ (PZT) based ferroelectric materials also exhibit high dielectric constant. But the dielectric properties of these materials are strongly temperature-dependent due to phase transition at Curie temperature, which restricts their application in high temperature range. Further Lead containing ceramics, being poisonous and eco-unfriendly, it is desirable to fabricate some non-ferroelectric materials with high dielectric constant to substitute these traditional ferroelectrics. Materials with the general formula $ACu_3Ti_4O_{12}$ (A = Ca, Ba, Sr) crystallize as body centred cubic perovskite-related structures in space-group *Im*3. One of the most recognised ceramics, $CaCu_3Ti_4O_{12}$ (CCTO) has been studied since the last one decade owing to its giant dielectric constant, $\varepsilon_r \sim 10^4$ for bulk and 10^5 for single crystal and moderately low dielectric loss (~ 0.10) despite its centrosymmetric nature with BCC structure (lattice parameter, a =7.391Å.) CCTO has got

---

*Corresponding author. *E-mail address:* kdmandal.apc@itbhu.ac.in (K. D. Mandal)

advantage over $BaTiO_3$ because its dielectric constant remains nearly temperature independent (100 – 600 K) and frequency independent (10^2-10^6Hz) without any structural phase transition [4]. Such type of temperature independent behaviour of CCTO is technically appealing alternative for the currently used ferroelectric materials and for the development of future microelectronic devices, with the same time suggesting itself a potential for technological applications, especially in memory devices specifically in dynamic random access memory (DRAMs), capacitive components or microwave devices, due to the possibility to reduce the dimensions needed in microelectronic equipments [5].

The correlation between the crystal structure and intrinsic dielectric properties of the unusual cubic perovskite based compounds of $ACu_3Ti_4O_{12}$ family has been a controversial issue for several years, exclusively for $CaCu_3Ti_4O_{12}$ (CCTO) family, the most studied member. Different mechanisms have been proposed to interpret the origin of giant dielectric response of CCTO. The giant dielectric response in CCTO may be supposed either due to disorder caused at Ca/Cu-site in CCTO crystal or due to an extrinsic mechanism caused by internal barrier layer capacitance (IBLC) model [6]. The IBLC mechanism has been widely accepted but is still controversial. Some researchers correlate this phenomenon to origin of local dipole moments associated with the displacement of some ions or due to relaxor behaviour like dynamical slowing down of dipolar fluctuations in nano-size domains or off-centric displacement of Ti^{4+} ions within TiO_6 octahedra within CCTO molecule [7]. Various research papers revealed that the dielectric response of CCTO is significantly influenced by factors like processing routes and conditions applied during its fabrication like sintering temperature and its duration, cooling rate and partial pressure of the sintering atmosphere. Apart from this, doping schemes as well as nature of dopants and their stoichiometric variations affecting the microstructure of the material and the measuring frequency and temperature range are quite important [8]. Various papers describing substitution effect of dopants on dielectric properties of CCTO has been reported earlier are scattered in literature.

The structural isomorphs of $ACu_3Ti_4O_{12}$ materials with a general formula $A_{2/3}Cu_3Ti_4O_{12}$ (A = $Y_{2/3}$, $Bi_{2/3}$, $La_{2/3}$) also show similar dielectric properties and are also reported in literature. Structural isomorphs of $A_{2/3}Cu_3Ti_4O_{12}$ (ACTO) where A is a rare earth metals have gained tremendous research attraction due to its various possible applications in microelectronics, high energy density storage applications and high performance dielectric devices. Such compounds could also be a promising candidate to replace relaxors as dielectrics. Apart from this, a wide variety of doping or partial substitution may be carried out at A, Cu or Ti-site of ACTO, which may be suitable for different industrial applications such as dynamic and versatile memory components, capacitors, transducers and sensors etc. The choice of substituent to modify the physical properties of the material is based on many factors including charge neutrality, tolerance factor, the ionic radius and solubility. As we know partial isovalent substitution of metal cations in different interstices can improve properties associated with ferroelectricity and dielectric response in these materials due to the partial modification of mixed-valent structure. Further, structural flexibility and chemical versatility of the materials make them more suitable for device applications [9] and it is quite inspiring to investigate systematically new $ACu_3Ti_4O_{12}$ like materials. A detail literature survey shows that not much work has been reported on $Y_{2/3}Cu_3Ti_4O_{12}$ (YCTO) compound which is isostructural to CCTO which may offer a large database to search for high dielectric constant materials [10-12]. Being non-ferroelectric, lead-free and eco-friendly too, YCTO material exhibits a high dielectric constant with good thermal stability. However, the dielectric loss (tan δ) of this material is high which is the main obstacle limiting its technical use. But due to advantage of flexible structure, properties can be modified by changing the chemical compositions through different possible substitutions at Y^{3+}, Cu^{2+} or

Ti^{4+} sites. Till date not much work was done on $Y_{2/3}Cu_3Ti_4O_{12}$ system [13-14]. To our best knowledge, no work on the effects of substituting impurity in YCTO at a lower processing temperature has been reported. Presently, we have worked out to control the dielectric loss factor by doping of Fe^{3+}at Ti^{4+} site in continuation to a detail research work on YCTO system.

In the present communication, we reported the effect of cationic substitution of Ti^{4+} site by doping with a heterovalent Fe^{3+} ion in $Y_{2/3}Cu_3Ti_{4-x}Fe_xO_{12}$ (where x = 0.00, 0.05, 0.15) ceramic. The dopent was introduced into YCTO system at different concentration level to synthesize the single phase iron doped $Y_{2/3}Cu_3Ti_{4-x}Fe_xO_{12}$ (where x = 0.00, 0.05, 0.15) samples by the semi-wet route at a relatively lower sintering temperature 950 °C.

2. MATERIALS AND METHODOLOGY

Iron doped $Y_{2/3}Cu_3Ti_4O_{12}$ ceramic, with the regular composition $Y_{2/3}Cu_3Ti_{4-x}Fe_xO_{12}$ (where x = 0.00, 0.05, 0.15), were prepared by semi-wet route. In the current synthesis, analytical grade chemicals $Y(NO_3)_3.6H_2O$ [99%, HiMedia, India], $Cu(NO_3)_2.3H_2O$ [99.5%, Merk, India], $Fe(NO_3)_3.9H_2O$ [99.8%, Merk, India], TiO_2 [99.5%, Merk, India], and glycine [99.5%, Merk, India], were used as a raw materials. The semi-wet route is somewhat a modified chemical route in terms of its simple processing steps. It involves mixing of solutions of a metal precursor and an organic poly -functional acid possessing at least one hydroxyl and one carboxylic acid group such as citric, glycine, tartaric and glycerol, which results in complex formation of the metal with the poly-carboxylic acid. It is more convenient for achieving chemical homogeneity where the individual constituents intermix at ionic level under the controlled wet chemical conditions. In this chemical method, nitrate solutions of yttrium and copper ions, in stoichiometric ratio are mixed along with solid TiO_2 and glycine of comparatively low cost solid is used by avoiding expensive titanium iso-propoxide $Ti(OR)_4$ in solution form and got single phase of YCTO at comparatively lower sintering temperature in short durations. Glycine, as a complexing agent, can complex with metal cations at both the carboxylic end and the amino group end; provides fuel for the ignition step. It helps in increasing the ignition temperature as it is oxidized readily by the nitrate ions to form a very fine crystalline precursor powder at very low temperature [15].

For the preparation of different composition $Y_{2/3}Cu_3Ti_{4-x}Fe_xO_{12}$ (where x = 0.00, 0.05, 0.15), The standard solutions of the required metal nitrates were prepared using double distilled water. Solutions of metal nitrates in stoichiometric amount of Y^{3+}, Cu^{2+} and Fe^{3+} ions, each as per $Y_{2/3}Cu_3Ti_{4-x}Fe_xO_{12}$ composition (where x = 0.00, 0.05, 0.15), were mixed in a beaker and a calculated amount of TiO_2 and glycine, equivalent to metal ions, were added to the solution. The solution was heated on a hot plate with constant stirring using a magnetic stirrer (REMI, 2 MLH) at 70-80 °C to evaporate water till blue gel was obtained. During combustion process a high viscous mass is formed by evaporation of a solution of metal nitrates and while glycine is ignited to produce a loose mass of very fine crystalline powder, particle size less than tens of nanometers is obtained after ignition, so, in contrast to the Pechini's method, no grinding is required. Glycine, an amino acid, forms complexes with the metal ions in the solution which increases the solubility and prevents the precipitation of the metal ions as water is evaporated. A good chemical homogeneity is, therefore, achieved in contrast the Pechini method. YCTO ceramic powder thus obtained is crystalline and, of course, very fine size. It may be probably be due to a direct but short exposure to high temperature during the ignition step. Glycine-nitratre process therefore offers a relatively an inexpensive route for the preparation of simple oxides as well as complex oxides.

Thermal studies were carried out for the dry powder using TG/DTA (Pyris Diamond TG/DTA, Perkin Elmer Instrument, USA) with heating rate 10 °C per minute from ambient to1000 °C. The dry powder was calcined in air at 500 °C and 800 °C for 5h and 8h respectively in a muffle furnace. The calcined powder was ground into fine powder and then pressed into cylindrical pellets of dimension 9 mm in diameter and 2 mm in thickness using a hydraulic press by applying uniaxial pressure of 4 ton. Poly vinyl alcohol (PVA) was used as a binder to reduce the brittleness of pellets. The pressed pellets were again heated to 500 °C to remove the binder by the burn-out process and then sintered in air at 950 °C for 12h in air. Prior to the structural characterization and physical property measurements, the pellets were sintered at 950 °C for 12h in air to get maximum density.

The crystalline phase structure of all compositional variant were identified using an X-ray diffractometer (D/max-2550/PC, Rigaku, Tokyo, Japan) employing Cu-K_α radiation (λ=1.5414 Å)) in the range of 2θ = 20-80°. The morphology of the fractured surfaces of different compositional variant YCTFO ceramic were characterized by scanning electron microscopy (SEM, Model JEOL JSM5410) and transmission electron microscopy (TEM, FEI Tecnai-20G^2) with an accelerating voltage of 200 kV. For TEM analysis, powder form of sintered ceramic of each composition was dispersed in acetone differently and 1-2 drops of each were mounted on a carbon coated copper grid. The solvent was subsequently allowed to escape in air at room temperature. Density and porosity of all the five compositions of $Y_{2/3}Cu_3Ti_{4-x}Fe_xO_{12}$ (where x = 0.00, 0.05, 0.15) sintered at 950 °C were measured with the help of Archimedes Principle. For dielectric measurements, both the flat and parallel surfaces of one pellet of each compositional variant was polished and coated by air drying conducting silver paint. To form electrodes, required for further electrical characterizations, those pellets was dried at 200 °C for 30 minutes to remove moisture, if any, and then cooled at room temperature, before taking electrical measurements. Dielectric measurement data were taken by LCR meter (PSM1735-NumetriQ, Newton 4th Ltd, U.K.) in a wide range of frequency 2Hz – 5MHz at different temperature 35-227 °C. The value of dielectric constant was evaluated using the standard relation, $\varepsilon_r = Cd/\varepsilon_0 A$, where C is capacitance, d is the thickness of the pellet, A is the effective cross sectional area of the pellet and $\varepsilon_0 = 8.854\times10^{-12}$ F/m.

3. RESULTS AND DISCUSSION

To ascertain the ideal temperature required for the thermal decomposition of precursor dry powder of $Y_{2/3}Cu_3Ti_{4-x}Fe_xO_{12}$ ceramic (where x = 0.00, 0.05, 0.15), TG/DTA was carried out. Fig. 1 shows simultaneous TG/DTA/DTG plots of precursor powder of YCTFO (where x = 0.05) ceramic. Two major stages of weight loss in the temperature range from 100 °C to 1000 °C was observed in TGA plot, the first major weight loss at 250 °C is due to an exothermic reaction which may be attributed to dehydration and combustion of the gel while the second weight loss at 800 °C is due to the formation of an intermediate compound. Further, an additional but very small weight loss at 950 °C was also observed which is due to an exothermic addition reaction of the intermediate compound to give the final product $Y_{2/3}Cu_{3-x}Ti_{4-x}Fe_xO_{12}$ where (x = 0.05). The corresponding DTA and DTG plot also show two strong exothermic peaks near 250 °C and 850 °C besides a small peak at 950 °C. It was also observed that DTG peaks are more intense than DTA peaks.

XRD pattern of $Y_{2/3}Cu_3Ti_{4-x}Fe_xO_{12}$ (where x = 0.00, 0.05, 0.15) sintered at 950 °C for 12 h is illustrated in Fig. 2. It clear from Fig 2 that all the compositional variants show distinct diffraction peaks corresponding to (211), (220), (013), (222), (321), (400), (422), (440) orientations which were found to have a good matching with $CaCu_3Ti_4O_{12}$ (JCPDS card no. 75-2188), demonstrating thus a polycrystalline characteristic of YCTFO ceramics. In addition to a dominant pure $CaCu_3Ti_4O_{12}$ (CCTO) phase, Fig 2 indicates the presence of small amounts of CuO as a minor, secondary phase of CuO (JCPDS 80-1917). The curves of sample with doping x = 0.05, 0.15 were almost identical with that of x = 0.00 i.e the undoped one and with the same

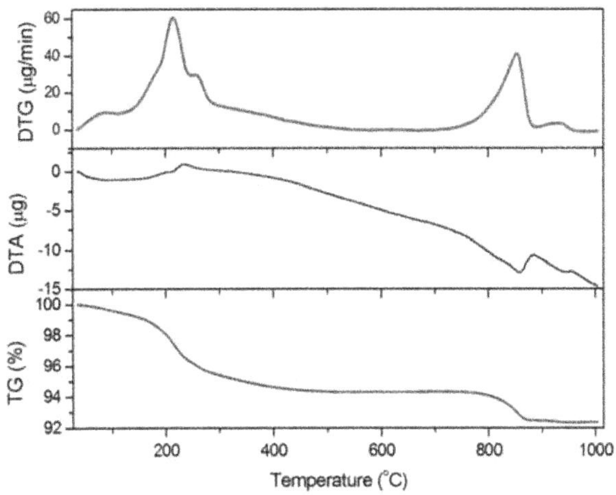

Figure 1 : DTA/TGA curve for precursor powder of $Y_{2/3}Cu_3Ti_{4-x}Fe_xO_{12}$ ceramic (where x = 0.05).

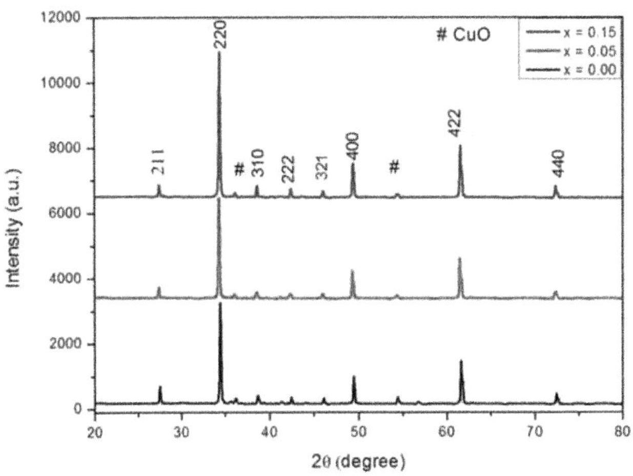

Figure 2 : XRD patterns of $Y_{2/3}Cu_3Ti_{4-x}Fe_xO_{12}$ ceramic (where x = 0.00, 0.05, 0.15) sintered at 950 °C for 12 h.

time, not a single Fe^{2+}/Fe^{3+}-related phase was detected in the XRD patterns which indicating that Fe^{3+} hadn't influenced the crystalline nature of YCTFO ceramic. At the same time, no significant secondary phase could be detected in the XRD data, for these samples probably because their quantities may be lower than the detection limits of the technique. It is also clear from the Table 1 that the lattice constant increases monotonically with increasing of x values. It indicates that the Fe^{3+} may have taken the place of Ti^{4+} as expected. The increase of the lattice constant is with doping concentration is may be either due to the difference in ionic radii of the dopant Fe^{3+} (0.785Å) to that of parent Ti^{4+} (0.745Å) ion and thus causing a decrease in defect density due to doping.

XRD data were indexed on the basis of a cubic unit cell similar to CCTO. The lattice parameter and unit cell volume of different compositional YCTFO ceramic were determined by least squares refinement method using 'Cel' software and are listed in Table 1. The presence of split peaks in XRD pattern for the reflections 400, 422 and 440 may be due to the presence of Cu-Kα_2 along with Cu-Kα_1 in the X-ray radiations used for diffraction. It is supported by the fact that in all these reflections, the intensity of peaks due to Cu-Kα_2 is close to 50% of intensity of peak due to Cu Kα_1 as expected [16].

The crystallite size (D) of the different YCTFO ceramic were estimated using the Debye-Scherrer's formula, as shown in equation (1)

$$D = \frac{k\lambda}{\beta \cos \theta} \tag{1}$$

where k is the crystal shape coefficient (k = 0.89), λ is the wave length of X-ray radiation, θ is the diffraction angle and β is the corrected full width at half maximum (FWHM). The calculated values of crystallite size for the different YCTFO ceramic were found to be 68±6 nm, 65±10 nm, 60±7 nm, rwespectively which is comparable to the earlier value reported in literature [13-14].

Table 1. Crystal system, lattice parameter and unit cell volume for $Y_{2/3}Cu_3Ti_{4-x}Fe_xO_{12}$ ceramic (where x = 0.00, 0.05, 0.15) sintered at 950 °C for 12 h.

System	Composition	Pellet Dimension Thick. (mm)	Diameter (mm)	Bulk Density	Porosity Bulk	Apparent	Crystal system	Lattice Parameter (Å)	Unit cell Volume (Å)^3
$Y_{2/3}Cu_3Ti_{4-x}Fe_xO_{12}$	x =0.00	3.5	12.5	1.195	80.06	22.02	Cubic	7.4983 ± 0.1037	421.5833 ± %17.5032
	x = 0.05	2	9	1.167	80.38	11.19	Cubic	7.5149 ± 0.1013	424.3867 ± %17.1629
	x = 0.15	2	9	1.291	78.21	18.67	Cubic	7.5173 ± 0.0994	424.7931 ± %16.8559

The scanning electron micrograph (SEM) of the fractured surfaces of $Y_{2/3}Cu_3Ti_{4-x}Fe_xO_{12}$ (where x = 0.00, 0.05, 0.15) ceramic sintered at 950 °C for 12h are shown in Fig 3(a-c). The SEM micrograph exhibits non-uniform grain size distribution consisting of bimodal nature with frequent segregation. The microstructure is dominated by small smooth surfaced grains with some pores. The grains are still in growing condition which is suggested by the variation in sizes of spherical, cylindrical and cubical structures as well. The bimodal grain size distribution contains regions of fine grains about ~0.4-1 μm and regions of extremely large grains about ~1.5-3.5 μm for all compositions with high degree of segregation. The variation in grain size thereby leading to abnormal grain growth may be due to the presence of CuO rich phase at the grain boundary of the ceramic. Fang *et al* also reported the presence of CuO along the grain boundary which transforms into the liquid phase during sintering. It may induce discontinuous grain growth [17] which is in accordance with the SEM micrograph of Fig.3. The CuO phase may contribute significant effects to promote the grain growth and densification of YCTFO samples. Fig 3(a) shows a discontinuous grain growth for the undoped sample, however, a continuous grain growth can be seen clearly for all the samples with doping of iron with a decrease in the average grain size from 0.61-0.52 μm for smaller grains and 2.30-3.14μm for larger grains. The doping of iron has effectively prevented the growth of large grains and average grain sizes are approximately 0.5~3.5 μm which is independent of Fe^{3+} concentration, as shown in Fig.3 b~c. It indicates that the adding of Fe^{3+} with variable concentration can reduce the grain size of YCTFO ceramic and improve the density of the ceramic. The bulk density of the samples was measured and found to be 1.195. 1.167 and 1.151 g/cc for x = 0.00, 0.05 and 0.15 respectively, considering the theoretical density of 5.995 g/cc for YCTO [14]. Such variation in density of these samples may be due to the presence of some pores as well as some oxygen vacancies in YCTFO ceramics which in turn influences the grain and grain boundary dimensions and dielectric properties of the ceramic material.

Figure 3: SEM micrograph of fractured surface of sintered $Y_{2/3}Cu_3Ti_{4-x}Fe_xO_{12}$ ceramic (where x =0.00, 0.05, 0.15).

It is also established that increase in sintering temperature significantly promote the grain growth and microstructural densification. It is quite obvious that with decrease in average size to those of smaller and larger grain, porosity decreases leading to more compactness of YCTFO ceramics. Further, it is reported in literature that below 1μm, the dielectric constant decreases significantly with grain size. As the grain size approaches to 0.5- 1.5 μm, the various types of stress operating in the grain suppress the domain and the dielectric constant falls to approximately 1000 [1]. Such type of relation between the grain size and dielectric constant value are reflected as well for different YCTFO ceramic samples. Besides grain size in the range of 0.5-1.5 μm, doped YCTFO samples (x =0.05 and 0.15) exhibits a low dielectric constant value (ε_r ~380, 326) than x =0.00 i.e. the undoped YCTO ceramic (ε_r ~2198) with grain size in the range of 1-3μm which may be due to porosity, compactness of grain and the structural quality of the material. It is also clear from the Fig. 3 that for different YCTFO ceramic samples exhibits high degree of porosity. The calculated values of porosity are also listed in the Table 1. The high porosity may be attributed to the formation of oxygen vacancies during sintering. It may be also due to lower sintering temperature. Increase in sintering temperature may lead to decrease in porosity owing to the grain growth phenomena. Moreover, the porosity also depends upon processing route, samples prepared by Pechini method and co-precipitation method exhibit a lower extent of porosity than those prepared by other processes at the same sintering temperature [18]. It was also observed that doping of Fe in YCTO ceramic causes decrease in density and increase in porosity. Density of different samples of YCTFO samples are in accordance with the reported value of undoped YCTO ceramic [13].

Fig. 4 shows EDX spectra of grain of different compositions of $Y_{2/3}Cu_3Ti_{4-x}Fe_xO_{12}$ (where x = 0.00, 0.05, 0.15) which indicates the presence of Y, Cu, Ti, Fe and O elements. Quantitative data for atomic and weight percentage of the elements present in the grain of YCTFO obtained from EDX data which is shown in Table 2. The atomic percentage of Y, Cu, Ti, Fe and O in different compositions of $Y_{2/3}Cu_{3-x}Ti_{4-x}Fe_xO_{12}$ ceramic was as per expected stoichiometry.

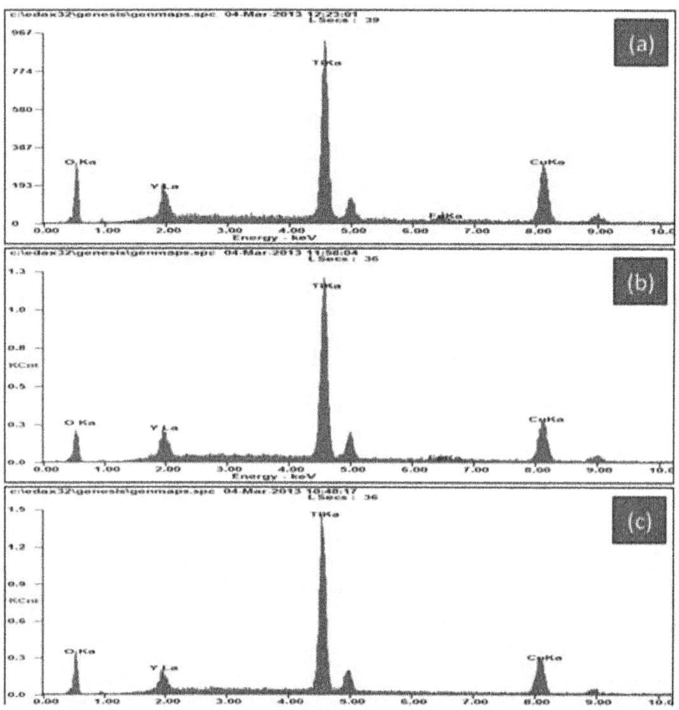

Figure 4: EDX spectra of grain surface of sintered $Y_{2/3}Cu_3Ti_{4-x}Fe_xO_{12}$ ceramic,where x = (a) 0.00, (b) 0.05, (c) 0.15.

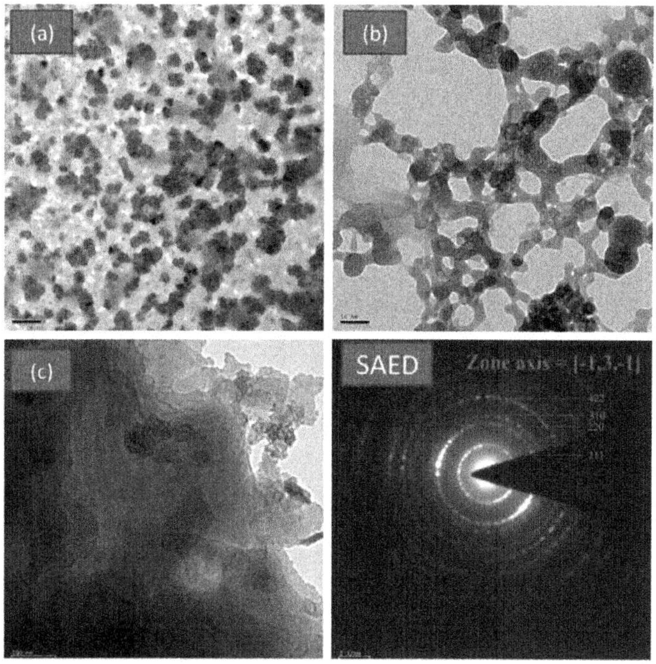

Figure 5: TEM bright field image of sintered (a) $Y_{2/3}Cu_3Ti_4O_{12}$ (b) $Y_{2/3}Cu_3Ti_{3.90}Fe_{0.10}O_{12}$ (c) $Y_{2/3}Cu_3Ti_{3.85}Fe_{0.15}O_{12}$ ceramic (d) corresponding SEAD pattern for $Y_{2/3}Cu_3Ti_{4-x}Fe_xO_{12}$ ceramic (where x = 0.05).

Table 2. Atomic percentage and weight percentage of elements for $Y_{2/3}Cu_3Ti_{4-x}Fe_xO_{12}$ ceramic (where x = 0.00, 0.05, 0.15) sintered at 950 °C for 12 h.

System	Composition	Atomic %					Wt %				
		Y	Cu	Ti	Fe	O	**Y**	Cu	Ti	Fe	O
$Y_{2/3}Cu_3Ti_{4-x}Fe_xO_{12}$	x =0.00	03.56	18.80	24.65	----	52.99	08.94	33.76	33.36	----	23.95
	x = 0.05	03.60	19.74	30.14	01.14	45.37	08.40	32.94	37.91	01.68	19.06
	x = 0.15	03.36	20.80	22.53	01.28	20.80	08.28	36.67	29.95	01.99	36.67

The bright field TEM image of $Y_{2/3}Cu_3Ti_{4-x}Fe_xO_{12}$ ceramic (where x = 0.00, 0.05, 0.15) sintered at 950 °C for 12h is represented in Fig. 5(a-c). The particles of different compositions of $Y_{2/3}Cu_3Ti_{4-x}Fe_xO_{12}$ are well dispersed with high extent of agglomeration. The average particles are in the range of 60±20 nm, 48±22 nm and 23±7 nm, respectively for x = 0.00, 0.05, 0.15 compositions. The lattice parameter calculated from the electron diffraction pattern is also in agreement with that obtained from the XRD data. However, the particle size obtained by TEM analysis is lesser than XRD results because a crystallite may consist of several particles. The corresponding selected area electron diffraction (SAED) pattern of TEM image for YCTFO ceramic is shown in Fig. 5(d). TEM diffraction pattern was indexed on the basis of body centred cubic perovskite structure and the zone axis of the selected area electron diffraction (SAED), which is a common crystal plane shared by the two planes (211) and (310) which is found to be $[\bar{1}\,3\,\bar{1}]$. As the corresponding selected area electron diffraction (SAED) pattern of TEM image for different samples are more or less same, SAED picture for composition x = 0.05 only is reproduced here. The presence of a few clear rings in SAED pattern again confirms the formation of polycrystalline phase of thermodynamically stable YCTFO ceramic.

Temperature dependence of dielectric properties of YCTO ceramic at 100 Hz, 1, 10 and 100 kHz is shown in Fig 6 exhibits the temperature dependence of dielectric (ε) and dielectric loss (tan δ) measured at a few frequencies for $Y_{2/3}Cu_3Ti_{4-x}Fe_xO_{12}$ (where x = 0.00, 0.05, 0.15) ceramic. It is clearly observed from Fig 6(a) that the YCTFO ceramic, with high concentration of Fe^{3+} exhibits step like increase in dielectric constant with increasing measuring temperature. Along with this, a broad dielectric peak is also observable for almost all Fe^{3+} doped YCTFO compositions with x = 0.00, 0.05, 0.15 near temperature 350-60 K, but the dielectric peak appears to be suppressed one for compositions with doping concentration, x = 0.05 and 0.15. These curves exhibit almost temperature independent behaviour after temperature 425-500 K. Further, with increasing measuring frequency, increase in dielectric constant suppressed with amplitude which indicates that the existence of at least one Debye like relaxation in different YCTFO ceramics. In general, one Debye like relaxation is accompanied by the appearance of dielectric loss peak or hump in tan δ-T plot. As shown in Fig 6(b), only one dielectric loss peak is observed around 360 K for YCTFO ceramic for all concentration. It is also noticeable that with increase in concentration of Fe, the hump in the low temperature range gets intensified in both ε-T and tan δ-T plots which may be attributed to increase in the grain boundary resistance therefore it can be inferred that the observed low temperature dielectric relaxation supports the grain boundary effects.

Fig. 7(a-c) illustrates the variation of real and imaginary part of dielectric constant (ε) and dielectric loss with frequency for Fe^{3+} doped YCTFO ceramic sintered at 950° C for 9 h within the frequency range 2Hz to 5MHz at different concentration. The values of dielectric constant (ε') of YCTFO ceramic with different compositions are 100837, 43065 and 29958, respectively. All the synthesized compositions exhibit a constant dielectric constant value 10^4-10^3 at the frequency 2 Hz to 5 MHz with a continuous decrease in the value of dielectric constant with frequency, however, this decrease is rapid in lower frequency range (2 Hz-100 Hz) and very slow in higher frequency range (1 kHz to 5 MHz). A large value of dielectric constant under frequency 2-100Hz may be associated to electrode contact while the drastic decrease in the value of dielectric constant at high frequency greater that f = 100 Hz thus suggests the presence of conductive grains. Furthermore, the rapid decrease in dielectric constant with frequency may be due to the contribution of space charge accumulation at the interface which leads to polarization of the ionic medium and hence the value of ε' increases. In high frequency regions, the periodic reversal of the field takes place so rapidly that there is no charge accumulation at the interface, resulting into a constant ε′ value [19].

The variation of imaginary part of dielectric constant (ε″) with frequency at room temperature for all compositions exhibit an appreciable decrease in the loss magnitude as shown in Fig 7(b). At high frequencies, the losses are much lower than the one occurring at low frequencies. Such kind of dependence of the dielectric loss with frequency is associated with losses due to the conduction mechanism.

A rapid decrease in dielectric constant of $Y_{2/3}Cu_3Ti_{4-x}Fe_xO_{12}$ with different Fe^{3+} concentration at higher frequency may be due to inability of dipoles to keep stepping with the change in frequency. This is the reason for high frequency dielectric relaxation of YCTFO ceramic which gets induced by a mixed valent structure of Cu^{+}/Cu^{2+} and Ti^{3+}/Ti^{4+} while the low frequency dielectric relaxation of YCTFO ceramic may be regarded as a result of oxygen vacancies created during sintering process. The two factors, the mixed valent structure and oxygen vacancies may be significantly contribute for the origin of the high dielectric constant of YCTFO ceramic. Along with these two factors interfacial polarisation and formation of barrier layer at grain and grain boundaries interface significantly contribute to the high dielectric constant of YCTFO ceramic.

The room temperature frequency dependent dielectric loss (tanδ) of $Y_{2/3}Cu_3Ti_{4-x}Fe_xO_{12}$ ceramic (where x = 0.00, 0.50, 0.15 as a function of concentration of Fe^{3+}) is shown in Fig. 8. The value of tan δ decreases rapidly over the entire frequency range which may be attributed to the presence of small interfacial polarization in the YCTFO ceramic [8]. No constant variation in tan δ with frequency is observed for all compositions. Dielectric loss for all compositions show a sharp decreasing trend up to 100 Hz and then a small rising at 1 kHz for undoped ceramic but after 1 kHz the dielectric losses decreases sharply, however the dielectric losses for YCTFO

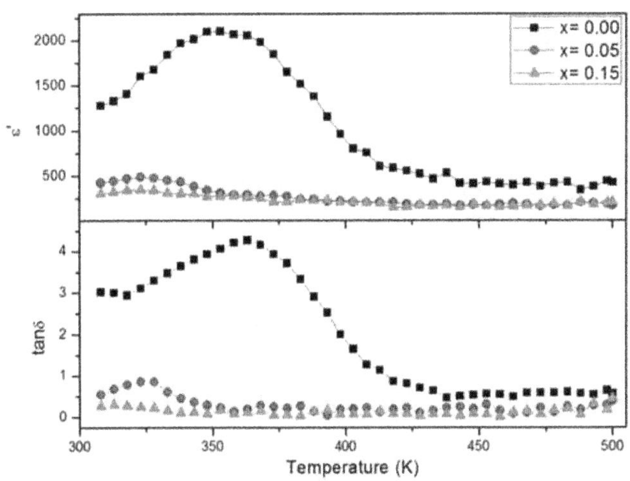

Figure 6: Plots of dielectric constant (ε′) and loss tangent (tan δ) vs temperature at few selected frequencies for sintered $Y_{2/3}Cu_3Ti_{4-x}Fe_xO_{12}$ ceramic (where x = 0.00, 0.05, 0.15).

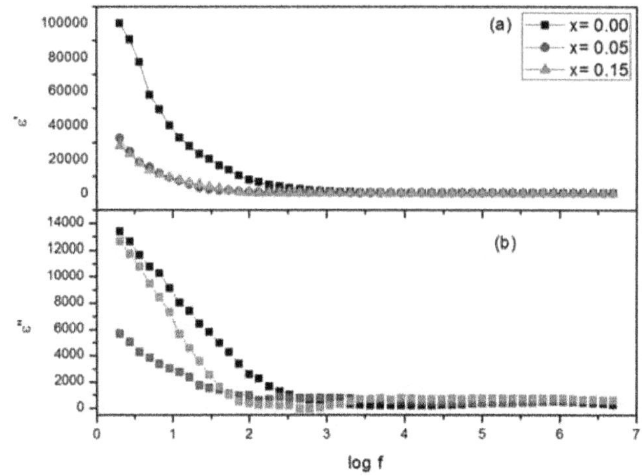

Figure 7: Plots of (a) real dielectric constant (ε') and (b) imaginary dielectric constant (ε'') with frequency at different temperature for $Y_{2/3}Cu_3Ti_{4-x}Fe_xO_{12}$ ceramic (where x = 0.00, 0.05, 0.15).

ceramic are higher at low frequency below 10 kHz. Further the tan δ value decreases significantly at a higher frequency form 10 kHz-5MHz which may be due of interfacial polarization originating in the YCTFO ceramic. Higher concentration of Fe^{3+} doping may introduce more oxygen vacancies and space charges producing thereby dielectric relaxation peaks which cause more dielectric loss.

It is also clear from Fig. 8 that the curve passes through a minimum tan δ value of 2.5, for x = 0.00, and 0.74, 0.51 for x = 0.05, 0.15 respectively at a frequency, $f = 10^{3.5}$Hz. In all, Fe^{3+} doping in YCTO ceramic decreases the dielectric loss at high frequency ranges. The undoped YCTO ceramic shows a decrease in tan δ value with a clear hump at or below 10 k Hz but the curve for doped one i.e. for x= 0.15, 0.15 got relatively flat, with no hump, and these were similar to some extent to those of variation in dielectric constant value with increasing frequency. The reason for such anomalous behavior of variation in dielectric loss of Fe may be due to bimodal distribution of grain in YCTFO ceramic for different x values. It may also be due to the grain sizes, its distribution and grain growth phenomena of both small and larger sized grain as already discussed in microstructure section. Such anomalous trend were also reflected in variation of ε-T and tan δ-T (as shown in Fig. 6) supporting thereby the fact that a proportionate introduction of Fe^{3+} flattens the dielectric constant appreciably with a significant reduction in dielectric loss (tan δ) simultaneously in the YCTFO ceramic.

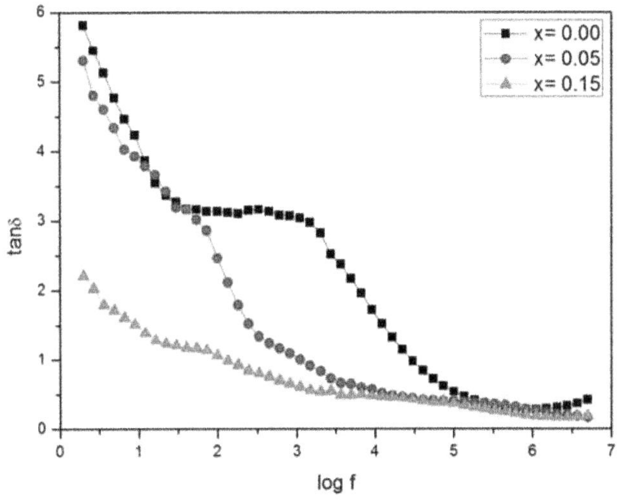

Figure 8: Plots of loss tangent (tanδ) vs frequency at few selected temperature for $Y_{2/3}Cu_3Ti_{4-x}Fe_xO_{12}$ ceramic (where x = 0.00, 0.05, 0.15).

The decreasing value of both, ε_r and tan δ with increasing frequency are due to interfacial polarization which arises as discussed earlier, due to micro-heterogeneities arisen in the ceramic with different extent of doping. These may be present because of the slow diffusion controlled thermochemical process used in their preparation. Fe^{3+} substitution in $Y_{2/3}Cu_3Ti_4O_{12}$ (YCTO), lower downs the dielectric loss (tan δ) and it shows stronger dependence on frequency as well as temperature. It is expected that using different atmosphere during sintering and post sintering, the above properties may be improved showing a positive direction for clues to decrease the dielectric loss to an appreciable extent for potential application of YCTFO ceramic, the next task to the researcher.

4. CONCLUSIONS

Nano-sized $Y_{2/3}Cu_3Ti_{4-x}Fe_xO_{12}$ ceramic (where x = 0.00, 0.05, 0.15) were synthesized by semi-wet route using metal nitrate solution and solid TiO_2 powder in stoichiometric ratio. XRD characterization confirmed the regular phase formation along with the presence of minor phase of CuO. The crystallite size of the YCTFO ceramic was found to be 68±6 nm, 65±10 nm, 60±7 nm, respectively by XRD while TEM studies revealed the size of agglomerated particles to be 60±20 nm, 48±22 nm and 23±7 nm respectively for x = 0.00, 0.05, 0.15 compositions. Dielectric constant and tangent loss of the ceramic are temperature independent in higher frequency region but temperature dependent in low frequency region. Low temperature dielectric relaxation is due to the grain boundary effects. The mixed valent structure and oxygen vacancies may be significantly contribute for the origin of the high dielectric constant of YCTFO ceramic.

ACKNOWLEDGEMENTS

Authors are very thankful to Prof. R K Mandal, Department of Metallurgical Engineering, IIT (BHU) for extending TEM and SEM facility.

REFERENCES

[1] A. J. Moulson, J.M. Herbert. Electroceramics, Second edition. Wiley, England, 2003.
[2] D.E. Kotecki, Integr. Ferrorlectrics 16 (1997) 1.
[3] C.C. Homes, T. Vogt, S.M. Shapiro, S. Wakimoto, A.P. Ramirez, Science 293 (2001) 673.
[4] M.A. Subramanian, A.W. Sleight, Solid State Sci. 4 (2002) 347.
[5] CM Garner, G Kloster, G Atwood, L Mosley, AC Palanduz. Microelectr. Reliab. 45 (2005) 919.
[6] D.C. Sinclair, T.B. Adams, F.D. Morrison, A.R. West, Appl. Phys. Lett. 80 (2002) 2153.
[7] Y. Liu, R. L. Withers, and X. Y. Wei, Phys. Rev. B **72**, (2005) 134104.
[8] T. B. Adams, D. C. Sinclair, A. R. West, Decomposition Reactions in $CaCu_3Ti_4O_{12}$ Ceramics, *J.* Am. Ceram. Soc., **89**, 2833 (2006).
[9] K.D. Mandal, Alok Kumar Rai, D. Kumar, Om Parkash, J. Alloy. Comp. 478 (2009) 771.
[10] L. Jianjun, D. Chun-Gang, W. N. Mei, R. W. Smith, J. R. Hardy, J. Appl. Phys., **98**, (2005) 093703.
[11] L. Pengfei, Y. Zupei, C. Xiaolian, L. Zonghua, J. Am. Ceram. Soc., 95 (2012) 2218.
[12] L. Junwei, L. Pengfei, Y. Jing, C. Xiaolian, Y. Zupei, J. Am. Ceram. Soc., 98-3 (2015) 795.
[13] Sunita Sharma, Shiv Sundar Yadav, M. M Singh., K.D Mandal, J. Adv. Dielectr., 4 (2014) 1450030-1-8 DOI: 10.1142/ S2010135X14 500301.
[14] S. Sharma, M.M Singh., U.S., Rai, K.D. Mandal, Mater. Sci. Semicond. Process. 31(2015)720.
[15] L.A. Chick, L.R. Pederson, G.D. Maupin, J.L. Bates, L.E. Thomas, G.J. Exarhos, Mater. Lett. 10, (1990) 6.
[16] B. D. Culity, S.R. Stock. Elements of X-ray Diffraction, Prentice Hall, New Jersey, Chapt 1, 2001.
[17] Fang TT, Liu CP. Chem Mater 17 (2005) 5167.
[18] E. Niwa, C.Uematsu, T.Hashimoto, J. Am. Ceram. Soc. 95 (2012), 3802-3806.
[19] B. Tareev. Physics of Dielectric Materials, Mir Publication, Moscow, (1975).

CALCIUM COPPER TITANATE BASED HIGH DIELECTRIC CONSTANT MATERIALS FOR ENERGY STORAGE APPLICATIONS

Disna P. Samarakoon, Nirmal Govindaraju, and Raj N. Singh
School of Materials Science & Engineering, Helmerich Advanced Technology Research Center, Oklahoma State University, Tulsa, OK 74106.

ABSTRACT

Effective use of alternative energy requires high capacity energy storage technology for many applications. This requires the development of materials with high dielectric constant combined with low dielectric losses. Calcium copper titanate ($CaCu_3Ti_4O_{12}$; CCTO) high-k dielectrics are attractive for energy storage applications because of high dielectric constants in the range of 10^4 - 10^5. Solid-state processing techniques to prepare $CaCu_3Ti_{4-x}Al_xO_{12-x/2}$ (x=0, 0.06, 0.1, and 0.5), which will enable effective mitigation of dielectric losses will also be described. X-ray diffraction (XRD) studies of the synthesized samples show that it is possible to synthesis phase pure CCTO by simple solid-state method at relatively low calcination temperature (850 °C). Therefore, Al-doping was done in the form of aluminum nitrate into calcined pure-CCTO and re-calcined at 850 °C. Microstructure observations by scanning electron microscopy (SEM) showed relatively homogeneous microstructure for the sintered samples in the range of 1050 – 1080 °C. Density measurements of pure-CCTO showed the highest density at 1060 °C and hence Al-doped samples were also sintered at the same temperature. No Al-containing phase could be seen by XRD in Al-doped CCTO calcined powders as well as ceramics (as-sintered and polished) indicates substitution of Al^{+3} for Ti^{+4} as there is no Al containing phases in the Al doped CCTO up to x=0.5. However, small CuO phase seen by XRD patterns of as-sintered specimen indicated partial re-oxidation of CuO/Cu_2O.

INTRODUCTION

High dielectric materials are capable of use as higher energy storage devices such as capacitors. Maximum stored energy density per unit volume (U) of a dielectric is given by $U = k\varepsilon_0E^2/2$ where, E is the applied field, k is the dielectric constant, and ε_0 = vacuum permittivity ($8.854x10^{-12}$ F/m) [1]. The capacitance (C) of a parallel plate capacitor is given by $C=k\varepsilon_0A/d$, where, A and d are the area and the space between the plates respectively. Hence, the larger the dielectric constant (k), the more charge can be stored leading to miniaturization of energy storage devices. Solid $CaCu_3Ti_4O_{12}$ (CCTO) emerged as a colossal dielectric constant ($k=10^4$ - 10^5 at room temperature) ceramics that is capable of use in supercapacitors [2, 3]. CCTO is stable dielectric over a wide temperature range of temperature 100 – 600 K [4, 5]. Furthermore, CCTO is a lead free dielectric material.

The dielectric properties of CCTO are closely related to its microstructure. CCTO has an electrically heterogeneous ABO_3 type perovskite structure in which grain boundaries are believed to be the origin of giant dielectric constant (k) [4-14]. Nevertheless, since the discovery of CCTO by Subramanian et al. [4], the origin of its huge dielectric response has been the subject of some controversy [15]. An internal barrier layer capacitor model (IBLC) has been used to explain the origin of an abnormally high dielectric constant in CCTO [4-14]. In IBLC model, dielectric properties of CCTO are treated as electrically heterogeneous structures composed of semi-conducting grains and insulating grain boundaries [4-14].

In spite of the high dielectric constant, high dielectric loss in CCTO prevents its use as a capacitor. So far, a dielectric constant as high as $3.5x10^4$ and tan δ (dielectric loss) as low as 0.014 at 1 kHz at room temperature were achieved by sol-gel synthesis of pure-CCTO [16]. Therefore, many studies have been mainly attempting to minimize the dielectric loss as well as conductivity in CCTO by means of altering the chemistry and structure of insulating grain boundaries via changing the processing conditions and by introducing element doping. Influence of processing conditions such as sintering can enhance the densification and microstructure, which ultimately promotes optimized dielectric properties [17-22]. Doping has been the area of interest for both theoretical and application study of CCTO for many researchers for over a decade [23-37]. Much work have been carried out in the recent past on partial substitutions of Ti sites by cations such as Al [22, 31], Cr [32], Fe [33], Mn [34], V [35], Hf [36], and Ga [37], etc.

Many research studies implied the possible Al-doping to modify the CCTO compositions thereby enhance the electric performance of capacitors [22-31]. Substitution of Ti sites by Al is an acceptor substitution, in which valence of the impurity ion Al (+3) is less than the valence of the cation Ti (+4), and hence the resulting structure can create holes or oxygen vacancies [38]. Studies [22-31] proposed an enhanced boundary resistance that eventually resulted in lower dielectric loss in CCTO. For instance, Choi et al [25] studied the doping of Al_2O_3 to prepare $CaCu_3Ti_{4-x}Al_xO_{12-x/2}$ (x=0, 0.06, and 0.1) and reported increased grain boundary resistivity by an order of magnitude and decreased dielectric loss 0.034 at x=0.06 and dielectric constant of 41000 at 10 kHz. Very recently, several groups [27-29] demonstrated that $CuAl_2O_4$ in CCTO increase the breakdown field while inhibiting the grain growth, which is associated with the formation of insulating $CuAl_2O_4$ in the intergranular regions.

Despite implementing different Al-doping strategies, it is difficult to make a good comparison of Al-doping content in each of their systems due to the variations in the preparation methods [23-31]. Moreover, synthesizing of pure and Al-doped CCTO ceramics aiming at low dielectric loss while retaining its high k is a challenge. Current work addresses this challenge by synthesizing pure and Al-doped $CaCu_3Ti_{4-x}Al_xO_{12}$ (x=0, 0.06, 0.1, and 0.5) using simple solid-state route. Reduced dielectric loss in CCTO is expected by increase of grain boundary resistance due to insulating $CuAl_2O_3$ phase or due to partial substitution of Ti and hence ensure low power loss when it is used as a dielectric material for a capacitor.

EXPERIMENTAL PROCEDURE

Powder Processing

Pure CCTO powder was synthesized by solid-state reaction using stoichiometric ratio of the high-purity $CaCO_3$ (VWR, 99.95%), CuO (VWR, 99.7%), and TiO_2 (Sigma-Aldrich, Anatase 99.8%) raw materials. These raw materials were mixed in anhydrous ethanol and ball-milled for 24 h in a polyethylene bottle containing zirconia media. The mixed slurry was dried and then calcined at the temperature range of 550 - 850 °C for 6 h in air in order to observe the phase evolution and to select suitable calcination temperature. Calcined powders were mixed in ethanol and then attritor-milled for 2 h into fine powders in zirconia media. Finely ground dried-slurry was then sieved using 120 µm sieve. For Al-doped CCTO, appropriate amounts of aluminum nitrate ($Al(NO_3)_3.9H_2O$) (Sigma-Aldrich, 99.997%) dissolved in anhydrous ethanol was added to calcined pure CCTO powders as per the the composition given by $CaCu_3Ti_{4-x}Al_xO_{12-x/2}$ (where, x = 0, 0.06, 0.1, and 0.5). Al-added dried mixtures were further re-calcined at 850°C for 6 h. Calcined CCTO powders were then mixed with 1 wt.% polyvinyl alcohol (PVA) and uniaxially pressed at 165 MPa into 13 mm diameter and 2.5 - 3 mm thickness cylindrical pellets. Samples are prepared by heating the pellets at different temperatures (1050 – 1080 °C) with the holding time of 5 h in order to select a suitable sintering temperature. Debinding of the pressed pellets was performed at 500 °C for 2 h followed by pre-sintering at 950 °C for 5 h and a constant heating and cooling rate of 2 °C/min was maintained. Based on the density measurements of the ceramics using Archimedes method, highest density was achieved for samples sintered at 1060

°C. In order to optimize the holding time at this temperature, sintering was performed for another set of samples at 1060 °C for 3, 10 and 15 h, and was shown that a 5 h holding was optimal.

Characterization

Thermogravimetric analysis (TGA, Netzsch STA 449 F1 Jupiter) was used to identify the thermal stability of the pre-calcined CCTO powders at a heating rate of 5°C/min in static air from room temperature to 850°C. Particle size was analyzed using Malvern Zetasizer Ver. 7.11 instruments. X-ray diffraction (XRD, Bruker AXS D8 Discover) with Cu-Kα radiation used to characterize the phases and structural variations in the precursor powders and sintered ceramics with the wave length 1.541 Å and the scan rate of 0.01°/s, with 2θ varying from 20° to 80°, and a count time of 60 s. Samples sintered at the chosen optimum temperature were examined by Scanning Electron Microscope (SEM, Hitachi S-4800), XRD, and subjected to the determination of lattice parameters. Pure and Al-doped CCTO lattice parameters were computed based on XRD peaks from the (220) and (420) planes. Powder morphology of precursors and the microstructure of sintered samples were also collected at room temperature (RT) by SEM.

RESULTS AND DISCUSSION

Figure 1 shows the TGA curve for decomposition of pre-calcined pure CCTO precursors. Almost negligible weight loss beyond 700 °C indicates that crystalline oxides slowly reacted and transferred to polycrystalline pure-CCTO. Weight loss up to this temperatures can be seen as nearly as 9%, which may be due to decomposition of organic groups.

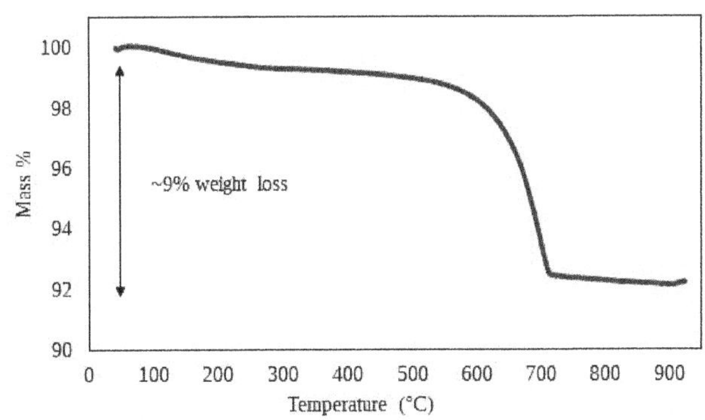

Figure 1. TGA for pure-CCTO pre-calcined precursors at a heating rate of 5 °C/min in static air from room temperature to 900 °C show no further loss beyond 700 °C.

XRD patterns for the evolution of CCTO phases from as-dried to calcined powders in the temperature of 650 – 950 °C for 6 h are shown in Figure 2 (a). Secondary phases are diminished upon heating above 750 °C indicating a higher crystallinity. The lowest temperature at which single-phase CCTO powders with a cubic structure are formed is at a calcination temperature of 850 °C for 6 h. Therefore, for all further experiments (including Al-doping) the calcination temperature was fixed at 850 °C. Figure 2 (b) shows the XRD patterns for Al-doped CCTO, for the composition $CaCu_3Ti_{4-x}Al_xO_{12-x/2}$, where x = 0.06, 0.1, and 0.5 calcined at 850 °C for 6 h. No secondary Al phases such as $CuAl_2O_4$ were found up to x=0.5.

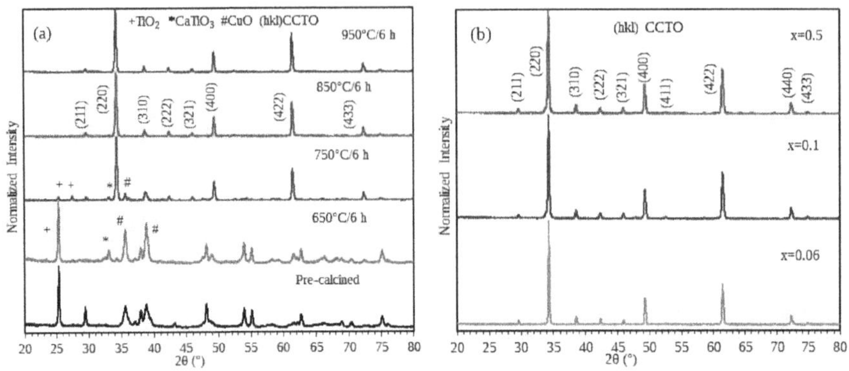

Figure 2. XRD patterns of CCTO with the composition of $CaCu_3Ti_{4-x}Al_xO_{12-x/2}$, where, x=0, 0.06, 0.1, and 0.5 for (a) pure CCTO pre-calcined and calcined powders (at 650 – 950 °C for 6 h) forming phase pure CCTO at 850 °C and (b) Al-doped CCTO (x=0.0.6, 0.1, and 0.5) calcined at 850 °C for 6 h respectively.

Using Malvern Zeta-sizer, particle size distribution was measured for pure-CCTO dried and calcined powders. Average particle size for all the calcined and sieved powders was measured as 500 nm, similar particle size range (250 – 500 nm) was observed in SEM micrographs. Figure 3 (a) and (b) are two different magnifications of homogeneously distributed CCTO calcined powder morphologies.

Figure 3. SEM micrographs of pure-CCTO powders calcined at 850 °C for 6 h showing the particle size as small as 250 - 500 nm. Different magnifications of the calcined powders are shown in (a) and (b) respectively.

SEM micrographs in Figure 4 (a) - (f) show the evolution of microstructure of polished samples sintered at different temperatures (1050 – 1080 °C). Larger magnifications of the images (a), (b), and (c) are indicated as (d), (e), and (f) respectively. As can be seen from the sintered images in Figure 4 (d) - (f), grains are not connected with clear boundaries indicating incomplete grain growth until temperature 1080 °C. Figure 4 (d) - (f) showed the relatively homogeneous microstructure of sintered samples at 1060 °C.

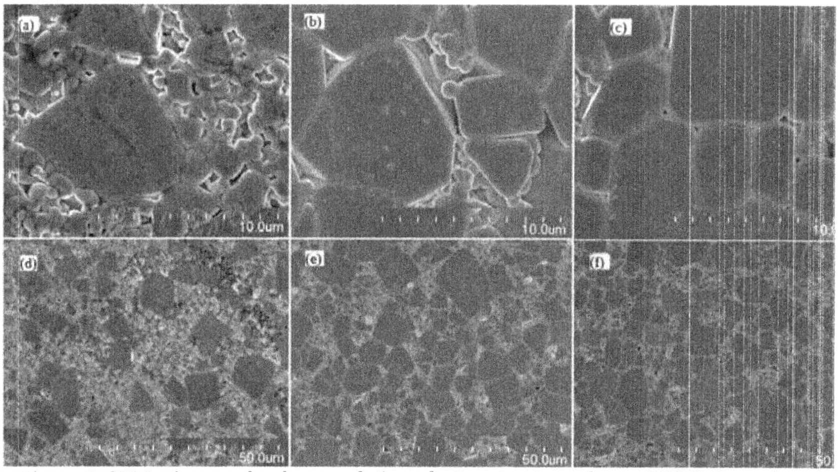

Figure 4. SEM micrographs show evolution of microstructure of pure-CCTO sintered ceramics prepared at different sintering temperatures of (a) 1050 °C, (b) 1060 °C, and (c) 1080 °C for 5 h. Figure 3 (d), (e), and (f) are the different magnifications of images (a), (b), and (c) respectively.

Later, sintering was carried out at 1115 °C for 4 h and then SEM observations on such a polished surfaces showed the completed grain growth with clear grain boundaries (see Figure 5 (a)-(b)). Increasing temperatures can cause an increase in liquid phase amount of CuO, which acted as sintering aid with increased porosity due to its high volatility during cooling [16, 39-43].

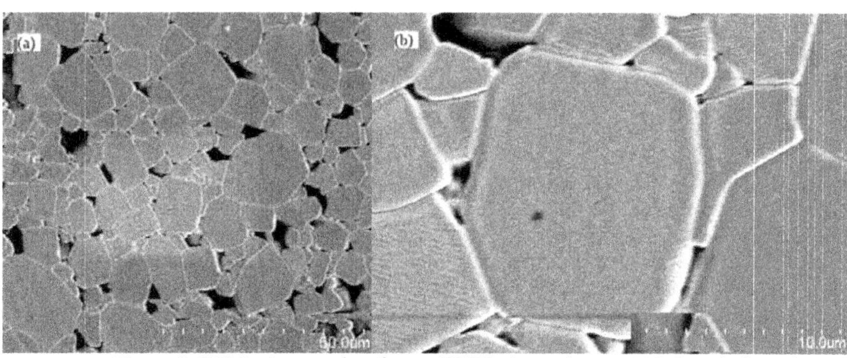

Figure 5. SEM micrographs of pure-CCTO sintered at 1115°C for 4 h show the completed grain growth with clear grain boundaries, where (a) and (b) are two different magnifications of the sintered sample.

Sintering behavior was further evaluated using density measurements by the Archimedes principle for the CCTO ceramics as a function of various temperatures and holding times. Figure 6 (a) is the graphical representation of the measured bulk density of sintered ceramics at 1050 – 1080 °C for an equal sintering cycle in terms of heating/cooling rates (2 °C/min) and holding time (5 h). When the sintering temperature is increased from 1050 °C to 1080 °C, relative density increases to a maximum value of 4.691 g/cm^3 at 1060 °C and then decreases to 4.563 g/cm^3 at 1080 °C. Samples sintered at 1060 °C showed the highest density of 4.691 g/cm^3, which is about 93 % of the theoretical density (reported by ICDD data as 5.053 g/cm^3) with the standard deviations of ±0.025. Once the holding time is varied from 3 to 15 h at this temperature (1060 °C), maximum values of densities as 4.691 g/cm^3 and 4.691 g/cm^3 were confirmed at the holding time of 5 h and 10 h respectively (See Figure 6 (b)). CCTO particles grow at an optimum temperature (1060 °C) resulting in highest density, while low and high sintering temperatures result in insufficient sintering and recrystallization leading to lower densification [16].

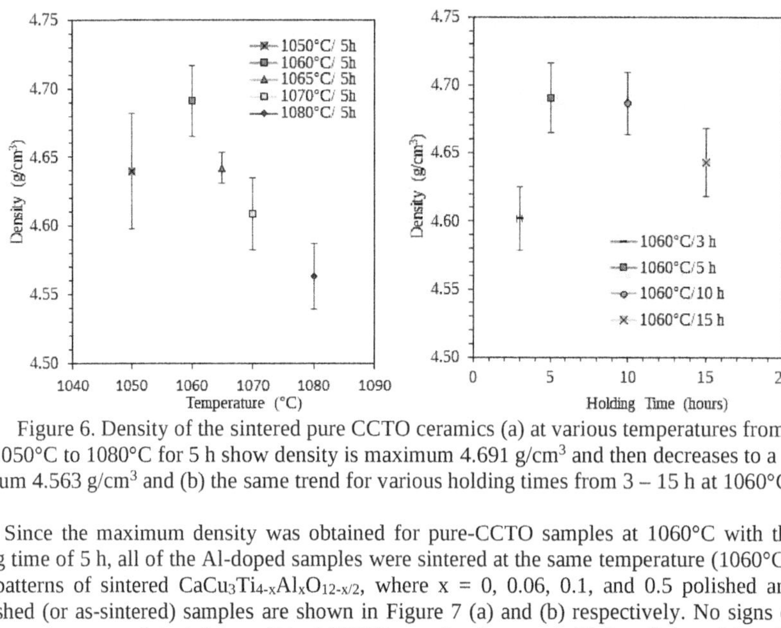

Figure 6. Density of the sintered pure CCTO ceramics (a) at various temperatures from (a) 1050°C to 1080°C for 5 h show density is maximum 4.691 g/cm^3 and then decreases to a minimum 4.563 g/cm^3 and (b) the same trend for various holding times from 3 – 15 h at 1060°C.

Since the maximum density was obtained for pure-CCTO samples at 1060°C with the holding time of 5 h, all of the Al-doped samples were sintered at the same temperature (1060°C). XRD patterns of sintered $CaCu_3Ti_{4-x}Al_xO_{12-x/2}$, where x = 0, 0.06, 0.1, and 0.5 polished and unpolished (or as-sintered) samples are shown in Figure 7 (a) and (b) respectively. No signs of secondary phases are detected in polished CCTO samples (pure and Al-doped) by raising the sintering temperature to 1060°C. Conversely, a weak CuO peak observed in unpolished samples at around 2θ = 35.6°, which was consistent with the reported observation [39-41]. Many reports on sintering studies of CCTO ceramics suggested the reason of such an observation is as a result of segregation of liquid CuO and/or Cu_2O at grain boundaries via the reduction of Cu^{+2} to Cu^{+1} at high temperature (1000 - 1120°C) [39-43]. During cooling re-oxidation of Cu_2O convert to CuO, whereas evaporation of CuO can be high. If the cooling rate is rapid, inadequate re-oxidation of Cu_2O to CuO can lead to inconsistent microstructures in the surface and interior of the ceramics [39-43].

On the other hand, experimental results showed [25] the substitution of titanium by aluminum in the TiO_6 octahedra of CCTO, if the doping level is relatively small (x=0.06 and

0.1). some other reports [27-28] suggested that presence of minor TiO_2 phase is due to the combined effect of substitution and the formation of $CuAl_2O_4$, if the doping level is relatively large (x=0.7, 1, and 1.4). Therefore, some traces of CuO in as-sintered CCTO surface in Figure 7 (a) can be attributed to two phenomena. Firstly, it can be a result of the formation of reduction and oxidation states of CuO/Cu_2O. Secondly, it can also be due to the substitution of Ti^{+4} by Al^{+3} without consumption of CuO to produce $CuAl_2O_4$ phase. However, no Al containing secondary phases observed for doping level up to x=0.5 indicates that Ti may have been replaced by Al in its lattice positions without distorting the crystal structure. Regardless of polishing, CCTO phases are the largest in all of the pure and Al-doped CCTO samples.

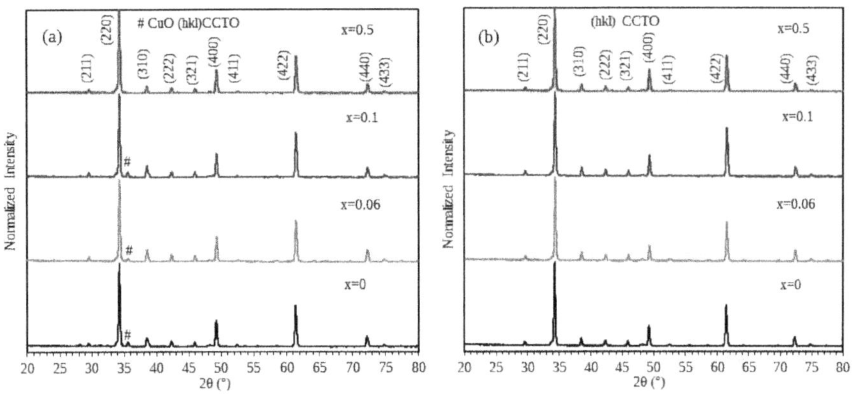

Figure 7. XRD for sintered $CaCu_3Ti_{4-x}Al_xO_{12-x/2}$, where, x = 0, 0.06, 0.1, and 0.5 ceramics at 1060°C for 5 h show no Al containing phase up to doping x=0.5 with (a) CuO traces in as-sintered samples up to x=0.1 and (b) no such traces in polished samples.

Lattice parameter for pure and Al-doped sintered ceramics were computed based on the two strong XRD Bragg reflection by CCTO planes (220) and (422). The average lattice constant of pure CCTO was determined to be 7.396 Å with the standard deviation of ±0.001, which is consistent with the reported value for standard sample 7.391 Å (JCPDS # 75- 2188) [25, 26]. The lattice constants of Al-doped CCTO compositions for x=0.06, 0.1, and 0.5 were 7.391, 7.389, and 7.388 Å respectively, which were nearly consistent with the reported values [25, 26]. The lattice constant decreased with increasing the Al-doping up to 0.5, which could be attributed to the ionic size effect. Ionic radius of Ti^{+4} (0.605 Å) is slightly larger than Al^{+3} (0.535 Å), and hence lattice parameter decreases with increasing the amount of Al incorporation into CCTO lattice [25, 27].

CONCLUSION

In conclusion, pure and Al-doped CCTO precursor powders and sintered ceramics successfully prepared with the composition of $CaCu_3Ti_{4-x}Al_xO_{12-x/2}$, where x = 0, 0.06, 0.1, and 0.5 using a solid-state chemical reaction approach. XRD patterns showed phase pure CCTO powders were successfully prepared at relatively low calcination temperature at 850 °C. SEM micrographs show homogeneous calcined powders with the average particle size of 500 nm. Density measurements proved the highest density at 1060 °C for 5 h holding time. SEM images for sintered samples at 1060 °C show relatively homogeneous microstructure, while sintering at high temperature (1115 °C) showed completed grain growth with clear grain boundaries. No Al-containing phase for the doping level of $CaCu_3Ti_{4-x}Al_xO_{12-x/2}$, where x = 0, 0.06, 0.1, and 0.5 was seen by XRD, can be attributed to possible substitution of Ti^{+4} sites by Al^{+3}. Furthermore, traces of CuO phases also seen by XRD for as-sintered samples can be a result of the formation of reduction and oxidation states of CuO/Cu_2O. The lattice constant decreased, when the Al doping is increases, which could be attributed to the ionic size effect of Ti^{+4} and Al^{+3}. Further investigation is needed to explore more on above observations and to link those results to their dielectric measurements.

REFERENCE

[1]Y. Bai, Z. Y. Cheng, V. Bharti, H. S. Xu, and Q. M. Zhang, High-dielectric-constant ceramic-powder polymer composites, *Appl. Phys. Lett.*, **76**, 25, (19 June 2000).

[2]J. W. Lee and J. H. Koh, Grain size effects on the dielectric properties of $CaCu_3Ti_4O_{12}$ ceramics for super capacitor applications, *Ceramics International*, **41**, 10442–10447 (2015).

[3]R. K. Pandey et al, Applications of CCTO supercapacitor in energy storage and electronics, *AIP Advances*, **3**, 062126 (2013).

[4]M. A. Subramanian, D. Li, N. Duan, B. A. Reisner, and A. W. Sleight, High Dielectric Constant in $ACu_3Ti_4O_{12}$ and $ACu_3Ti_3FeO_{12}$ Phases, *Journal of Solid State Chemistry*, **151**, 323-325 (2000).

[5]D. C. Sinclair, T. B. Adams, F. D. Morrison, and A. R. West, $CaCu_3Ti_4O_{12}$: One-step internal barrier layer capacitor, *Applied Physics Letters*, **80**, 2153 (2002).

[6]P. Lunkenheimer et al, Origin of apparent colossal dielectric constants, *Physical Review B*, **66**, 052105 (2002).

[7]P. Lunkenheimer, R. Fichtl, S. G. Ebbinghaus, and A. Loidl, Nonintrinsic origin of the colossal dielectric constants in CaCu3Ti4O12, *Physical Review B*, **70**, 172102 (2004).

[8]R Schmidt, S. Pandey, P. Fiorenza, and D. C. Sinclair, Non-stoichiometry in ''$CaCu_3Ti_4O_{12}$'' (CCTO) ceramics, *RSC Adv.*, **3**, 14580–14589 (2013).

[9]P. Thongbai, J. Jumpatam, T. Yamwong, S. Maensiri, Effects of Ta5+ doping on microstructure evolution, dielectric properties and electrical response in $CaCu_3Ti_4O_{12}$ ceramics, *Journal of the European Ceramic Society*, **32**, 2423–2430 (2012).

[10]W. X. Yuan, Impedance and electric modulus approaches to investigate four origins of giant dielectric constant in $CaCu_3Ti_4O_{12}$ ceramics, *Solid State Sciences*, **14**, 330-334 (2012).

[11]W. X. Yuan and S.K. Hark, Investigation on the origin of the giant dielectric constant in $CaCu_3Ti_4O_{12}$ ceramics through analyzing $CaCu_3Ti_4O_{12}$–HfO_2 composites, *Journal of the European Ceramic Society*, **32**, 465–470 (2012).

[12]J. Y. Li, X. T. Zhao, S. T. Li, and M. A. Alim, Intrinsic and extrinsic relaxation of $CaCu_3Ti_4O_{12}$ ceramics: Effect of sintering, *Journal of Applied Physics*, **108**, 104104 (2010).

[13]J. Li, A.W. Sleight, and M. A. Subramanian, Evidence for internal resistive barriers in a crystal of the giant dielectric constant material: $CaCu_3Ti_4O_{12}$, Solid State Communications, **135**, 260–262 (2005).

[14]R. Schmidt, M. C. Stennett, N. C. Hyatt, J. Pokorny, J. Prado-Gonjal, M. Li, and D. C. Sinclair, Effects of sintering temperature on the internal barrier layer capacitor (IBLC) structure in $CaCu_3Ti_4O_{12}$ (CCTO) ceramics, *Journal of the European Ceramic Society*, **32**, 3313–3323 (2012).

[15]Lei Zhang, Electrode and grain-boundary effects on the conductivity of $CaCu_3Ti_4O_{12}$, *Applied Physics Letters*, **87**, 022907 (2005).

[16]Y. Li, P. Liang, X. Chao, and Z. Yang, Preparation of $CaCu_3Ti_4O_{12}$ ceramics with low dielectric loss and giant dielectric constant by the sol–gel technique, *Ceramics International*, **39**, 7879–7889 (2013).

[17]Q. Zheng and H. Fan, Influence of Fabrication Parameters on the Phase Formation and Dielectric Properties of $CaCu_3Ti_4O_{12}$ Ceramics, *J. Mater. Sci. Technol.*, **28** (10), 920–926 (2012).
[18]S. D. Hutagalung, O. L. Ying, and Z. A. Ahmad, Effect of Sintering Temperature on the Properties of Modified Mechanical Alloyed $CaCu_3Ti_4O_{12}$, *IEEE transactions on ultrasonics, ferroelectrics, and frequency control*, 54, 12 (December 2007).
[19]B. S. Prakash and K. B. R. Varma, Influence of sintering conditions and doping on the dielectric relaxation originating from the surface layer effects in $CaCu_3Ti_4O_{12}$ ceramics, *Journal of Physics and Chemistry of Solids*. **68**, 490–50 (2007).
[20]S. K. Jo and Y. H. Han, Sintering behavior and dielectric properties of polycrystalline $CaCu_3Ti_4O_{12}$, *J Mater Sci: Mater Electron*, **20**, 680–684 (2009).
[21]P. K. Patel and K. L. Yadav, Dwell time effect on the barrier layer capacitor structure in $CaCu_3Ti_4O_{12}$ ceramic, *Ceramics International*, **41**, 12386–12392 (2015).
[22]R. Yu, H. Xue, Z. Cao, L. Chen, and Z. Xiong, Effect of oxygen sintering atmosphere on the electrical behavior of CCTO ceramics, *Journal of the European Ceramic Society*, **32** , 1245–1249 (2012).
[23]Y. Hu, Y. Pu, P. Wang, and T. Wu, Effects of Kaolinite additions on sintering behavior and dielectric properties of $CaCu_3Ti_4O_{12}$ ceramics, *J Mater Sci: Mater Electron*, **25**, 546–551 (2014).
[24]A. R. Darvishi, R. Bayati, O. S. Bishe, L. D. Wang, W. L. Li, J. Sheng, and W. D. Fei, Giant dielectric response and low dielectric loss in Al_2O_3 grafted $CaCu_3Ti_4O_{12}$ ceramics, *Journal of Applied Physics*, **117**, 094103 (2015).
[25]S. W. Choi and S. H. Hong, Effect of Al Doping on the Electric and Dielectric Properties of $CaCu_3Ti_4O_{12}$. *J. Am. Ceram. Soc.*, **90**, 4009–4011 (2007).
[26]L. Shengtao, W. Hui, L. Chunjiang, Y. Yang, and L. Jianying, Dielectric Properties of Al-doped $CaCu_3Ti_4O_{12}$ Ceramics by Coprecipitation Method, Conference Proceedings of ISEIM (2011).
[27]R. Jia, X. Tang, J. Li, and X. Zhao, Large Breakdown Field and Dielectric Performance of $CaCu_3Ti_4O_{12}$ Ceramics Modified by Al_2O_3, 2013 IEEE International Conference on Solid Dielectrics, Bologna, Italy, (June 30 – July 4, 2013).
[28]J. Li, L. Hou, R. Jia, L. Gao, K. Wu, S. Li, Influences of $CuAl_2O_4$ doping on the dielectric properties of $CaCu_3Ti_4O_{12}$ ceramics, *J Mater Sci: Mater Electron*, **26**, 5085–5091 (2015).
[29]C. T. Su, H.Y. Lee, B.-K. Wu, and M.Y. Chern, Development of phase-pure $CuAlO_2$ thin films grown on c plane sapphire substrates prepared by RF sputtering, *Journal of Crystal Growth*, **328**, 25–29 (2011).
[30]L. Liu, Y. Huang Y. Li, D. Shi, S. Zheng, S. Wu, L. Fang, C. Hu, Dielectric and non-Ohmic properties of $CaCu_3Ti_4O_{12}$ ceramics modified with NiO, SnO_2, SiO_2, and Al_2O_3 additives, *J Mater Sci*, **47**, 2294–2299 (2012).
[31]C. Warangkanagool and G. Rujijanagul, Improvement in dielectric and mechanical performance of $CaCu_{3.1}Ti_4O_{12.1}$ by addition of Al_2O_3 nanoparticles, *Nanoscale Research Letters*, 7, 68 (2012).
[32]Q. Zheng, H. Fan, C. Long, Microstructures and electrical responses of pure and chromium-doped $CaCu_3Ti_4O_{12}$ ceramics, *Journal of Alloys and Compounds*, **511**, 90– 94 (2012).
[33]C. Mu, H. Zhan, Y. He, and P. Liu, Influence of temperature on dielectric properties of Fe-doped $CaCu_3Ti_4O_{12}$ ceramics, *Physica B*, **405**, 386–389(2010).
[34]X.H. Zheng, C. Zhang, B. L. Liang, D. P. Tang, X. Huang, and X.L. Liu, An effective way to tune the microstructure and dielectric properties of $CaCu_3Ti_4O_{12}$ ceramics, *Journal of Alloys and Compounds*, **505**, L10–L14 (2010).
[35]A. Sen, U. N. Maiti, R. Thapa, and K. K. Chattopadhyay, Effect of vanadium doping on the dielectric and nonlinear current–voltage characteristics of $CaCu_3Ti_4O_{12}$ ceramic, *Journal of Alloys and Compounds*, **506**, 853–857 (2010).

[36]M. A. de la Rubia, P. Leret, A. del Campo, R.E. Alonso, A. R. López-Garcia, J. F. Fernández, and J. de Frutos, Dielectric behavior of Hf-doped $CaCu_3Ti_4O_{12}$ ceramics obtained by conventional synthesis and reactive sintering, *Journal of the European Ceramic Society*, **32**, 1691–1699 (2012).
[37]J. Jumpatam, B. Putasaeng, T. Yamwong, P. Thongbai, and S. Maensiri, Enhancement of giant dielectric response in Ga-doped $CaCu_3Ti_4O_{12}$ ceramics, *Ceramics International*, **39**, 1057–1064 (2013).
[38]L. Singh, U.S. Rai, K. D. Mandal, N. B. Singh, Progress in the growth of $CaCu_3Ti_4O_{12}$ and related functional dielectric perovskites, *Progress in Crystal Growth and Characterization of Materials*, 60 15–62 (2014).
[39]W. Yuan, Investigation on the decomposable process and the secondary liquid phase effect on the dielectric properties of $CaCu_3Ti_4O_{12}$ ceramics, *J. Phys. D: Appl. Phys.* **42**, 175401 (2009).
[40]W. X. Yuan, Z. Luo, and C. Wang, Investigation on effects of CuO secondary phase on dielectric properties of $CaCu_3Ti_4O_{12}$ ceramics, *Journal of Alloys and Compounds*, **562**, 1–4 (2013).
[41]S. Kwon and D. P. Cann, Relationship among the phase equilibria, microstructures, and dielectric properties of $CaCu_3Ti_4O_{12}$ ceramics via different sintering time, *J Mater Sci*, **44**, 4117–4123 (2009).
[42]S. Kwon & D. P. Cann, Influence of the processing rates and sintering temperatures on the dielectric properties of $CaCu_3Ti_4O_{12}$ ceramics, *J Electroceram*, **24**, 231–236 (2010).
[43]B. Wan, Y. P. Pu, H. D. Wu, K. Chen, and N. Xu, Influence of sintering atmosphere on dielectric properties and microstructure of $CaCu_3Ti_4O_{12}$ ceramics, *Ceramics International*, **39**, S525–S528 (2013).

Sintering and Related Powder Processing

SYNTHESIS, CHARACTERIZATION AND GIBBS ENERGY OF THERMOELECTRIC Mg_2Si

Mallikharjuna R. Bogala, Ramana G. Reddy
Department of Metallurgical and Materials Engineering, The University of Alabama
Tuscaloosa, Alabama 35487-0202, United States of America

ABSTRACT

A simple and rapid synthesis of magnesium silicide powders was carried out using pelletizing and sintering methods. A mixture containing stoichiometric quantities of magnesium and silicon powders was cold-pressed to form pellets. Sintering was conducted on the pellets from 300 to 550 °C and for 0 to 3 hours to obtain Mg_2Si powders. The product samples were characterized using X-ray diffraction (XRD), scanning electron microscopy (SEM), and energy dispersive spectroscopy (EDS) methods. The thermal stability of Mg_2Si powders was studied using differential scanning calorimetry (DSC) technique. The results show the formation of uniform microparticles of pure Mg_2Si powders at sintering temperature and time of 500 °C and 5 min respectively. Furthermore, the data from heat capacity (C_p) and change in Gibbs energy (ΔG) calculations suggests that the obtained Mg_2Si powders are thermally stable.

INTRODUCTION

In recent years, there is a growing concern over conventional electricity generation, which not only deplete natural resources by burning of fossil fuels (coal, oil and gas), but also cause global warming through significant emissions of CO_2 gas into the earth's atmosphere[1]. Research efforts are made to minimize the global warming via development and implementation of energy efficient renewable technologies for power generation from natural resources[2-3]. Thermoelectric (TE) technology is very attractive because of its unique ability to directly convert waste heat into electricity. The TE generator works on the principle of Seebeck effect[4] i.e., the temperature differences between the hot and cold sides of the TE device results in the diffusion of charge carriers (electrons and holes), which produces potential difference or electric voltage across the two terminals of TE device[5-6]. The performance of a TE device is determined from the dimensionless figure of merit (ZT) given by equation (1).

$$ZT = (S^2\sigma T)/\kappa \quad (1)$$

where S, σ, and κ are Seebeck coefficient (μVK^{-1}), electrical conductivity ($Ohm^{-1}m^{-1}$), and thermal conductivity ($Wm^{-1}K^{-1}$) of the TE material respectively, and T is the absolute temperature (K)[5-8]. The TE technology involve solid state devices that contain immovable, reliable, and small components with negligible CO_2 gas emissions[9]. Several TE materials have been developed for waste heat recovery application in automobile, aerospace, military and consumer-based industries[7-8]. Although commercially available Pb-Te and Bi-Te based TE materials exhibit higher ZT values at low and middle temperatures, they are also toxic, heavy, and less abundant in the earth's crust[10-11]. Therefore, it is important to develop sustainable TE materials for middle and high temperature applications.

Magnesium silicide (Mg_2Si) is a promising TE material with high ZT at middle temperature range of 500 – 800 K[12-16]. Mg_2Si is the only possible binary phase of Mg-Si system. Mg_2Si contain anti-fluorite CaF_2 type structure with $Fm\bar{3}m$ space group and cubic crystal lattice cell parameters (a = b = c = 6.351 Å and $\alpha = \beta = \gamma = 90°$)[8]. The Si atoms occupy the corners and face-centered lattice sites of Mg_2Si unit cell, while Mg atoms occupy eight tetrahedral positions in the interior region of the unit cell. The coordination number of Si and Mg are eight and four respectively. The cubic lattice of Mg_2Si transforms to hexagonal lattice at high pressure (> 25 kbar) and temperature

(> 900 °C). The density and narrow energy band gap of Mg_2Si are 1.99 gcm^{-3} and 0.77 eV respectively[17, 18]. In addition, Mg_2Si exhibits superior mechanical behavior with high young's modulus (120 GPa) and elevated compressional strength (1640 MPa)[17]. Mg_2Si is very attractive candidate due to its low cost, higher abundance, light weight, and non-toxic nature [17, 19]. Therefore, Mg_2Si based TE devices have huge potential to replace the existing Pb-Te based TE devices.

Several synthetic procedures have been used to improve mechanical and TE properties of Mg_2Si through refinement of grain size and lowering of thermal conductivity. Some of these techniques include mechanical alloying or ball milling[20-22], vacuum melting[15], high temperature solid state reaction[23], spark plasma sintering or plasma activated sintering[24,25], vertical Bridgeman (VB) method[26], and microwave irradiation[27]. Mechanical alloying is a time consuming process and the samples are subjected to easy contamination and oxidation during the synthesis. Vacuum melting is hard to control the synthesis of Mg_2Si. The structure, composition, and TE performance of Mg_2Si synthesized from these conventional methods are inconsistent because of (i) large difference in the melting points of Mg (650 °C) and Si (1414 °C), (ii) high volatility or evaporation of Mg at temperature (1090 °C) close to the melting point of Mg_2Si (1085 °C), and (iii) easy contamination or oxidation of Mg to MgO[17, 28]. Hence, it is important to develop low temperature, fast and energy efficient methods for synthesis of Mg_2Si.

In this paper, we describe a simple and rapid procedure for synthesis of Mg_2Si powder using cold pressing and sintering methods. The product samples were tested for phase structure, surface morphology and elemental analysis using XRD, SEM and EDS methods. The specific heat of Mg_2Si powder was determined from DSC experiments and the thermal stability of Mg_2Si was evaluated from change in Gibbs energy at 25 to 350 °C.

EXPERIMENTAL

Elemental powders of magnesium (-325 mesh, 99.8%) and silicon (-325 mesh, 99.5%) were purchased from Alfa Aesar, MA, U.S.A. Figure 1 shows the schematic drawing of simple and rapid procedure used for the synthesis of Mg_2Si alloy. Mg reacts with atmospheric oxygen to form a thin oxide surface layer, which further inhibits the reactivity of the metal. Hence, the mixing of stoichiometric amounts of starting materials i.e., Mg and Si powders (2:1 molar ratio) were carried out in a Labconco glove box (Kansas City, MO, U.S.A) under argon gas atmosphere. About 1 gram of this uniform mixture was placed in a dye of 1.3 cm inner diameter and then cold-pressed at 90 kN using a Carver press. The obtained silver-greyish cylindrical (2Mg+Si) pellet was placed in a glass vial and sealed under vacuum. Sintering of (2Mg+Si) pellets was carried out using Fischer Scientific furnace to obtain dark-bluish Mg_2Si powders. Table I summarizes the sintering parameters i.e., temperature (300 to 600 °C) and time (0 to 3 hrs) for synthesis of Mg_2Si powders.

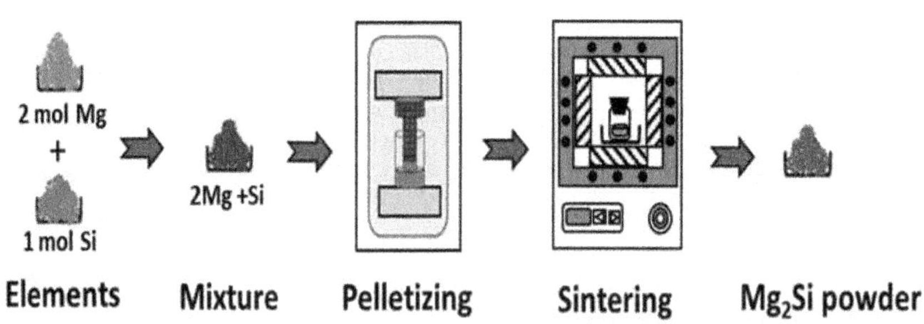

Figure 1. A schematic showing the experimental procedure for synthesis of Mg_2Si powder.

The Mg_2Si powders obtained from sintering process was characterized using XRD, SEM, and EDS methods. Small amounts of the Mg_2Si powders were used for obtaining X-ray diffraction spectrum using Phillips X'PERT MPD instrument. The XRD data was acquired by using Cu Kα (λ = 1.5405 Å) source at diffraction angles of 2θ (= 20° - 120°). Phase analysis of the samples was done by manual assignment and comparison of XRD peak reflections using standard ICDD diffraction data (Mg_2Si, Mg, and Si). The surface morphology of Mg_2Si powders was determined using 7000 JEOL FE SEM. Small amounts of the powder sample was adhered on to carbon tape of the holder. The sample was then inserted into the JEOL instrument for SEM imaging and EDS spectral analysis. A power of 20 kV was applied between the sample and SEM probe. The working distance was maintained at 10 mm. The SEM images of Mg_2Si sample were captured at higher magnification of 1500X. The Mg_2Si sample was later subjected to elemental analysis using EDS detector that is attached to the 7000 JEOL FE SEM.

Table I. Sintering Parameters (Temperature and Time) of Samples Tested for Synthesis of Mg_2Si.

Sample ID	Sintering		Observation
	Temperature	Time	
1	300 °C	5 min and 1 hr	No formation of Mg_2Si
2	350 °C	5 min	No formation of Mg_2Si
3	400 °C	5 min and 1 hr	No formation of Mg_2Si
4	450 °C	5 min	No formation of Mg_2Si
5	500 °C	5 min, 30 min, 1, 2 and 3 hrs	Formation of Mg_2Si
6	550 °C	5 min	Formation of Mg_2Si
7	600 °C	5 min	Formation of Mg_2Si

To determine the phase stability of Mg_2Si at higher temperatures, thermal analysis was conducted on the synthesized Mg_2Si powder (500 °C and 5 min) using DSC technique. About 8 mg of Mg_2Si powder was placed on the Al sample holder of PerkinElmer Diamond DSC instrument. Differential heat flow data was acquired using *iso-scan-iso* mode from 300 to 650 K using Pyris software. The heat flow data was used to determine the specific heat capacity of Mg_2Si from 300 - 650 K. The change in Gibbs energy of Mg_2Si was calculated from the C_p data, which is compared with thermal properties predicted using HSC 7.1 and FactSage 6.3 software at 25 °C.

RESULTS AND DISCUSSION

As shown in the Figure 1, the synthesis of Mg_2Si powder from (2Mg + Si) mixture was carried out using cold-pressing and sintering methods. Table I lists the sintering conditions (time and temperature) and the outcome of the experiments. The optimum sintering time and temperature

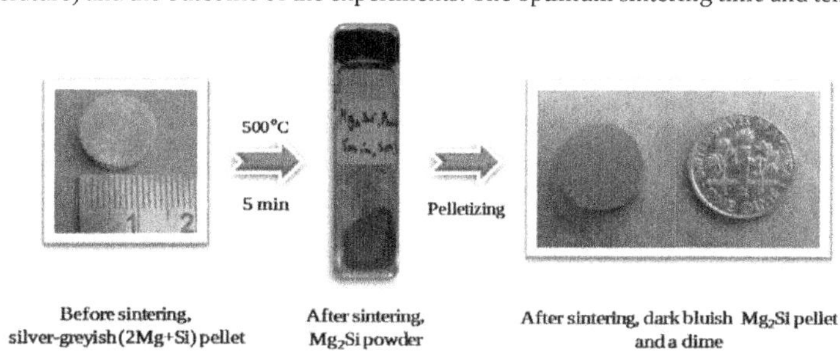

Figure 2. Images of (2Mg+Si) and Mg_2Si pellets before and after sintering at 500 °C and 5 min.

for synthesis of Mg_2Si powders are 500 °C and 5 min respectively. Figure 2 shows the image of (2Mg+Si) pellet, Mg_2Si powder and Mg_2Si pellet obtained before and after the sintering process.

The sample powders were analyzed for phase identification, surface morphology, and elemental composition using XRD, SEM and EDS techniques respectively. Figure 3 (a and b) compares the XRD spectra of 2Mg + Si mixture and Mg_2Si powder formed after sintering at 500 °C and 5 min. The disappearance of major reflections {Mg (101) and Si (111)} and appearance of major characteristic reflection of Mg_2Si (220) at 2θ ~ 40° in the XRD spectra shows the formation of single product phase at 500 °C. The lattice constant for the cubic lattice of the Mg_2Si powder is

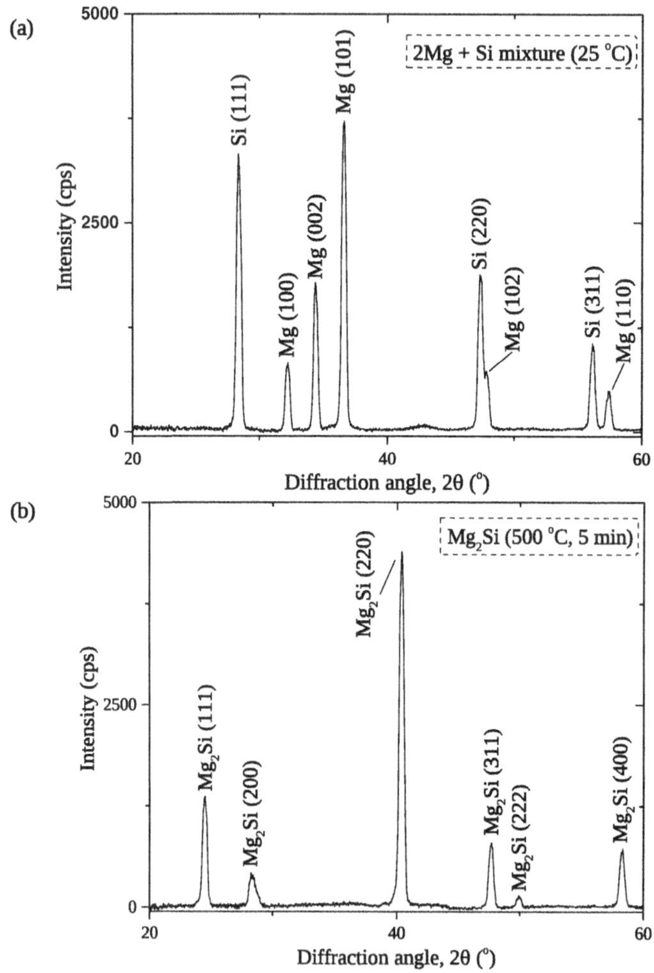

Figure 3. XRD spectra of (a) 2Mg + Si, & (b) Mg_2Si powders from sintering at 500 °C for 5 min.

calculated based on strongest peak (220) of the XRD spectrum (Figure 3b) by using equation (2).

$$a(\text{Å}) = d_{hkl}(\text{Å}) \times (h^2 + k^2 + l^2)^{1/2} = (\lambda/2 \sin\theta_{hkl}) \times (h^2 + k^2 + l^2)^{1/2} \quad (2)$$

The value of lattice constant for Mg_2Si powder synthesized is 6.278 Å, which is in good agreement with the lattice constant of 6.351 Å reported in standard ICDD database (Pdf card 00-034-0458).

Figure 4 displays the 1500X SEM image of Mg_2Si powder. The size and shape analysis of Mg_2Si sample indicates the formation of fine irregular sized microparticles of Mg_2Si with an average size < 10 µm. The Mg_2Si sample was later subjected to elemental analysis using the EDS detector that is attached to the 7000 JEOL FE SEM. Figure 5 corresponds to the EDS spectrum of

Figure 4. SEM image of Mg_2Si powder obtained from sintering at 500 °C for 5 min.

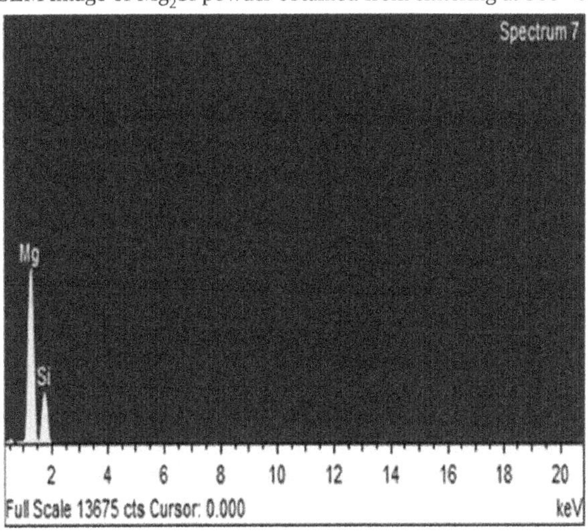

Figure 5. EDS spectrum of Mg_2Si powder obtained from sintering at 500 °C for 5 min.

total area of the SEM image. The elemental composition of the powder reveals the formation of ultra-high pure Mg_2Si at 500 °C and 5 min.

Thermal stability of Mg_2Si powder was studied using DSC technique. The differential heat flow data was acquired using *iso-scan-iso* mode of Pyris software from 300 to 650 K. The heat flow data was used to determine the specific heat capacity (C_p) of Mg_2Si sample from 300 - 650 K. Figure 6 shows the plot of C_p vs. temperature of Mg_2Si sample, and the expression for $C_p(T)$ is given by equation (3).

$$C_p(T), Jmol^{-1}K^{-1} = -0.0001T^2 + 0.1282T + 38.429 \quad (3)$$

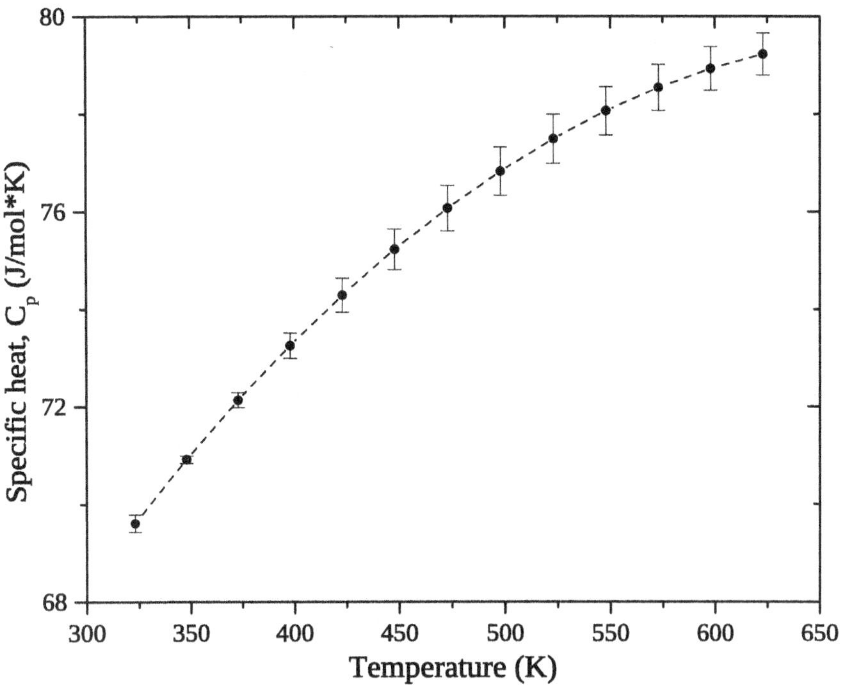

Figure 6. Specific heat vs. temperature of Mg_2Si obtained at 500 °C and 5 min.

Thermodynamic calculations were conducted, using $C_p(T)$, to estimate the change in enthalpy (ΔH), and the change in entropy (ΔS) of Mg_2Si sample. The expressions for ΔH(T), and ΔS(T) are given by equations (4 and 5) respectively.

$$\Delta H(T) = H_T^o - H_{298.15\,K}^o = \int_{298.15\,K}^{T} C_p(T)\, dT \quad (4)$$

$$\Delta S(T) = S_T^o - S_{298.15\,K}^o = \int_{298.15\,K}^{T} [C_p(T)/T]\, dT \quad (5)$$

The change in Gibbs energy of Mg_2Si was calculated from equation (6) using ΔH(T) and ΔS(T).

$$\Delta G(T), kJmol^{-1} = G_T^o - G_{298.15\,K}^o = (H_T^o - H_{298.15\,K}^o) - T(S_T^o - S_{298.15\,K}^o)$$
$$= -79.412 + 0.0085\,T \quad (6)$$

Figure 7 shows ΔG vs. temperature plot of Mg_2Si powder obtained from sintering at 500 °C and 5 min. The negative Gibbs energy values of Mg_2Si suggests that the obtained Mg_2Si powder is thermally stable at higher temperatures. Table II compares the experimental values of thermodynamic properties (ΔG(T), C_p(298.15 K), $\Delta_f H^o$(298.15 K), $\Delta_f S^o$(298.15 K), and $\Delta_f G^o$(298.15 K)) with corresponding values from the literature, and predictions using HSC 7.1 and FactSage 6.3 software tools.

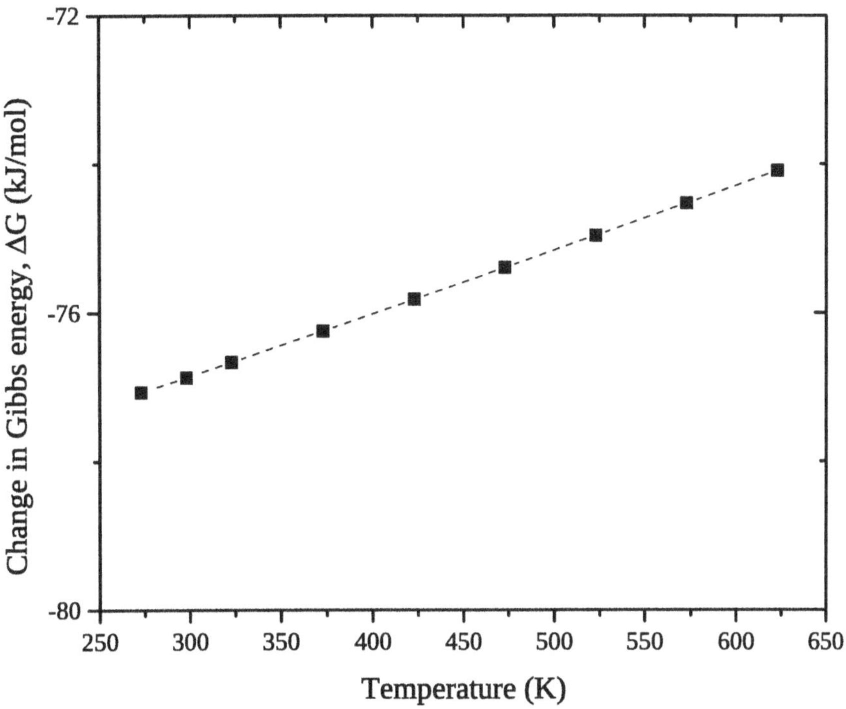

Figure 7. Change in Gibbs energy vs. temperature for Mg_2Si obtained at 500 °C and 5 min.

Table II. Change in Gibbs energy ΔG(T) and thermodynamic properties of Mg_2Si at 298.15 K

Mg_2Si	ΔG(T) (kJ/mol); T in K	At 298.15 K			
		C_p (J/mol*K)	$\Delta_f H^o$ (kJ/mol)	$\Delta_f S^o$ (J/mol*K)	$\Delta_f G^o$ (kJ/mol)
This study	-79.41 + 0.0085 T	67.76 ± 0.39	-79.41	-8.50	-76.87
Gerstein et al.[29]	-77.80 + 0.0080 T	68.43 ± 0.20	-77.80	-8.05	-75.40
Barin et al.[30]	-79.29 + 0.0084 T	67.83	-79.29	-8.40	- 76.78
HSC 7.1	-77.80 + 0.0081 T	67.84	-77.80	-8.12	-75.38
FactSage 6.3	-70.82 + 0.0015 T	67.77	-80.82	-15.06	-76.33

CONCLUSIONS

A simple and rapid synthesis of Mg_2Si powders was carried out using pelletizing, and sintering methods. Mg_2Si was produced from cold-pressing of 2:1 molar mixture of Mg and Si, followed by sintering of the pellet at optimum conditions of 500 °C for 5 min. The Mg_2Si samples were characterized using XRD, SEM, and EDS methods. Formation of uniform microparticles of pure Mg_2Si was observed. The thermal stability of Mg_2Si powders was studied using DSC technique. The heat capacity (C_p), and Gibbs energy (ΔG) data suggests that the Mg_2Si powders are thermally stable at higher temperatures.

ACKNOWLEDGEMENTS

The authors are thankful for the financial support from the National Science Foundation (NSF) agency, Grant No.DMR-1310072 and American Cast Iron Pipe Company (ACIPCO) and The University of Alabama during the course of the research project.

REFERENCES

[1]M. I. Hoffert, K. Caldeira, G. Benford, D. R. Criswell, C. Green, H. Herzog, A. K. Jain, H. S. Kheshgi, K. S. Lackner, J. S. Lewis, H. D. Lightfoot, W. Manheimer, J. C. Mankins, M. E. Mauel, L. J. Perkins, M. E. Schlesinger, T. Volk, and T. M. Wigley, Advanced technology paths to global climate stability: energy for a greenhouse planet, *Science*, **298**, 981-987 (2002).

[2]R. E. Sims, H. Rogner, and K. Gregory, Carbon emission and mitigation cost comparisons between fossil fuel, nuclear and renewable energy resources for electricity generation, *Energy Policy*, **31**, 1315-1326 (2003).

[3]I. Bhat, and R. Prakash, LCA of renewable energy for electricity generation systems—a review, *Renewable Sustainable Energy Rev.*, **13**, 1067-1073 (2009).

[4]T. J. Seebeck, Ueber die magnetische Polarisation der Metalle und Erze durch Temperaturdifferenz, *Ann. Phys.*, **82**, 253-286 (1826).

[5]F. Rosi, Thermoelectricity and thermoelectric power generation, *Solid-State Electron.*, **11**, 833-868 (1968).

[6]A. Minnich, M. Dresselhaus, Z. Ren, and G. Chen, Bulk nanostructured thermoelectric materials: current research and future prospects, *Energy Environ. Sci.*, **2**, 466-479 (2009).

[7]T. M. Tritt, H. Böttner, and L. Chen, Thermoelectrics: Direct solar thermal energy conversion, *MRS Bull.*, **33**, 366-368 (2008).

[8]D. M. Rowe, CRC handbook of thermoelectrics, *CRC press* (1995).

[9]E. J. Winder, A. B. Ellis, and G. C. Lisensky, Thermoelectric devices: solid-state refrigerators and electrical generators in the classroom, *J. Chem. Educ.*, **73**, 940 (1996).

[10]H. Kim, T. Oh, and D. Hyun, Thermoelectric properties of the p-type Bi_2Te_3–Sb_2Te_3–Sb_2Se_3 alloys fabricated by mechanical alloying and hot pressing, *J. Phys. Chem. Solids*, **61**, 743-749 (2000).

[11]Y. Gelbstein, O. Ben-Yehuda, E. Pinhas, T. Edrei, Y. Sadia, Z. Dashevsky, and M. Dariel, Thermoelectric properties of (Pb, Sn, Ge) Te-based alloys, *J. Electron. Mater.*, **38**, 1478-1482 (2009).

[12]R. G. Morris, R. Redin, and G. Danielson, Semiconducting properties of Mg_2Si single crystals, *Physical Review*, **109**, 1909 (1958).

[13]R. J. LaBotz, D. R. Mason, and D. F. O'Kane, The Thermoelectric Properties of Mixed Crystals of $Mg_2Ge_xSi_{1-x}$, *J. Electrochem. Soc.*, **110**, 127-134 (1963).

[14]T. Kajikawa, K. Shida, K. Shiraishi, T. Ito, M. Omori, and T. Hirai, Thermoelectric figure of merit of impurity doped and hot-pressed magnesium silicide elements, *XVII Int. Conf. Thermoelectrics Proc.*, **ICT 98**, 362-369 (1998).

[15]V. Zaitsev, M. Fedorov, E. Gurieva, I. Eremin, P. Konstantinov, A. Y. Samunin, and M. Vedernikov, Highly effective $Mg_2Si_{1-x}Sn_x$ thermoelectrics, Phys. Rev. B, 74, 045207 (2006).
[16]S. K. Bux, M. T. Yeung, E. S. Toberer, G. J. Snyder, R. B. Kaner, and J. Fleurial, Mechanochemical synthesis and thermoelectric properties of high quality magnesium silicide, *J. Mater. Chem.*, **21**, 12259-12266 (2011).
[17]L. Wang, X. Qin, W. Xiong, and X. Zhu, Fabrication and mechanical properties of bulk nanocrystalline intermetallic Mg_2Si, *Mater. Sci. Eng. A*, **459**, 216-222 (2007).
[18]N. Satyala, and D. Vashaee, Modeling of thermoelectric properties of magnesium silicide (Mg_2Si), *J. Electron. Mater.*, **41**, 1785-1791 (2012).
[19]S. Nakamura, Y. Mori, and K. Takarabe, Analysis of the Microstructure of Mg_2Si Thermoelectric Devices, *J. Electron. Mater.*, **43**, 2174-2178 (2014).
[20]Q. Zhang, J. He, T. Zhu, S. Zhang, X. Zhao, and T. Tritt, High figures of merit and natural nanostructures in $Mg_2Si_{0.4}Sn_{0.6}$ based thermoelectric materials, *Appl. Phys. Lett.*, **93**, 102109 (2008).
[21]W. Xiong, X. Qin, and L. Wang, Preparation and microstructural characterization of nanocrystalline Mg_2Si intermetallic compound bulk, *Chin. J. Nonferrous Met.*, **15**, 380-385 (2005).
[22]X. Niu, and L. Lu, Formation of magnesium silicide by mechanical alloying, *Adv. Perform. Mater.*, **4**, 275-283 (1997).
[23]H. Jiang, H. Long, and L. Zhang, Effects of solid-state reaction synthesis processing parameters on thermoelectric properties of Mg_2Si, *J. Wuhan Univ. of Technol. Mater. Sci. Ed.*, **19**, 55-56 (2004).
[24]J. Tani, and H. Kido, Thermoelectric properties of Sb-doped Mg_2Si semiconductors, *Intermetallics*, **15**, 1202-1207 (2007).
[25]L. Han, M. Yang, Q. Shen, and L. Zhang, Reaction sintering of magnesium silicide thermoelectric material by the spark plasma sintering technique, *J. Chin. Ceram. Soc.*, **36**, 337 (2008).
[26]M. Yoshinaga, T. Iida, M. Noda, T. Endo, and Y. Takanashi, Bulk crystal growth of Mg_2Si by the vertical Bridgman method, *Thin Solid Films*, **461**, 86-89 (2004).
[27]S. Zhou, and C. Bai, Microwave direct synthesis and thermoelectric properties of Mg_2Si by solid-state reaction, *Trans. Nonferrous Met. Soc. China*, **21**, 1785-1789 (2011).
[28]L. Wang, and X. Qin, The effect of mechanical milling on the formation of nanocrystalline Mg_2Si through solid-state reaction, *Scr. Mater.*, **49**, 243-248 (2003).
[29]B. Gerstein, F. Jelinek, M. Habenschuss, W. Shickell, J. Mullaly, and P. Chung, Thermal Study of Groups II—IV Semiconductors. Lattice Heat Capacities and Free Energies of Formation. Heat Capacity of Mg_2Si from 15°—300° K, *J. Chem. Phys.*, **47**, 2109-2115 (1967).
[30]B. Ihsan, Thermochemical data of pure substances, *VCH Publishers*, New York (1995).

MODELING DENSIFICATION DURING FAST FIRING OF YTTRIA-STABILIZED ZIRCONIA

Sergio Y. Gómez[(a)], Farshad Farzan[(a)], Ricardo H. C. Castro[(b)], Dachamir Hotza[(a)]

(a) Federal University of Santa Catarina (UFSC), Brazil
(b) University of California at Davis (UCDavis), USA

ABSTRACT

The fast firing process was studied in compacts of yttria-stabilized zirconia, both partially stabilized tetragonal phase (3YSZ) and the fully stabilized cubic phase (8YSZ). Fast firing consists in using a high heating rate until maximum temperature to produce relatively dense ceramics in shorter periods of time, avoiding grain coarsening during sintering. Experiments were carried out in a tube furnace with a heating rate of ~500 °C/min and the continuous density change of the YSZ compacts was compared to traditional sintering schedules. Moreover, a densification model was developed as a function of time, temperature and particle size for fast-fired samples. The experimental results disclosed a microstructural improvement with high heating rates. The model was able to predicting the achieved density at the sintering conditions with different particle size of the precursor powders.

1. INTRODUCTION

Fine-grained materials constitute a paramount research focus, stem from their enhanced properties [1-7]. One of the challenging scopes within the field is sintering planning, mainly for lack of tools to predict the density-time-temperature profile from available sintering data, and for simultaneous densification and grain growth during thermal cycles, where the last is often undesirable [8].

Simplified models in sintering are helpful to identify driving forces, mass transport paths, and geometric factors, but their practical applicability is restricted [9-11]. Efforts made to extend the models have been of a limited success [11].

The state-of-the-art approach to thermal schedule planning is the master sintering curve (MSC), proposed by Su and Johnson to characterize the sintering behavior of a given powder and green compact notwithstanding the heating profile [11]. The MSC approach presents the limitation of a necessary construction of a function-density relationship, so to retrieve data is necessary to compute the new data finding the corresponding ordinate value at that point.

To tackle the grain growth effect, Harmer and Brook proposed a thermal procedure called fast firing (FF), which consists in using a very high heating rate and a short holding time at elevated temperatures to bypass temperature regions that do not contribute to densification and promote grain growth [12-15].

In this work the sintering of ceramic materials was analyzed in two complementary approaches. Firstly, fast firing was compared to conventional sintering in ceramic compacts of zirconia partially stabilized tetragonal phase (3YSZ) and the fully stabilized cubic phase (8YSZ). Secondly, a simple model is proposed to predict the densification for fast-fired samples.

2. METHOD

Commercial, high purity yttria stabilized zirconia powders, ZrO_2–3 mol%Y_2O_3 (TZ-3Y), with powder density of 6.05 g/cm^3, and ZrO_2–8 mol%Y_2O_3 (TZ-8Y), with powder density of 5.95 g/cm^3, (both from Tosoh, Japan) were used for the experiments. Particle size measurements of the suspension of YSZ powders (0.5 wt% in deionized water) were performed (Nano/Zeta-sizer ZEN 3600, Malvern, USA).

Powders were compacted in a cylindrical die by uniaxial pressing to produce green compacts with 19.05 mm diameter, and ~0.5 mm height. No additives such as organic binders or dispersants were used for pellet forming to avoid carbon contamination. The green density was 45% of the pore-free (theoretical) density (%TD).

Sintering was performed in a tube furnace with a heating rate of ~500 °C/min (named fast firing, FF) and 10 °C/min (named conventional sintering, CS). In both cases, the samples were quenched in air to room temperature. The maximum temperature (T_{max}) used was setup at 1400 °C and the holding times at T_{max} were 1, 10 and 100 min (counted after reaching the isotherm in both FF and CS). The continuous density change of the YSZ compacts were measured by the Archimedes method from 3 samples of each experimental condition.

3. RESULTS AND DISCUSSION

Experimental

Table I shows the theoretical density achieved by 3YSZ and 8YSZ compacts for different holding times at 1400°C, for CS and FF. An initial difference of ~30% in densification is observed at holding time of 1 min. Such variance between methods is attributed to the densification occurred during the heating procedure before reaching maximum temperature. The difference in density becomes smaller with time, due to higher densification rate of FF samples at the holding temperature. Furthermore, the differences in relative density are negligible for holding time longer than 10 min. The data show no difference in final densities when comparing both procedures, but FF evidently takes a much shorter time to achieve the densities, which could result in decreased grain growth.

The reduction in grain growth for FF becomes evident when observing the average grain size: growth factor of 1.11 and 1.82 for FF and CS in 3YSZ, respectively. For 3YSZ samples, the differences in average grain size correspond to a decrease of 63% for 3YSZ-FF compacts as compared to 3YSZ-CS.

Table I. Theoretical density and average grain size for 3YSZ and 8YSZ compacts sintered at different holding times (t) at 1400°C.

t	% TD				Average grain size (nm)			
(min)	3YSZ-CS	3YSZ-FF	8YSZ-CS	8YSZ-FF	3YSZ-CS	3YSZ-FF	8YSZ-CS	8YSZ-FF
0	45.0	45.0	45.0	45.0	188	188	243	243
1	98.7	70.7	78.9	50.6	-	-	-	-
10	99.8	99.5	86.2	81.1	-	-	-	-
100	100	100	97.0	96.2	340	210	1100	245

The average grain size of 8YSZ-FF compacts reveal a limited grain growth during the process. It is interesting to observe that though starting with a submicron powder (243 nm), the grain size of 8YSZ-FF compacts are in the order of typical sizes achieved from powders with initial grain sizes below 50 nm for this same composition but using different densification techniques; i.e., 295 nm for TSS, 210 nm for SPS, and 370 for HP [5,16].

The overall results and the comparison with the literature suggest that FF procedure is advantageous for grain growth suppressing during sintering of zirconia. This effect is less pronounced for 3YSZ samples because 3YSZ is known to have slow grain growth kinetics [10]. That is, even for CS, coarsening was somehow limited. The grain sizes reported here for FF (~200 nm) are very similar to those corresponding to conventional sintering under similar conditions but starting with 75 nm powder, suggesting that submicrometric 3YSZ-FF can yield comparable results as those obtained with nanopowders [10].

Model

The linear-viscous version of the Olevsky continuum theory of sintering provides the porosity kinetics during free sintering as [17,18]:

$$\theta = \theta_0 e^{-\frac{3}{8}\tau_s} \tag{1}$$

where θ_0 is the initial porosity, θ is the porosity at any given time and τ_s is the specific dimensionless time of sintering, defined as:

$$\tau_s = 3\int_0^t \frac{\alpha}{\eta_0 G} dt \tag{2}$$

where t is the time of sintering: α the surface tension; η_0 the shear viscosity of the fully dense material; and G the average particle size. Thus, the porosity kinetics for free sintering is characterized by an exponential porosity decrease with an asymptotic approaching a fully dense state at $\lim_{t\to\infty}\theta = 0$. If isothermal sintering conditions and negligible grain growth are assumed, all the parameters in Eq. (2) become time invariant and the integral can be numerically solved. Using Eq. (1) the specific time of sintering can be determined based on the dilatometry data for a given material through isothermal sintering experimentation as [19]:

$$\theta = \theta_0 e^{-\frac{9\alpha}{8\eta_0 G}t} \tag{3}$$

The drawback of this approach is the apparent nonlinear function when used for ceramic materials, which results in poor fitting. To accurately fit on experimental isothermal data, in this work a phenomenological approach is proposed as:

$$\theta = \theta_0 e^{-F(T,t)} \tag{4}$$

where $F(T,t)$ is an unknown function depending on T and t, temperature and sintering time respectively. At isothermal conditions, the sintering process is a thermal activated process, thus, an Arrhenius type function is assumed. A power model for the sintering time was proposed to linearize the data. Thus, the isothermal function is set as:

$$F(t) = Ae^{-Q/RT}t^{1/n} \tag{5}$$

where A is a preexponential factor that includes the initial average grain size G_0 ($A=A'/G_0$); Q is the activation energy of the densification process; R is the gases constant (8.314 J/mol K); and n is a constant related to the sintering process. Taking natural logarithm of both sides of Eq (5) for the isothermal case, Eq. (6) is obtained:

$$\ln\left(\frac{\theta}{\theta_0}\right) = Ae^{-Q/RT}t^{1/n} \tag{6}$$

To find n, a generalized reduced gradient algorithm was used to optimize the fitting on porosity isothermal data, maximizing the square root regression coefficient to Eq. (6) of the experimental data at different temperatures simultaneously. The results are shown in Fig. 1, where the coefficients for linear fit are in average >0.99. The outcome for both materials (3YSZ and 8YSZ) was similar with values of $n_{3YSZ} = 2.37$ and $n_{8YSZ} = 2.36$.

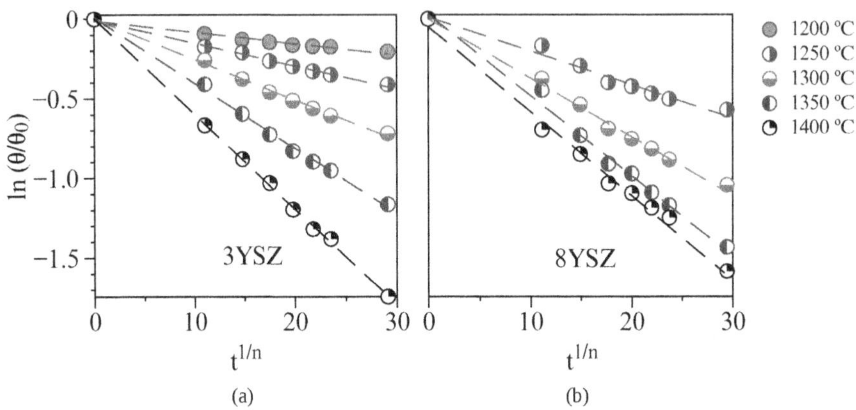

Fig 1. Isothermal sintering data for (a) 3YSZ and (b) 8YSZ with n_{3YSZ}=2.37 and n_{8YSZ}=2.36. Data from the literature[20].

Thus, the slope (z) for each experiment is function of the temperature as:

$$z = Ae^{-Q/RT} \tag{7}$$

Taking natural logarithm in both sides of Eq. (7), Eq. (8) is obtained:

$$\ln z = \ln A - \frac{Q}{RT} \tag{8}$$

Now it is possible to find the model parameters with a linear fitting of z *vs.* 1/T. The fitting results show that the proposed model describes the sintering from the data in the literature[20], for a particle size of 600 nm and 530 nm for 3YSZ and 8YSZ, respectively. The constants obtained for the studied materials are in Table II.

Table II. Parameters for the studied materials from model fit.

Parameter	3YSZ	8YSZ
ln(A)	12.92	6.61
Q (kJ/mol)	215	132
n	2.37	2.36
G (nm)	600	530

It is possible to obtain the relative density from other experiments, recalling that the relative density is related with the porosity with the relationship:

$$\rho = 1 - \theta \tag{9}$$

The model results are shown in Fig. 2.

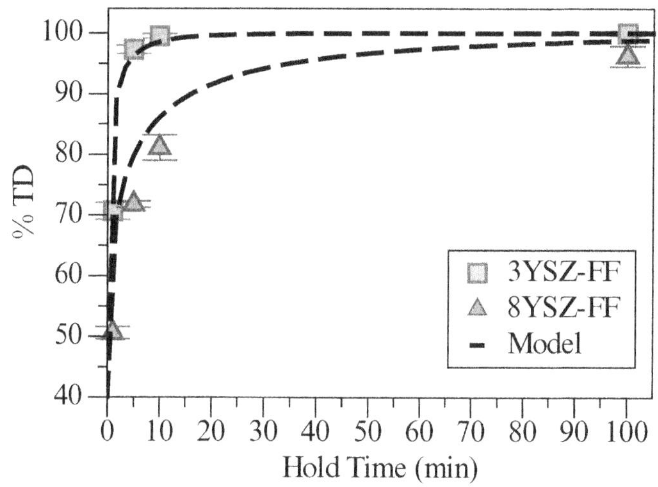

Fig. 2. Theoretical density *vs.* holding time for fast-fired YSZ samples at 1400°C in comparison to the model prediction.

The model describes the behavior of both materials with a maximum error of ±7%, starting with experimentally obtained constants for larger particle sizes, thus showing the model robustness.

CONCLUSIONS

This study examines the microstructure evolution of yttria-stabilized zirconia compacts comparing fast firing with conventional sintering. A model is also developed and validated with the experimental data.

The fast firing method saves energy and time in ceramic processing maintaining density values after sintering similar to those in the conventional process. Fast firing promotes lower grain coarsening than conventional sintering and the results are comparable to widely employed techniques such as spark plasma sintering, hot pressing and two-step sintering.

The model predicts the densification on 3YSZ and 8YSZ compacts, describing times and temperatures for fast-fired samples with different particle sizes by means of three experimental constants: This model may be a valuable tool to plan fast firing sintering schedules and optimize time-temperature profiles.

ACKNOWLEDGEMENTS

The Brazilian funding agencies CAPES and CNPq are gratefully acknowledged.

REFERENCES

[1]Bellino, M. G., Lamas, D. G. & Walsoe de Reca, N. E. Preparation and ionic transport properties of YDC-YSZ nanocomposites. *J Mater Chem* 18, 4537-4542 (2008).
[2]Suzuki, T., Kosacki, I. & Anderson, H. U. Microstructure–electrical conductivity relationships in nanocrystalline ceria thin films. *Solid State Ionics* 151, 111-121 (2002).
[3]Han, M., Tang, X., Yin, H. & Peng, S. Fabrication, microstructure and properties of a YSZ electrolyte for SOFCs. *J Power Sources* 165, 757-763 (2007).
[4]Razavi Hesabi, Z., Mazaheri, M. & Ebadzadeh, T. Enhanced electrical conductivity of ultrafine-grained $8Y_2O_3$ stabilized ZrO_2 produced by two-step sintering technique. *J Alloy Compd* 494, 362-365 (2010).
[5]Mazaheri, M., Valefi, M., Hesabi, Z. R. & Sadrnezhaad, S. K. Two-step sintering of nanocrystalline $8Y_2O_3$ stabilized ZrO_2 synthesized by glycine nitrate process. *Ceram Int* 35, 13-20 (2009).
[6]Tran, T. B., Hayun, S., Navrotsky, A. & Castro, R. H. R. Transparent nanocrystalline pure and Ca-Doped MgO by spark plasma sintering of anhydrous nanoparticles. *J Am Ceram Soc* 95, 1185-1188 (2012).
[7]Yang, H.-S., Bai, G. R., Thompson, L. J. & Eastman, J. A. Interfacial thermal resistance in nanocrystalline yttria-stabilized zirconia. *Acta Mater* 50, 2309-2317 (2002).
[8]Mazaheri, M., Zahedi, A. M. & Hejazi, M. M. Processing of nanocrystalline 8 mol% yttria-stabilized zirconia by conventional, microwave-assisted and two-step sintering. *Mater Sci Eng A* 492, 261-267 (2008).
[9]Kang, S.-J. L. in Sintering (ed Suk-Joong L. Kang) 39-55 (Butterworth-Heinemann, 2005).
[10]Mazaheri, M., Simchi, A., Dourandish, M. & Golestani-Fard, F. Master sintering curves of nanoscale 3Y-TZP powder compacts. *Ceram Int* 35, 547-554, (2009).
[11]Su, H. & Johnson, D. L. Master sintering curve: a practical approach to sintering. *J Am Ceram Soc* 79, 3211-3217 (1996).
[12]Harmer, M. P. & Brook, R. J. Fast firing-microstructural benefits. *Trans J Brit Ceram Soc* 5, 147-148 (1981).
[13]Landin, S. M. & Schulze, W. A. Rapid Sintering of stoichiometric zinc-modified lead magnesium niobate. *J Am Ceram Soc* 73, 913-918 (1990).
[14]Garcia, D. E., Klein, A. N. & Hotza, D. Advanced ceramics with dense and fine-grained microstructures through fast firing. *Revi Adv MateriSci* 30, 273-281 (2012).
[15]Hotza, D., García, D.E., & Castro, R.H.R. Obtaining highly dense YSZ nanoceramics by pressureless, unassisted sintering. *Int Mater Rev* 7, 1-23 (2015).
[16]Dahl, P. et al. Densification and properties of zirconia prepared by three different sintering techniques. *Ceram Int* 33, 1603-1610 (2007).
[17]Olevsky, E. A. Theory of sintering: from discrete to continuum. *Mater Sci Eng R* 23, 41-100 (1998).
[18]Olevsky, E. et al. Sintering of multilayered porous structures: Part I- Constitutive models. *J Am Ceram Soc* 96, 2657-2665 (2013).
[19]Ni, D. W. et al. Sintering of multilayered porous structures: Part II–Experiments and model applications. *J Am Ceram Soc* 96, 2666-2673 (2013).
[20]Suarez, G., Rendtorff, N. M., Scian, A. N. & Aglietti, E. F. Isothermal sintering kinetic of 3YTZ and 8YSZ: Cation diffusion. *Ceram Int* 39, 261-268 (2013).

MECHANISTIC STUDIES OF COMPACTED AND SINTERED ROCK SALT

Claudia H. Swanson[*,†], Susanne Böhme[‡], Jens Günster[‡]
[†] Department of Natural Sciences, University of Chester
Pool Ln., Ince, Chester, CH2 4NU, United Kingdom
[‡] Division: Ceramic Processing and Biomaterials, Federal Institute of Materials Research and Testing BAM
Unter den Eichen 44-46 12203 Berlin, Germany

ABSTRACT

This research addresses the densification via compaction and sintering of naturally occurring rock salt. In a saline environment, rock salt is thermodynamically stable and shows a mechanical behavior compatible to the surrounding host material. The sintering of rock salt powder compacts was systematically investigated using temperature and pressure as the process variables. The rock salt showed segregations of anhydrite, $CaSO_4$, at the grain boundaries of sodium chloride crystals. Powder compacts that were subjected to low pressures lost those anhydrite segregates over time, while high pressure treated compacts retained the anhydrite segregates at the grain boundaries. The maximum experimental density of the sintered rock salt samples was 2.008 g cm^{-3} which corresponded to 92.5 % of the theoretical density of pure sodium chloride. The increase in the density at low pressures (< 200 MPa) and high temperatures (> 550°C) makes the sintering process a promising technique for rock salt applications such as underground building blocks.

INTRODUCTION

Rock salt has been of interest for several different applications such as food ingredient and as storage medium. The ductile nature of rock salt under high pressure is noteworthy for its possible application as building bricks. For the manufacture of construction materials, elements made of rock salt are considered with different approaches. One possibility is cutting simple geometries, i.e. bricks, directly out of natural rock salt. This process does not automatically lead to a product with homogeneous and isotropic properties, but will translate all irregularities from the natural material to the building blocks. Therefore, a powder based approach is more attractive than conventional methods. In a typical powder processing route, the raw material provided as a powder, is compacted to one part by pressing and sintering.

In 1990s, early applications of rock salt as storage material for gas capture were published.[1, 2, 3] Pure sodium chloride and rock salt were compared as a model for compaction and isolation of gas from sodium chloride. The gas storage experiments showed that sodium chloride formed cracks through which the gas could move. Recent studies on addition of anhydrite added to halite resulted in synthetic rock salt that showed variations in mechanical properties.[4] Cosenza *et al.* performed field experiments which confirmed that rock salt is permeable to gas and brine, even relatively far from underground openings.[5, 6]

Besides halite (sodium chloride, NaCl), the naturally occurring rock salt, typically contains a variety of trace metals such as calcium, potassium and magnesium and anions such as chloride, sulfate and bromide. Its composition usually depends on the evaporation process connected to the formation of the rock salt deposits.[7] Several studies have been investigated on the properties of pure halite. The study of sodium chloride properties which were relevant to our research included the examination of variations in coefficient of thermal expansion from $\alpha = 40.3 \cdot 10^{-6}$ to $78.5 \cdot 10^{-6}\ K^{-1}$ at 0 to 750°C, respectively.[8] Marmottant *et al.*, Brinckmann *et al.* and Price examined the compaction of sodium chloride and synthetic rock salt using different conditions and techniques.[9, 10, 11] Sodium chloride exhibited 50 % higher contact for non-compacted particles than for pressure treated compacted particles which was further verified from comparison of computational and experimental compaction, resulting of pure sodium chloride. Interestingly, the high humidity experiments from Wu *et al.* shows that sodium chloride self-assembled into greatly ordered arrays under high humidity conditions i.e. at relative humidity higher than 40 %.[12]

Kingery *et al.* conducted first sinter experiments with pure sodium chloride and concluded that the sinter process follows an evaporation-condensation mechanism.[13, 14] No densification of the salt was observed using the evaporation-condensation mechanism. The particles changed from spheres to the ellipses at 700-750°C, and an increase in their size from 61.5 to 70 µm in addition to increase in the strength due to the sintering process. Goodall *et al.* analyzed the sintering kinetics at different particle sizes and temperatures for sodium chloride. In order to conduct water- and air-free experiments, argon was used as inert gas during the sinter process at 700-800°C.[15, 16] Moser *et al.* determined that the rate-controlling step in the material transport of the sinter process, occurring by evaporation-condensation mechanism, was the diffusion of sodium chloride vapor through a thin boundary layer near the condensing surface. Several research groups studied the sinter kinetics of sodium chloride and explained the sintering process *via* evaporation and condensation mechanism.[17, 18] Thompson *et al.* noticed a better densification using smaller particles with larger surface area and thus, resulted a more effective evaporation-condensation process. Sata *et al.* studied the expansion of sodium chloride during sintering at varied time and temperature settings.[19] Compaction pressures between 20 and 200 MPa resulted in green densities from 72 to 93 %, respectively. Sata *et al.* found a linear expansion of the salt with temperature, which was caused by an increase in closed pore volume and grain boundary relaxations. Morgan *et al.* observed a volume shrinkage of 18 % at 650°C for sodium chloride.[20] Ammar *et al.* conducted sinter experiments using sodium chloride under pressure and different particle sizes.[21] Ammar observed that powder compacts of particles smaller than 125 µm showed shrinkage during sintering.

Very few experiments have been conducted on sintering rock salt. Wawersik *et al.* conducted modeling and a mechanistic interpretation of creep in rock salt up to temperatures of 200°C.[22] The experiments were performed in triaxial creep tests by varying stress and temperature. The data were fitted with models for power-law creep, cross-slip, and dislocation glide. The results for the cross-slip model exhibited the best fit agreement. Other rock salt experiments included the study of viscosity on the material.[23] The dependency of the grain size and temperature on viscosity were significant and varied from 10^{17} Pa s for small grains and high temperatures to 10^{20} Pa s for large grains a low temperatures.

In the current study, we discuss different approaches for the densification of rock salt powders *via* compaction and sintering. The work focuses on the use of powder qualities obtained by crunching and milling of the natural rock salt. A detailed analysis on the effect of compaction pressure, time and temperature on sintering of rock salt is discussed.

EXPERIMENTAL SETUP

Natural rock salt samples (Gorleben) were crushed and milled in a disc mill (Fritsch, Pulverisette 14) to obtain fine powders. After milling, the powders were dried at 35°C for 8 hours. Particle size distributions were determined using a Master Sizer instrument from Malvern UK. For chemical characterization and phase analysis of the starting material and final samples, ion chromatography (Metrohm Type 761 Compact IC) and powdered X-ray diffraction (PXRD PANalytical X'Pert PW3020) were employed. Cylindrical powder compacts with a diameter of 5 mm and a height of 5 mm were prepared *via* uniaxial pressing and isocratic pressing. For different samples, the pressure for compaction varied from 50 to 560 MPa. The density of the obtained samples was determined by measuring their geometric volume and bringing it in relation to their weight. The as-prepared powder compacts were sintered for further densification. Sintering was performed in a hot stage microscope (EM-201, Hesse Inst., Germany) up to maximum temperature of 810°C at a heating rate of 10°C/min. The hot stage microscope allowed observing the *in-situ* volumetric shrinkage to be determined during the sintering process by measuring the area of the sample's projection in an optical setup by a CCD camera. Images were collected at a rate of 40 frames per minute. The structure of the samples was analyzed using environmental scanning electron microscopy (FEI XL30 ESEM) and field emission scanning electron microscopy (FEI FESEM Quanta 200 + TSL-EDAX).

RESULTS AND DISCUSSION

Characterization

Before the sintering experiments, a phase analysis of the natural rock salt material for its halite-anhydrite composition was carried out using IC and PXRD, as 98.3 % halite and 1.7 % anhydrite; (figure 1). In figure 2, a typical particle size distribution of the rock salt after milling is shown. The particle size varies between 5 to 200 µm. The mean particle size (d_{50}) is 30 µm. Thermogravimetric analysis was performed to verify the stability of the minerals and the absence of phase transitions in the temperature range of the sinter process (figure 3). At room temperature, rock salt exhibits hygroscopic properties and any weight gain due to moisture is lost immediately after the heating process started. At about 600-650°C, a small endothermic peak, most likely attributed to the release of water, was observed. Also, in this temperature regime, the onset of sintering and densification was observed using the hot stage microscope. At temperatures above 750°C, a significant weight loss of rock salt occurred due to endothermic melting of sodium chloride at 801°C and evaporation of the material.

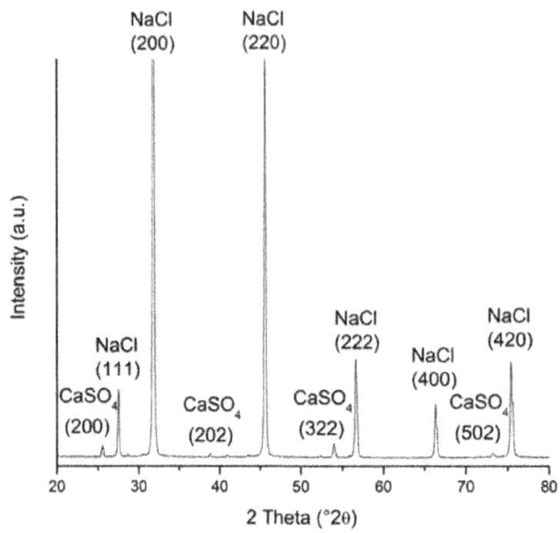

Figure 1: PXRD of rock salt from 20 to 80° 2Theta with the corresponding Miller indices.

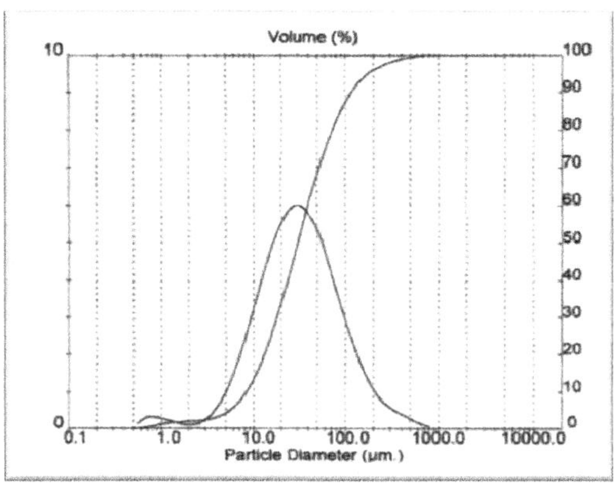

Figure 2: Particle size distribution of milled rock salt resulting in a mean particle size of 30 µm.

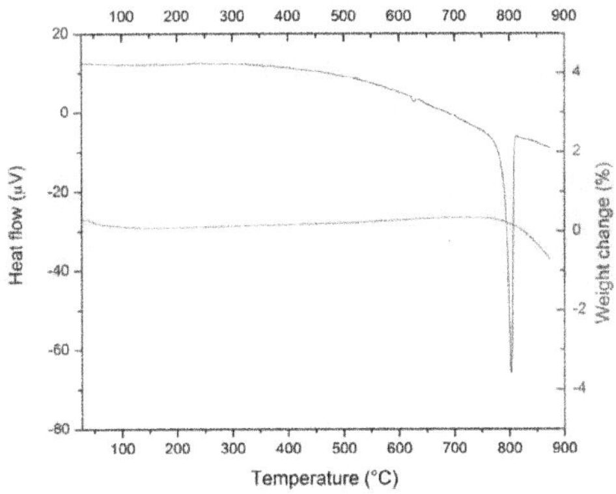

Figure 3: Thermogravimetric (TGA, red) and Differential Thermal Analysis (DTA, blue) of the rock salt.

Compaction results

Rock salt powders were compressed mainly *via* uniaxial pressing to obtain powder compacts. Isocratic pressing produced similar compression results and therefore, no further investigation was carried out on these samples. The density of the powder compacts varies with the applied pressure during compaction. Table 1 lists the obtained densities of the powder compacts as a function of applied pressure. The mineral composition of the naturally occurring rock salt with an expected density of 2.184 g cm^{-3} determined from PXRD, containing the two components: halite and anhydrite. At pressures of 560 MPa, the density of the compacted powder reaches a value of 2.145 g cm^{-3}, which is 98.2 % of the expected theoretical density. Figure 4 compares the microstructure of two powder compacts, prepared by uniaxial pressing at applied pressure of 100 and 560 MPa and densities of 1.776 and 2.145 g cm^{-3}, respectively. At a pressure of 100 MPa, larger and smaller particles are uniformly distributed over the entire imaged area. At 560 MPa, the cross section reveals areas of densely packed smaller particles and areas in which the particles are fused to larger domains. After the application of high pressures, the obtained samples are translucent and show the formation of macroscopic cracks. These observations suggest that, at pressures ~ 560 MPa the primary particles of the starting powder are fused to a dense structure. Typically, uniaxial pressing does not provide a uniform pressure distribution within the sample and, therefore, besides translucent regions. Staiger *et al.* also obtained similar experimental results that showed crystal orientations at the grain boundaries during the compaction of sodium chloride.[24]

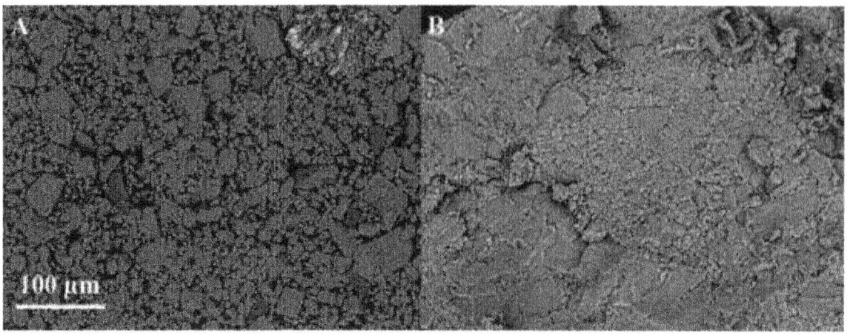

Figure 4: ESEM images as comparison of compressed rock salt samples at 100 MPa (A) and 560 MPa (B).

Table 1: Pressure applied for the compaction of rock salt and the respective density after compression.

Pressure [MPa]	Density [g cm^{-3}]	Density after sintering [g cm^{-3}]
60	1.640	1.928
80	1.737	1.843
100	1.776	1.949
110	1.811	1.936
120	1.872	2.008
200	1.908	1.485
300	2.038	1.526
400	2.090	1.513
500	2.100	1.552
560	2.145	-

Sinter results

Pure sodium chloride has a melting point of 801°C and a high thermal expansion coefficient of $40.\cdot 10^{-6}$ K^{-1} at 20°C). In order to examine the thermal expansion properties of naturally occurring rock salt, powdered and compacted samples (100 MPa) were heated in a hot stage microscope to 812°C. The corresponding sintering curve (change in dimension/area vs temperature) is illustrated in figure 5, which shows the behavior of the rock salt toward volume change while being heated. The rock salt sample increases in volume linearly with the temperature at a rate of $34.\cdot 10^{-6}$ K^{-1} (up to 300°C). Over 300°C, the thermal expansion varies and increases to $51.\cdot 10^{-6}$ K^{-1} above 500°C. Assuming the similarity of naturally occurring rock salt to pure sodium chloride, the thermal expansion coefficient should increase from about $40.\cdot 10^{-6}$ K^{-1} at 20°C to about $79.\cdot 10^{-6}$ K^{-1} at 750°C. Our measurements disagree with the literature; the thermal expansion is not close to the reported value from the literature. The step in the expansion curve (figure 5) at about 425°C belongs to a loss of crystalline water with needed adjustments on the fogged heating chamber, which was observed at about 600°C. Just before reaching the melting point, the volume of the sample was reduced by 7.5 %. From the hot stage microscope measurements, we also found that the melting point of rock salt is reduced, which is possibly due to the presence of calcium sulfate as impurity in sodium chloride. The melting point determined by hot stage microscopy was found to be at 796°C, which is 5°C below the melting point of pure sodium chloride as reported in the literature.

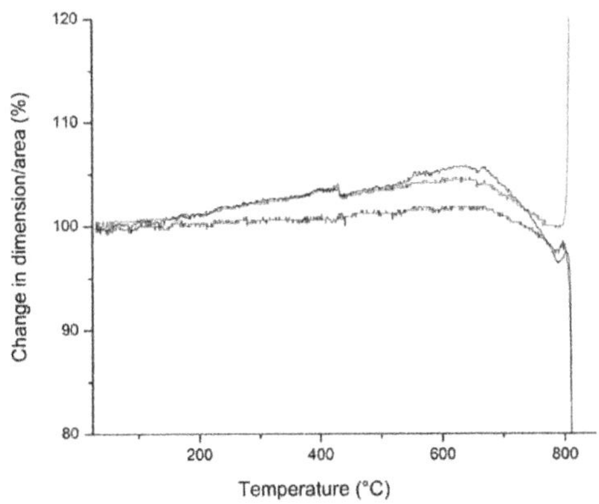

Figure 5: Hot stage microscopy of rock salt, with the height change (blue), the width change (red) and the area change (black).

Temperature and pressure dependence in the sinter process

In order to study the sintering kinetics of rock salt, powder compacts prepared by uniaxial pressing, at different maximum pressures, were annealed at a constant heating rate (10 K min^{-1}) to different maximum temperatures. These experiments were performed in a conventional sintering furnace under ambient atmosphere. After sintering, the samples were characterized by measuring their density and comparing their microstructures. In table 2, the investigated powder compacts and the respective sintering conditions are summarized. The densities after compaction by uniaxial pressing and by sintering are listed in table 1. Below

550°C, neither a visible change nor a change in density for the material could be observed. At 550°C, the rock salt samples showed a visible change of their outer contour. The crystalline boundaries of the cylindrical samples lost their sharp edges, which refer to the onset of the sintering. At 650°C, a change in density could be observed in addition to the visible change in geometry. All samples have a residual porosity. Figure 6 discloses the microstructure of the samples (fracture surface) sintered at temperatures of 550, 640, 700 and 750°C. In figure 6(B,C), between 640 and 700°C, significant changes in the microstructure are clearly noticeable. Typically during sintering, smaller particles disappear at the expense of larger ones. The larger particles show a relatively smooth surface. This trend is confirmed by figure 6D (750°C), where only large particles are present in the sample.

Pressures higher than 100 MPa were applied to powder compacts of rock salt in order to achieve higher densities. At higher compaction pressure, the risk of forming cracks in the powder compact increased. On the other hand, in the previous section, it was shown that longer sintering times can heal the material. Rock salt was compacted at 100, 200, 300, 400, and 500 MPa. Afterwards, the samples were sintered at 750°C for 20 h. The sintering often resulted in cracks and lower densities than the unsintered samples (table 1).

Figure 7 shows micrographs of sintered samples as a function of different compaction pressures applied to the powder compacts. In high resolution (right), the microstructures are not very different but in the overview (left), the formation of cracks becomes noticeable for samples compacted at pressures higher than 300 MPa. These cracks explain why samples compacted at pressures higher than 120 MPa reveal lower densities than samples subjected to low pressure (< 120 MPa) after the sintering, see table 1. Likely, gaseous species entrapped in the powder compact cannot escape when the samples are highly compacted prior to sintering.

Table 2: Compacted samples (100 MPa) at different temperatures of the sinter process.

Temperature [°C]	Density before sinter process [g cm^{-3}]	Green density [%]	Density after sinter process [g cm^{-3}]	Density after sintering [%]
350	1.779	81.4	1.808	82.8
450	1.811	82.9	1.771	81.1
550	1.776	81.3	1.789	81.9
640	1.799	82.4	1.794	82.2
700	1.790	82.0	1.836	84.1
750	1.802	82.5	1.916	87.7

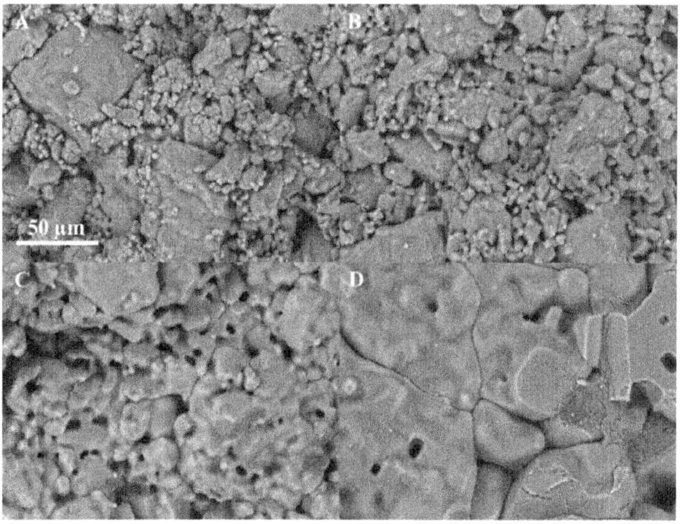

Figure 6: ESEM images of rock salt at different sinter temperatures of T=550°C (A), 640°C (B), 700°C (C), and 750°C (D).

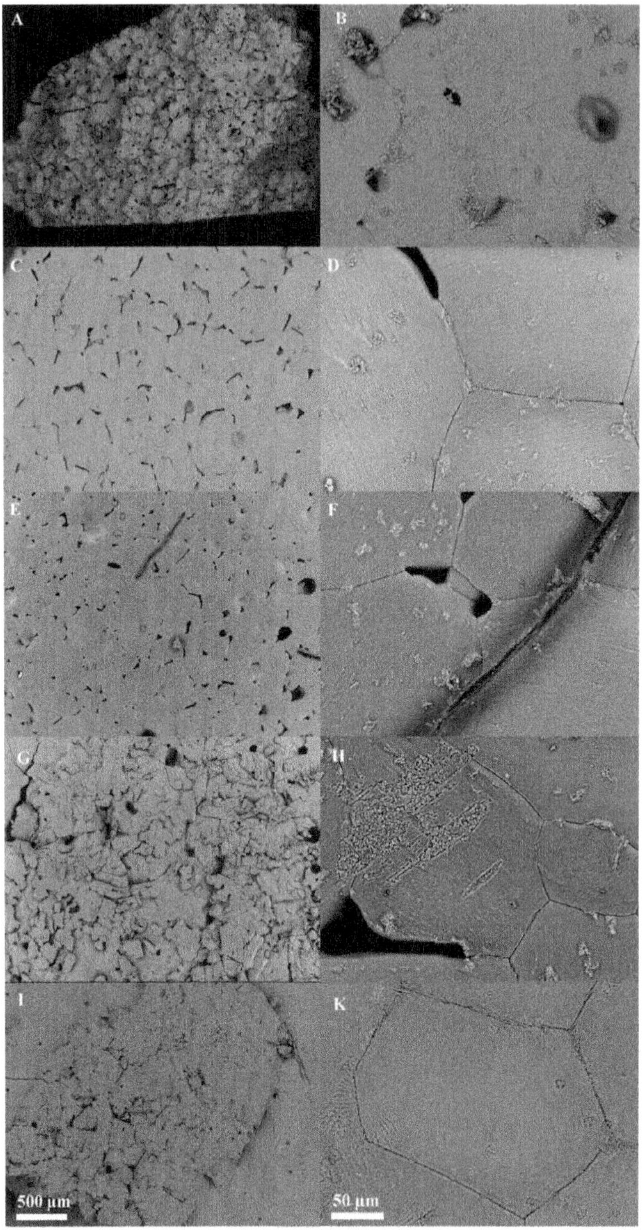

Figure 7: SEM images of sintered rock salt samples with the powder compacted at p = 100 MPa (A,B), 200 MPa (C,D), 300 MPa (E,F), 400 MPa (G,H) and 500 MPa (I,K).

Time dependence in the sinter process

Figure 8 illustrates the microstructure of the samples surface after sintering at 700°C for different dwell times. Both images reveal large particles that are very well interconnected and contain a closed outer surface within the sample. The sample sintered for 30 min showed elongated chip-like particles at the crystal boundaries. Those structures disappeared after sintering the sample for an extended period of 20 h. FESEM images were taken to analyze these elongated particles. The sample was vertically cut between two crystal grain boundaries. Figure 9 shows the FESEM of the chosen particle and the cross-cut of the elongated particle between crystal boundaries. The material runs like a wedge into the rock salt. This wedge contains a larger amounts of calcium sulfate compared to the main material. As the FESEM beam was focused on the sample, the wedge changed its appearance. This confirms the ductile property of salt and also explains the disappearance of the elongated particles after sintering for 20 h.

The small holes that are visible on the salt surface indicate that the interior volume of the samples is not dense, but is also porous. Previous micrographs shown in figure 6 (700 and 750°C) reveal the presence of these holes inside the volume of the samples. As shown in figure 7 (100 MPa), these holes are cross sections of worm-like cavities in three dimensions. With these structural features, the rock salt samples predicate a structure that is quite unique after the sintering process. In a previous study, similar structures have been reported to occur in pure NaCl sintered powder samples.[19] T. Sata attributed the formation of closed pores during sintering *via* pore coalescence and grain boundary relaxation in conjunction with a rapid grain growth. Thus, the formation of the observed pore structure is most likely not a feature exclusively related to the intrinsic impurities of the natural mineral, but associated with the rapid crystal growth during sintering. Pores similar to the one reported in this study were also observed by S. Olivella *et al.* and cited by Spiers *et al.*[6, 25]

We conclude that the microstructure of the compacted and sintered samples is formed by two phases, which are separated during sintering. One phase is anhydrite that is segregating to the majority of the sample particles surface and the second phase is halite forming the particles volume. For the formation of the worm-like cavities, a conclusive model cannot be offered, but a rapid grain growth of halite crystals is a decisive factor. A release of gas from the sample during sintering could not be confirmed in this context. The anhydrite forms plate-like structures, which are accommodated between the halite particles; the halite particles are interconnected by halite channels penetrating through the anhydrite plates. Only at the outer surface of the sample, the anhydrite plates can form a closed surface, which is pierced by openings of the worm-like vacancies.

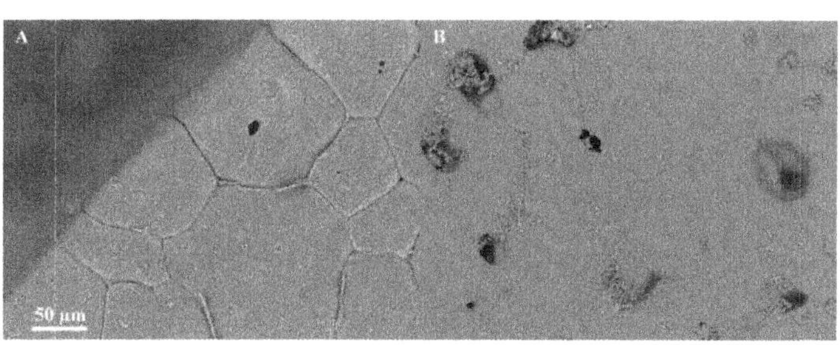

Figure 8: ESEM images of rock salt for different sinter times; 30 min (A) and 20 h (B).

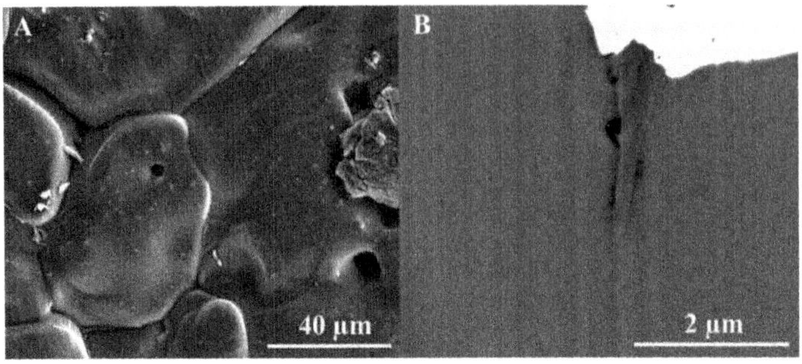

Figure 9: FESEM of cut through the elongated material running along the crystal boundaries.

CONCLUSION

For the first time, the sintering of powder compacts made from natural rock salt has been systematically examined. The major impurity contained in the investigated rock salt samples is anhydrite (1.7 %) and causes the main difference to pure rock salt. During sintering, the anhydrite segregates to the grain boundaries between individual sodium chloride crystals and thus blocking the complete consolidation of the powder sample. Compaction and sintering at different temperatures does not result in completely dense samples. The maximum density reached in this sintering study is 2.008 g cm^{-3} that corresponds to the 92.5 % of the theoretical density. The anhydrite impurities are not responsible for the incomplete densification, but rather results in distinct grain separation. The sintered rock salt forms a complex pore structure. Naturally occurring rock salt is not feasible through the sintering because of the formation of cracks and tunnels that occur during the sintering process. The samples from compaction at 560 MPa show the most promising density for high density required applications. An interesting future study is to explore in more detail the interaction of anhydrite and halite during the sinter process at different ratios of the two components and also the interaction of other inorganic salts towards their sinter behavior with sodium chloride.

REFERENCES

1. Y. Cinar, G. Pusch, and V. Reitenbach, "Petrophysical and Capillary Properties of Compacted Salt," *Transport in Porous Med.*, 64[2] 199-228 (2006).
2. J. Stormont and J. Daemen, "Laboratory study of gas permeability changes in rock salt during deformation," pp. 325-42. Vol. 29.
3. D. Zeuch, "Isostatic hot-pressing mechanism maps for pure and natural sodium chloride--applications to nuclear waste isolation in bedded and domal salt formations," pp. 505-24. Vol. 27.
4. L. Hunter, "The Variation with Temperature of the Principal Elastic Moduli of NaCl near the Melting Point," *Phys. Rev.*, 61[1-2] 84-90 (1942).
5. P. Cosenza, M. Ghoreychi, B. Bazargan-Sabet, and G. De Marsily, "In situ rock salt permeability measurement for long term safety assessment of storage," *Int. J. Rock Mechanics and Mining Sci.*, 36[4] 509-26 (1999).
6. S. Olivella and A. Gens, "A constitutive model for crushed salt," *Int. J. Numer. Anal. Met.*, 26[7] 719-46 (2002).
7. O. Braitsch, "Salt deposits their origin and composition," Vol. 4. Springer Science & Business Media, (2012).

8. P. D. Pathak and N. G. Vasavada, "Thermal expansion of NaCl, KCl and CsBr by X-ray diffraction and the law of corresponding states," *Acta Crystallogr. A,* 26[6] 655-58 (1970).
9. S. Brinckmann, G. Gao, and T. Siegmund, "A combined experimental-numerical study of the compaction behavior of NaCl," *Powder Technol.,* 194[3] 197-206 (2009).
10. A. Marmottant, L. Salvo, C. Martin, and A. Mortensen, "Coordination measurements in compacted NaCl irregular powders using X-ray microtomography," *J. European Ceram. Soc.,* 28[13] 2441-49 (2008).
11. R. H. Price, "Effects of anhydrite and pressure on the mechanical behavior of synthetic rocksalt," *Geophys. Res. Lett.,* 9[9] 1029–32 (1982).
12. J. H. Wu, S. G. Ang, and G. Q. Xu, "Atomic Force Microscopy Study of Self-Assembled Sodium Chloride Nanocrystallites and Their Morphology Transitions," *J. Phys. Chem. C,* 112[20] 7605-10 (2008).
13. R. Goodall, J. F. Despois, and A. Mortensen, "Sintering of NaCl powder: Mechanisms and first stage kinetics," *J. European Ceram. Soc.,* 26[16] 3487-97 (2006).
14. W. D. Kingery and M. Berg, "Study of the Initial Stages of Sintering Solids by Viscous Flow, Evaporation Condensation, and Self Diffusion," *J. Appl. Phys.,* 26[10] 1205-12 (1955).
15. J. Moser and D. Whitmore, "Kinetics of sintering of sodium chloride in the presence of an inert gas," *J. Appl. Phys.,* 31[3] 488-93 (1960).
16. A. Verma, K. Biswas, C. S. Tiwary, A. K. Mondal, and K. Chattopadhyay, "Combined Cryo and Room-Temperature Ball Milling to Produce Ultrafine Halide Crystallites," *Metall. Mater. Trans. A* 1-11 (2011).
17. R. J. Thompson and Z. Munir, "Influence of particle size on the sintering kinetics of ultrapure sodium chloride," *J. Am. Ceram. Soc.,* 65[6] 312-16 (1982).
18. W. Tomlinson and G. Astle, "Sintering kinetics of pure and CdCl 2—doped NaCl in air," *J. Mater. Sci.,* 11[11] 2162-65 (1976).
19. T. Sata, "Expansion during sintering of NaCl powders," *Ceram. Int.,* 20[1] 39-47 (1994).
20. C. S. Morgan, L. L. Hall, and C. S. Yust, "Sintering of Sodium Chloride," *J. Am. Ceram. Soc.,* 46[11] 559-60 (1963).
21. A. Ammar and D. Budworth, "Pressure sintering of NaCl and other ionic materials," *Trans. Brit. Ceram. Soc,* 66[3] 93-105 (1967).
22. W. Wawersik and D. Zeuch, "Modeling and mechanistic interpretation of creep of rock salt below 200 C," *Tectonophysics,* 121[2-4] 125-52 (1986).
23. P. E. van Keken, C. J. Spiers, A. P. van den Berg, and E. J. Muyzert, "The effective viscosity of rocksalt: implementation of steady-state creep laws in numerical models of salt diapirism," *Tectonophysics,* 225[4] 457-76 (1993).
24. M. Staiger, I. Kolbeinsson, J. Newman, T. Woodfield, and T. Sato, "Orientation imaging microscopy of polycrystalline sodium chloride," *Mater. Charact.,* 61[4] 413-19 (2010).
25. C. Spiers, "The Influence of Fluid-rock Interaction on the Rheology of Salt Rock: Final Report." Office for Official Publications of the European Communities, (1986).

CORRESPONDING AUTHOR

* Claudia H. Swanson, c.swanson@chester.ac.uk

ACKNOWLEDGMENTS

We like to thank Dr. Ralf Müller, BAM, for the discussions and the utilization of the facilities for the sintering process and Dr. Martin Sabel, BAM, for his help with the ESEM images. Dipl.-Ing. Reinhard Görke, we would like to acknowledge for his help with the FESEM images.

SINTERING OF NANOSTRUCTURED ZIRCONIA: A MOLECULAR DYNAMICS STUDY

Yi Zhang, Jing Zhang*
Indiana University-Purdue University Indianapolis, Indianapolis, Indiana, USA
*Corresponding author: jz29@iupui.edu

ABSTRACT

Zirconia (ZrO_2) is widely used as a thermal protective material in high temperature or corrosive conditions due to its refractory properties. Powder metallurgy is the primary processing route to fabricate solid zirconia components. In this work, sintering of zirconia is investigated using molecular dynamics calculations with a three-particle atomistic model. The simulated sintering activation energy is 318.4 kJ/mole in the particle core, and 224.0 kJ/mole on the particle surface. Diffusion mechanisms during the sintering process are also discussed.

1. INTRODUCTION

Zirconia is a ceramic material that has many different applications. It shows some unique properties such as high bulk modulus and fracture toughness[1], high dielectric constant[2], chemical stability[3] and low thermal conductivity[4]. Due to its properties, zirconia is widely used for functional materials in the fields including dentistry, fuel cells, corrosive protective coatings and thermal barrier coatings.

In order to achieve high densification, sintering, a powder metallurgy (P/M) process, is used for the fabrication of zirconia materials. The PM process compacts small powders together by heating or pressure without reaching the melting point. Several experiments have been conducted, with the considerations of sintering method and effect of particle size and sintering conditions [5]. Besides, some computational work on sintering of zirconia are reported. Ding [6] developed sintering simulation models at both continuum mechanics and molecular dynamics scales. Through comparing the two models, molecular dynamics has the ability to capture detailed scenarios than the continuum solid state mechanics method. Xu et al. [7] studied the sintering of nickel-YSZ (yttria-stabilized zirconia) system in solid oxide fuel cell structure. The work is focused on the empirical potential model fitting and the diffusion of nickel in the porous YSZ structure. Few computational work has been reported to investigate the pure zirconia sintering process with the analysis of sintering diffusion mechanisms.

In this paper, a molecular dynamics model has been developed to simulate the sintering process in zirconia nanoparticles. The configurational change of zirconia particles in the sintering process is analyzed. Sintering mechanisms, including surface mass transport and bulk mass transport mechanisms are discussed.

2. SIMULATION METHOD

2.1 Pairwise potential model

All calculations are using Molecular Dynamics (MD) method. Several interatomic potential functions for Zr-O compounds have been reported. The Lewis-Catlow (LC) potential[8], based on the Born-Mayer function, is one that widely used for lattice structural and dynamics calculation. However, this model is not preferable for sintering simulation due to the over prediction of melting point[9]. Pairwise potentials with partial charges developed by Pedone et al. [10] are implemented in this paper, due to its capability of high temperature simulation. This potential includes a Morse potential with an additional repulsive term, and long range Coulomb interaction. The expression for the potential is given by:

$$U(r)=\frac{z_i z_j e^2}{r}+D_{ij}\left[\left\{1-e^{-a_{ij}(r-r_0)}\right\}^2-1\right]+\frac{C_{ij}}{r^{12}} \quad (1)$$

The potential energy U(r) is calculated using the combination of three terms: (1) the long range Coulomb potential, where z are charges for each atom. (2) the short range Morse potential, and D, a, r_0 are its coefficients, and (3) the additional repulsive term C/r^{12}. Since for high temperature/pressure simulation, the distance r between each atoms are so small that makes the Morse function to an infinite value, so that the third term is necessary. All the coefficients, including D, a, r0 and C are based on empirical fitting. Table 1 shows the pair coefficient parameters used in this calculation.

Table 1: Pair coefficient parameters used in this calculation

	D_{ij} (eV)	a_{ij} ($Å^{-2}$)	r_0 (Å)	C_{ij} (eV $Å^{12}$)
$Zr^{2.4}$-$O^{-1.2}$	0.206237	2.479675	2.436997	1.0
$O^{-1.2}$-$O^{-1.2}$	0.042395	1.379316	3.618701	22.0

2.2 Atomic model

The three-particle atomic model is built by cutting three spheres out of a large supercell. The supercell is composed of 17×16×8 cubic zirconia unit cells with a unit cell side length of 5.07 Å. Each particle has a radius of 2 nm, or 4 nm in diameter. The three particles are placed with an initial particle center-to-center distance of 4.3 nm, so that the particle surface distance is 3 Å. In order to maintain charge neutrality and avoid the surface dipole, extra atoms are manually removed from the particle surfaces. The initial atomic model is shown in Figure1, where the gray atoms are zirconium and the red atoms are oxygen. Furthermore, for each particle, a surface layer of atoms in 0.7 nm thick is defined as "particle shell" and the atoms inside the sphere with a radius of 1.3 nm is defined as "particle core". This definition is for the analysis of diffusion of inner and outer regions of the particles.

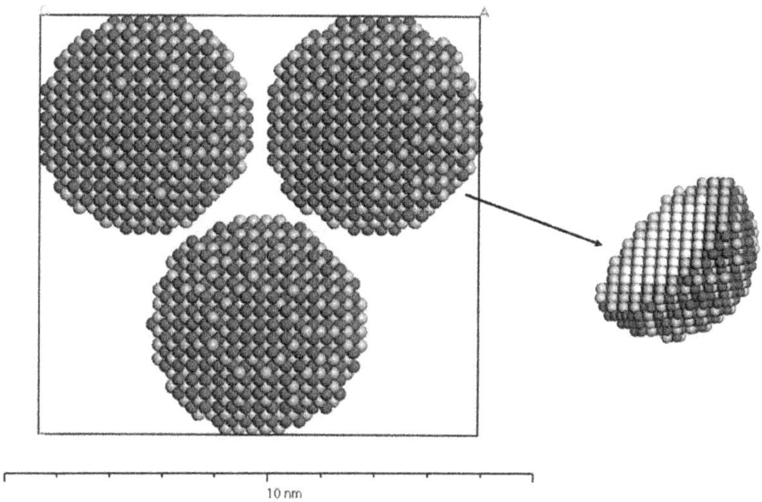

Figure 1: Cubic zirconia 3-particle atomic model. The cut half particle shows the "particle core" region in yellow. Red atoms are oxygen, and blue atoms are zirconium.

2.3 Computational details

The sintering simulation is conducted using LAMMPS [11] code. Initial zirconia particles are simulated in the canonical ensemble (NVT) with Nose-Hoover thermostat at constant temperatures. The Newton's equations of motion are integrated by using the Varlet's leapfrog algorithm with the integration time step of 2 fs. Total simulation time is 100 ps with 50000 time steps. During the simulation, the mean square displacements for the outside "shell" of the particle and the inside "core" of the particle are tracked. The diffusion coefficient for the sintering is calculated by the mean square displacement (MSD) results using the following:

$$D = \frac{1}{6}\frac{\Delta MSD}{\Delta t} \tag{2}$$

where D is the diffusion coefficient, MSD is the mean square displacement, and t is time. The diffusion coefficient is the slope of mean square displacement versus time curve. Constant 6 refers to a three dimensional diffusion.

In order to calculate the sintering activation energy, simulation cases with different temperatures from 2700K to 2900K are conducted. The sintering activation energy can be obtained by calculating diffusion coefficients at different temperatures (Equation 3):

$$\ln D = \ln D_0 - \frac{Q_d}{R}\left(\frac{1}{T}\right) \tag{3}$$

where Q_d is the activation energy, R is the gas constant, T is the temperature, and D_0 is a pre-exponential factor. In this equation, the activation energy can be obtained by linear fitting of the diffusion coefficient and the reciprocal of temperature.

3. RESULTS AND DISCUSSION

3.1 Configurational change

Three sintering initial conditions based on constant temperature (2700 K, 2800 K, and 2900 K) are simulated in the canonical assemble based on (NVT) Nose-Hoover thermostat for 100 ps, with a time step of integration of 2 fs. The atom positons are stored every 5 ps during the calculations, in order to observe the configurational change of the simulation model. Configurations for the 2800 K sintering at several sintering stages are shown in Figure 2.

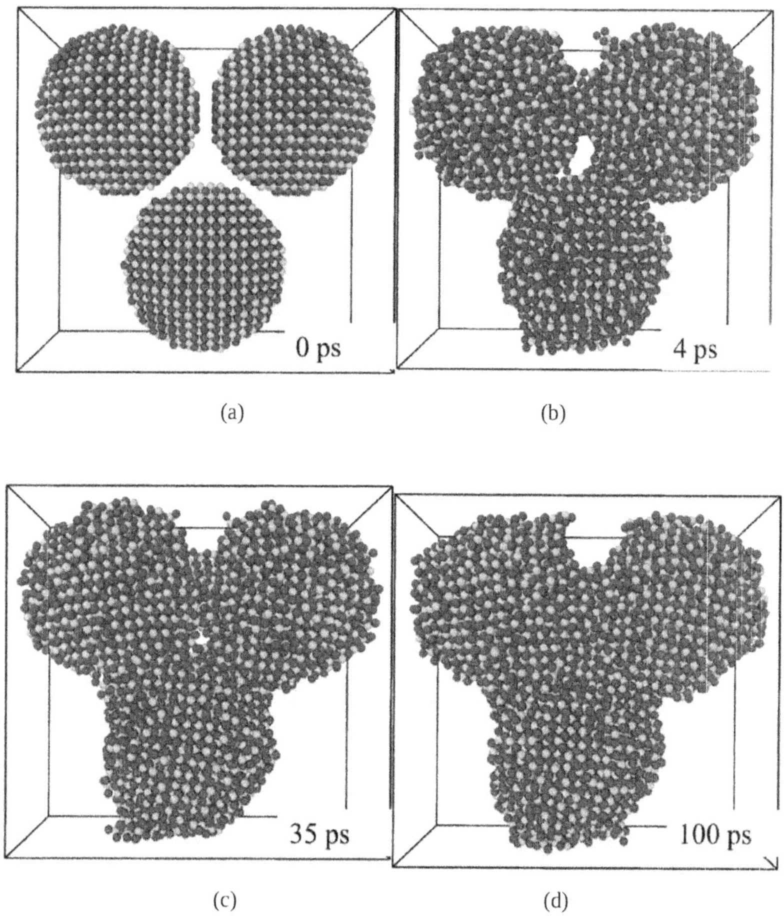

Figure 2: Configurational evolutions of the three-particle model sintered at 2800 K. (a) 0 ps, (b) 4 ps, (c) 35 ps, (d) 100 ps. Red atoms are oxygen, and blue atoms are zirconium.

At initial stage (0 ps), the particles in the three-particle model are slightly separated with a surface distance of 3 Å. Each particle is in spherical shape with perfectly crystalized structure. At the beginning stage of sintering (4 ps), atoms on the surface of the particles move closer to each other, due to the high kinetic energy induced by the high temperatures, as well as the tendency of lower surface energy near the particle gaps. The movement of surface atoms filling the initial gaps occurs in a very short time (4 ps), which can be described as the initial stage of sintering [12]. Figures 2 (c) and (d) show the intermediate stage [12] with sintering at 35 ps and 100 ps, respectively. During this stage, surface diffusion keeps driving atoms to minimize curvatures and the vacancies between each particles, which leads to the densification of atoms to the center of the system. At the end of this simulation (100 ps), the gap between the particle center has been totally filled up.

Temperature is an important factor for sintering. A higher temperature results in a higher kinetic energy, and a better diffusion of the atoms. Figure 3 shows the configurations of sintering at different sintering temperatures, 2700, 2800 and 2900 K, respectively, at the end of the simulation (100 ps). Comparing these figures, the 2700 K is less sintered since there are still vacancies in the center between three particles. For the 2800 K case, all the gaps and vacancies are filled up, which indicates it is further sintered than 2700 K. At temperature of 2900 K, there is no any vacancy, and the curvature between particles is smother than the 2800 K case. The temperature dependency of sintering is further discussed in the next session by means of diffusion analysis.

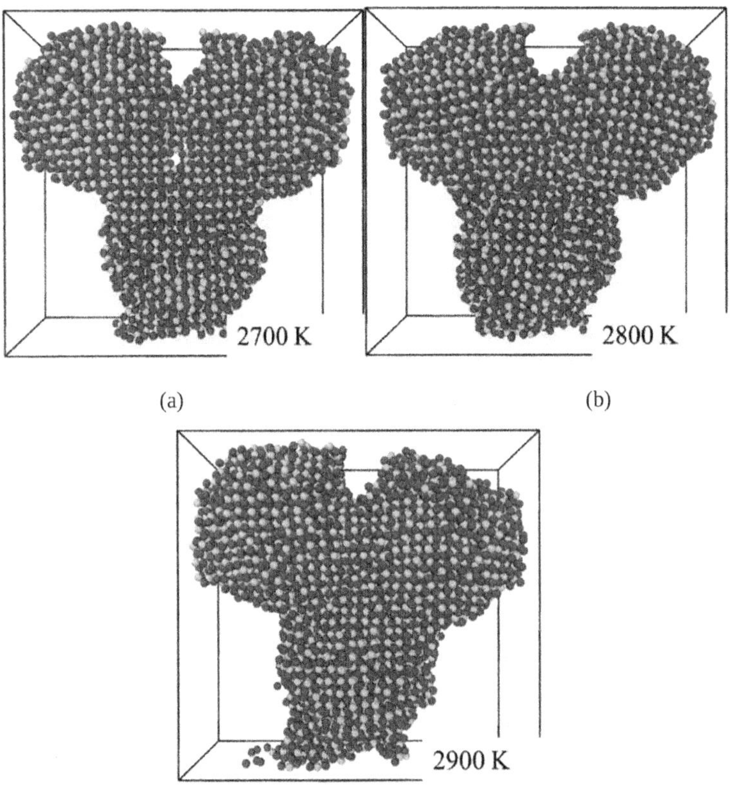

Figure 3: Configurations at temperature of (a) 2700 K, (b) 2800 K, and (c) 2900 K. Red atoms are oxygen, and blue atoms are zirconium.

Point defects are important for sintering process. In order to analyze the point defects during sintering process, a slice in the middle plane of the three-particle model is shown in Figure 4. This is along the (110) plane of cubic zirconia crystal as shown in Figure 4 (a), Zr atoms in the initial structure is perfectly located without any vacancies. As the sintering process proceeds, vacancies are generated in nanoparticle as shown in Figure 4 (b), which leads to the lattice diffusion inside the particle. Oxygen atoms on the surface of the nanoparticle fill up interstitial positions, resulting higher diffusion on the particle surface.

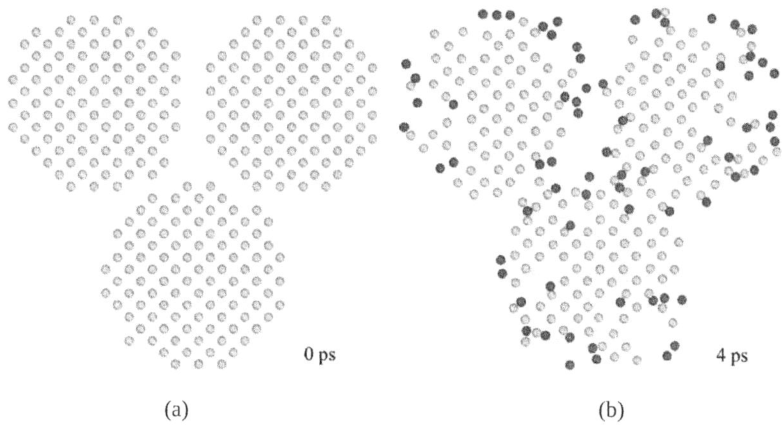

Figure 4: Point defects during the sintering process at (a) 0 ps, (b) 4 ps. Red atoms are oxygen, and blue atoms are zirconium.

3.2 Diffusion mechanisms

There are two major mechanisms driving mass transportation in sintering process: the bulk transport which contributes to both neck growth and densification; and the surface transport which contributes to neck growth only[12]. In order to investigate these two mechanisms, the nanoparticles in this work have been separated to two regions: a 0.7 nm thickness shell, and a 1.3 nm radius core. By calculation of the diffusion of the surface (shell) and the bulk (core) of the particles, the dominant driving force for sintering can be analyzed.

During the simulations, the MSDs of both the core and shell parts of the particles are tracked and plotted in Figures 5 and 6. It shows that higher temperature leads to higher MSD at the same amount of time, due to high kinetic energy. The MSD shows a larger slope at the time from 0 to 10 ps, which indicates the initial stage of sintering with a fast atom movement to form the initial neck between particles. After 10 ps, the MSD shows a linear relationship with time with less slope. This indicates the intermediate sintering stage, and it has a steady neck growth and densification.

According to Equation 2, the diffusion coefficient D at varying temperatures can be obtained by calculating the slope of linear fitting data of MSD versus time. The calculated diffusion coefficients are listed in Table 2. The particle shells show larger diffusion coefficients than the inner cores. For the particle core, bulk diffusion dominants the mass transport, and the mass transport is achieved through the movement of particles toward each other. On the other hand, in the surface shell region, both bulk diffusion and surface diffusion play important roles. As shown in the model, a large neck size is formed.

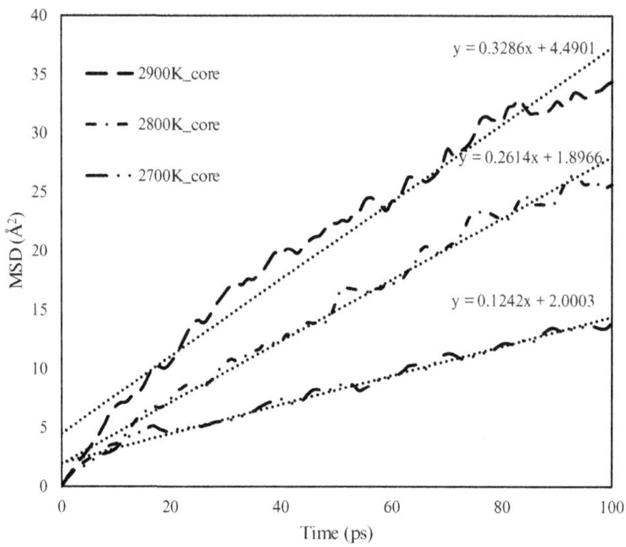

Figure 5: Mean square displacements of particle core at different temperatures during the 100 ps sintering simulation

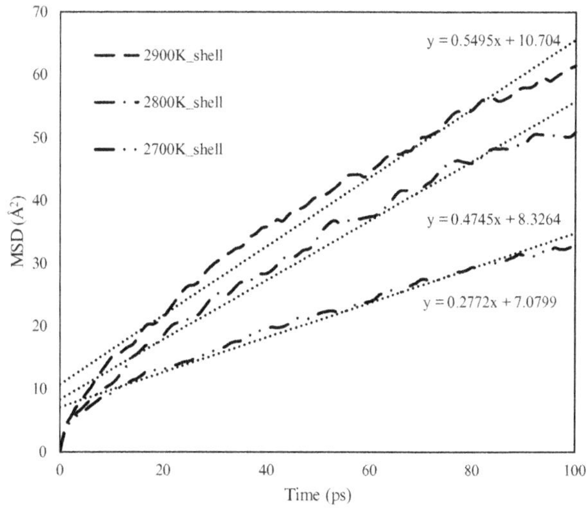

Figure 6: Mean square displacement of particle shell at different temperatures during the 100 ps sintering simulation

Table 2: Diffusion coefficients of the core and shell of zirconia particles

Sintering temperature (K)	D_{core} (m^2/s)	D_{shell} (m^2/s)
2700	2.07×10^{-10}	4.62×10^{-10}
2800	4.36×10^{-10}	7.91×10^{-10}
2900	5.48×10^{-10}	9.16×10^{-10}

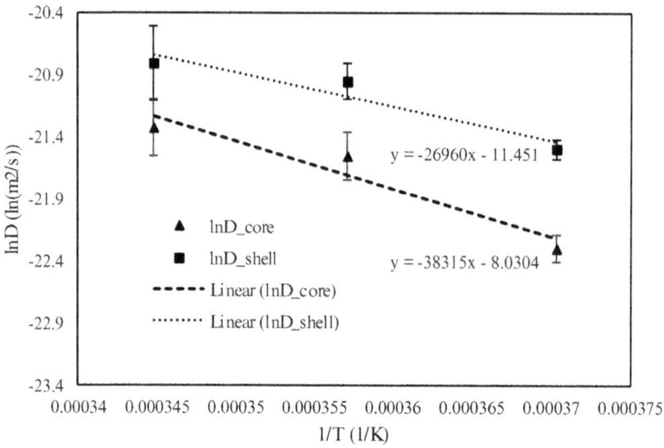

Figure 7: Temperature dependence of diffusion coefficients

Figure 7 shows the temperature dependence of diffusion coefficients, natural logs of diffusion coefficients are plotted together with the reciprocal of temperature. The activation energy for the diffusion in the core and shell parts is calculated by using Equation 3. The resulting diffusion activation energies for particle core and shell are 318.4 kJ/mole and 224.0 kJ/mole, respectively.

4. CONCLUSION

Sintering of nanostructured zirconia is studied by molecular dynamic modeling of a three-particle model. The results are summarized below.

1. The configurations of zirconia particles at selected sintering temperatures are determined. Two early stages of sintering, initial stage and intermediate stage, have been observed.
2. Diffusion mechanisms of sintering are discussed by analyzing the mean square displacement for inner and outer parts of the nanoparticle. Diffusivity on the surface of the particle is greater than the inner part. The particle surface has both surface diffusion and bulk diffusion, whereas the inner part has bulk diffusion only. The calculated particle shell diffusivity is approximately twice as the particle core.
3. By calculating the sintering diffusivity under different temperatures, the diffusion activation energies are 318.4 kJ/mole for the particle core, and 224.0 kJ/mole for particle shell.

ACKNOWLEDGEMENT

J.Z. acknowledges the financial support provided by the United State Department of Energy (Grant No. DE-FE0008868, program manager: Richard Dunst) and Indiana University-Purdue University Indianapolis Research Support Funds Grant (RSFG).

REFERENCES

[1] G. Fadda, L. Colombo, and G. Zanzotto, "First-principles study of the structural and elastic properties of zirconia," *Physical Review B,* vol. 79, p. 214102, 2009.

[2] X. Y. Zhao and D. Vanderbilt, "Phonons and lattice dielectric properties of zirconia," *Physical Review B,* vol. 65, Feb 15 2002.

[3] A. Medevielle, F. Thévenot, and D. Tréheux, "Wear resistance of stabilized zirconias," *Journal of the European Ceramic Society,* vol. 15, pp. 1193-1199, // 1995.

[4] T. Arima, S. Yamasaki, K. Yamahira, K. Idemitsu, Y. Inagaki, and C. Degueldre, "Evaluation of thermal conductivity of zirconia-based inert matrix fuel by molecular dynamics simulation," *Journal of Nuclear Materials,* vol. 352, pp. 309-317, 6/30/ 2006.

[5] G. Suárez, Y. Sakka, T. S. Suzuki, T. Uchikoshi, X. Zhu, and E. F. Aglietti, "Effect of starting powders on the sintering of nanostructured ZrO2 ceramics by colloidal processing," *Science and Technology of Advanced Materials,* vol. 10, p. 025004, 2009/04/01 2009.

[6] L. Ding, R. L. Davidchack, and J. Pan, "A molecular dynamics study of sintering between nanoparticles," *Computational Materials Science,* vol. 45, pp. 247-256, 4// 2009.

[7] J. Xu, R. Sakanoi, Y. Higuchi, N. Ozawa, K. Sato, T. Hashida, *et al.*, "Molecular Dynamics Simulation of Ni Nanoparticles Sintering Process in Ni/YSZ Multi-Nanoparticle System," *The Journal of Physical Chemistry C,* vol. 117, pp. 9663-9672, 2013/05/16 2013.

[8] F. Shimojo, T. Okabe, F. Tachibana, M. Kobayashi, and H. Okazaki, "Molecular Dynamics Studies of Yttria Stabilized Zirconia. I. Structure and Oxygen Diffusion," *Journal of the Physical Society of Japan,* vol. 61, pp. 2848-2857, 1992/08/15 1992.

[9] S. Davis, A. B. Belonoshko, A. Rosengren, A. C. van Duin, and B. Johansson, "Molecular dynamics simulation of zirconia melting," *Central European Journal of Physics,* vol. 8, pp. 789-797, 2010.

[10] A. Pedone, G. Malavasi, M. C. Menziani, A. N. Cormack, and U. Segre, "A new self-consistent empirical interatomic potential model for oxides, silicates, and silica-based glasses," *The Journal of Physical Chemistry B,* vol. 110, pp. 11780-11795, 2006.

[11] S. Plimpton, "Fast Parallel Algorithms for Short-Range Molecular-Dynamics," *Journal of Computational Physics,* vol. 117, pp. 1-19, Mar 1 1995.

[12] R. German, *Sintering: From Empirical Observations to Scientific Principles*: Butterworth-Heinemann, 2014.

Processing and Performance of Materials using Microwaves, Electric, and Magnetic Fields

RAPID SYNTHESIS OF NANOSTRUCTURED TITANIUM BORIDE (TiB) BY ELECTRIC FIELD ACTIVATED REACTION SINTERING

K. S. Ravi Chandran, A P. Sanders, J. Du
Department of Metallurgical Engineering, The University of Utah, Salt Lake City, UT 84112

L.S. Walker
Thermal Technology LLC, Santa Rosa, CA

ABSTRACT

A relatively new ceramic hard material, nano titanium boride (TiB), containing Fe and Mo in solid solution and which has a very uniform nanostructure, has been synthesized in bulk form by reaction sintering under electric field induced heating. X-ray diffraction and SEM analyses indicate that this material is almost 100% TiB and 100% dense. The titanium boride forms in situ as bundles of whiskers during reaction sintering. It was found that the size, morphology, and distribution of TiB whiskers as well as the residual titanium phase were largely independent of the electrical heating path (powder/die) used in the synthesis. In the sintered material, Fe appears to be non-uniformly distributed within the TiB. However, an interesting finding is that hardness values, measured at random throughout the microstructure do not appear to be greatly affected by the inhomogeneous Fe distribution. This suggestes that the TiB synthesized by reaction-sintering under electric field induced heating is largely similar to the TiB material synthesized using a conventional hot-pressing technique. A comparison of the physical and mechanical properties of this material with the Cerbec silicon nitride indicates that the present nanostructured TiB ceramic is superior in performance and could be useful in engineering applications.

I. INTRODUCTION

The present work demonstrates that rapid synthesis of fully dense and nanostructured titanium boride can be achieved, in a shorter processing time than conventional hot pressing, through electric field activated sintering (EFAS). The process is a relatively new technique for synthesizing ceramics, and it has been increasingly employed in the synthesis of a variety of ceramic materials[1]. The process consists basically of heating the powdered constituent materials using electrical current directed through the powders or their containment die, and simultaneously pressurizing them via a ram. Three modes of heating are possible: (i) direct heating (DH) where all current is forced through the powder; (ii) indirect heating (IH), where all the current is forced through the dies, and (iii) a combination of these two (CH). At least in the direct powder heating, the heating is more rapid than that provided by the established approaches such as pressureless sintering, hot pressing, and hot isostatic pressing. This increased heating rate could translate into reduced process time and a more refined microstructure, and, because of shorter exposure time at high temperature, cause less grain growth.

Our recent research has shown that titanium boride ceramic with finely distributed nanoscale whiskers can be made in a one step process in a hot press, by reaction sintering a mixture of Ti, TiB_2, and Fe-Mo powders. This produced a fully dense ceramic consisting of >99% nanoscale TiB whiskers and <1% residual Ti. Because the starting powders are electrically conductive, it is of high interest to see whether TiB can be synthesized in a short time using EFAS. The titanium boride material exhibits excellent mechanical properties including good fracture toughness and high wear resistance, which may be largely attributed to its interesting nanostructure. Using the Ti-B system as a model, several works have demonstrated the feasibility of synthesizing TiB-Ti metal-matrix composites and cermets via EFAS[2,3,4,5,6,7]. However, the formation of nanostructured TiB and its densification using the Ti and TiB_2 source powders has not been demonstrated by EFAS processing.

The commonly known ceramic materials such as alumina[8], silicon nitride[9], and zirconia[10], all of which are insulating, have been sintered to full density and have much refined microstructures, which are attributed to the rapid heating inherent in EFAS. Experimental and analytical studies of EFAS processing have shown that with both insulating and conductive powders, there can exist thermal gradients, which are affected by properties of the powders and the design of the current path, including the use and the adequacy of electrical and thermal insulations[11,12,13]. Thus, a potential disadvantage of EFAS processing is the risk of producing inhomogeneous material with non-uniform microstructure or properties, if the thermal gradients are excessive. One study demonstrated that an electrically conductive ceramic (TiN) exhibited greater thermal gradients during EFAS than an insulating one (ZrO_2).[14] On the other hand, electrically conductive powders may be advantageous because of the simple fact that the electrical current can be directly routed though the powders. Any effect of current path, or the tools, or a combination of both, can then be investigated to determine if these parameters have an effect on the final product and the microstructure.

The objective of this work was to synthesize, for the first time, monolithic, fully dense, nanostructured TiB ceramic using an EFAS process, and to compare it with a benchmark material of the same composition produced by hot pressing (HP). We investigated the effects of different electrical current paths by directing direct the current through 1) powder mixture only, or 2) die only or 3) both. The variations in process voltage, current, pressure, temperature and ram displacement were recorded during the reaction sintering of TiB to assess the material variations, if any, caused by the current path variations. The density, microstructure, and mechanical properties of these compacts were examined.

II. EXPERIMENTAL PROCEDURE

II.1. Powder

The powder mixture consisted of Ti, TiB_2, and Fe-Mo powders whose sizes were in the micron range. The details of the powders and their compositions are as follows. The Ti powders were 99.95% pure Ti (α -Ti) with an average particle size of 45 μm and had the composition (in wt.%) of 0.23% O, 0.02% N, 0.01% C, 0.04% Fe, and 0.024% H (supplied by Alfa Aesar, Ward Hill, MA). The TiB_2 powder had an average particle size of 3 μm and had the composition (in wt. %): 30.3% B, 0.67% Zr, 0.01% C, 0.04% Fe, and 0.022% H (supplied by Advanced Ceramics Corp., Cleveland, OH). The Fe-Mo alloy powders had a nominal stoichiometry of Fe-59Mo with 59.8% Mo and average particle size of 10 μm (supplied by F.W. Winter Inc. & Co., Camden, NJ). The composition of the mixture is the same that previously yielded monolithic TiB ceramic via hot press for 2 h at 1350 °C and with a pressure of 20 MPa[15]. The powders were loaded in a stainless steel jar, which was evacuated after powder loading and backfilled with argon. The mixture was blended for 10 h on a jar mill, with 12 mm diameter niobium slugs as milling media to help break down the agglomerates.

II.2. SPS pressing

The processing by EFAS was performed in a 25 ton press (Model SPS 25-10, Thermal Technology, Santa Rosa, CA), applying up to 20 MPa pressure and 10 V, 10,000 A electrical current to produce disks of 13 mm thickness and 60 mm diameter. Three trials were performed under identical sintering conditions other than the current path. The current path was modified by placing insulating sheets (graphite foil spray-coated with boron nitride) as needed. In direct hearing, the current was forced through the powder by insulating the interior of the die walls. In indirect heating, the graphite punch ends were insulated to prevent the current path through the powder mixture. In the combined heating process, neither insulation approach was used, allowing free current flow through the powder and the die. The powder mixture was poured into the graphite die and tapped to level the surface. The die assembly including the graphite punches, powder, and insulators was placed under the rams. The chamber was then evacuated and backfilled with argon.

The electric power was controlled to provide a 100 °C/min heating rate and to reach 1350 °C and 20 MPa pressure was applied upon reaching the holding temperature. These conditions were maintained for 20 min, followed by cooling to approximately 300 °C. At this point, the chamber was vented to laboratory air and opened for room temperature cooling. The following parameters were recorded during the process: current and voltage, pressure, ram position and temperature, punch temperature, and die wall temperature. For process control, the temperature, as measured from a thermocouple embedded in the punch was used. The die wall temperature was recorded by a pyrometer focused at the bottom of a small diameter hole drilled through the external die wall to a depth of about 5 mm from the internal wall.

II.2. Microstructure Characterization and Mechanical Testing

For microstructural study, samples were cut by electric-discharge-machining (EDM). The cut surfaces were ground, metallographically polished, and etched to reveal the microstructure. Multiple locations from the specimen's center to its edge were investigated using SEM/EDS tools to characterize the uniformity of microstructure and the composition. Micro-hardness measurements were made on metallographically polished surfaces using a Vickers diamond indenter at the load of 9.81 N.

The fracture toughness was determined using samples machined out of the center area of each specimen. For this purpose, single-edged pre-cracked beam (SEPB) technique, following ASTM C 1421, was employed. SEPB specimens with dimensions: 3x4x50 mm were cut by wire EDM after slicing away a top layer of about 1 mm. The cut was made such that the 4 mm direction of the specimen was parallel to the EFAS pressing direction. This is to ensure that the crack plane and crack propagation directions were both parallel to the pressing axis. All the cracks were created within a material zone that had originated within 10 mm of the disk center. The test technique followed the standard steps to create a pre-crack from a Vickers indentation crack. With suitable pre-crack lengths (a/w about 0.5), the specimens were fractured in four-point bending.

III. RESULTS AND DISCUSSION

III.1. Process Parameters

Among the process measurements, the die wall temperature (recorded by the pyrometer), electrical power (product of voltage and direct current), and ram position showed notable variation between trials (Figure 1), while the computer-controlled hold temperature (1,350 °C) and pressure (20 MPa) were indistinguishable between trials.

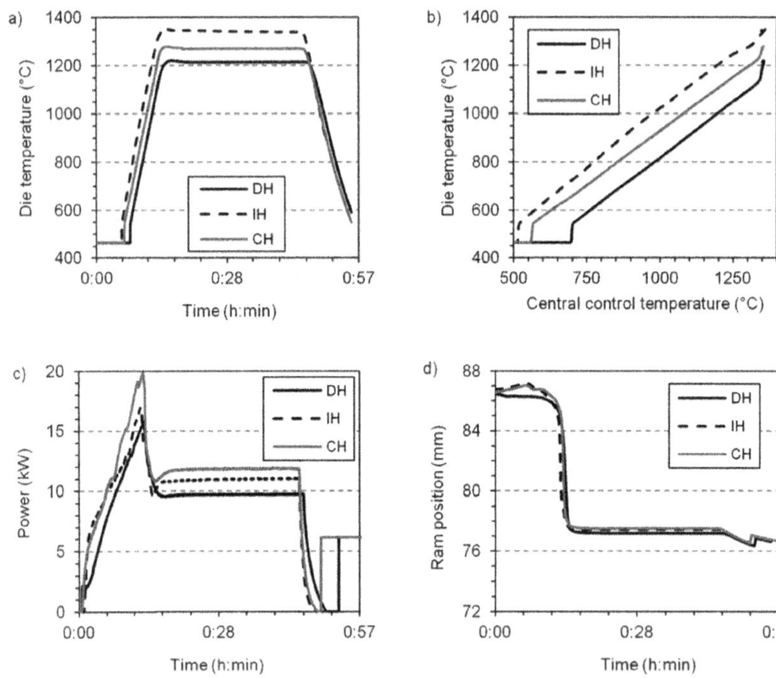

Figure 1: Process measurements for the three trials: a) Die wall temperature, b) Die vs. central temperature (through the end of hold stage), c) Electrical power, d) Ram position.

The temperature measurements show that the DH trial had the largest thermal gradient between the center and edge of the specimen, with a difference of ~200 °C during the ramp stage, diminishing to ~140 °C through the hold stage (Figure 1a and 1b). The CH trial showed a smaller thermal gradient, and the IH trial showed nearly equal temperatures at the central and die wall locations. In the DH conditions, the high gradient results from heat lost to the surroundings and from thermal insulation provided by the die wall. Also, in the DH case, the control thermocouple and the current path are both centrally located, so the configuration provides little control over the temperature at the specimen's edge. In the IH conditions, the current path was distant from the control thermocouple, yet the central temperature was nearly equal to that of the die, probably because at the center, heat loss was practically absent because there was no external boundary. The electrical power required during the hold stage was lowest in the DH trial, probably as a direct effect of having the lowest die temperature and thus lowest heat loss to the surroundings in addition to the highest electrical resistance. The IH trial had the highest die temperature, yet it had the second lowest power requirement during the hold stage. The maximal power requirement of the CH trial is somewhat anomalous since this trial had only the second highest die temperature; however, this trial also had the highest temperature in the steel ram (412 °C vs. 392 °C in the IH trial, during the hold stage), and highest current (~4,010 A vs. ~3,590 A in the IH trial), which indicate that heat losses were higher outside the specimen/tool assembly in the CH trial.

III. 2. Observations on the microstructure

The SEM micrographs are presented in Figures 2-4, for the direct, indirect, and combined heating conditions, respectively. The titanium boride actually forms in situ as bundles of whiskers during reaction sintering. It was found that the size, morphology, and distribution of TiB phases as well as the residual titanium phase were largely independent of the electrical heating path (powder/die). In general, the microstructural characteristics largely resemble the material made by hot pressing. However, a subtle difference may be noted. In the indirect heating process, the TiB whiskers within a colony appear to be oriented nearly in the same direction (Figure 3b). Such preferred orientation was not seen in the other two heating methods (Figures 2b, 4b) where the TiB whiskers appear to be randomly oriented with respect to each other.

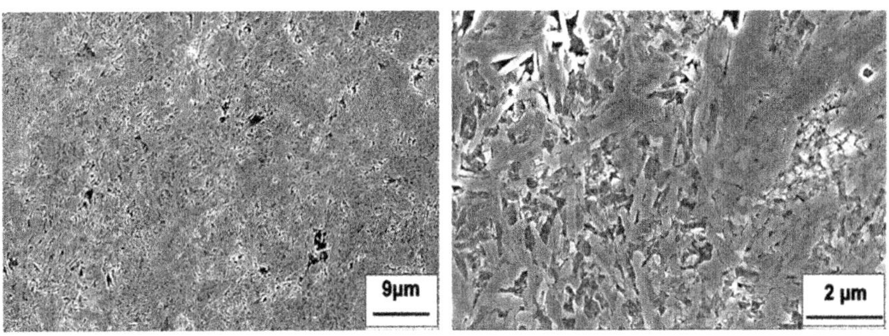

(a) (b)

Figure 2. Microstructures of TiB ceramic sintered by direct current heating: (a) low and (b) high magnifications.

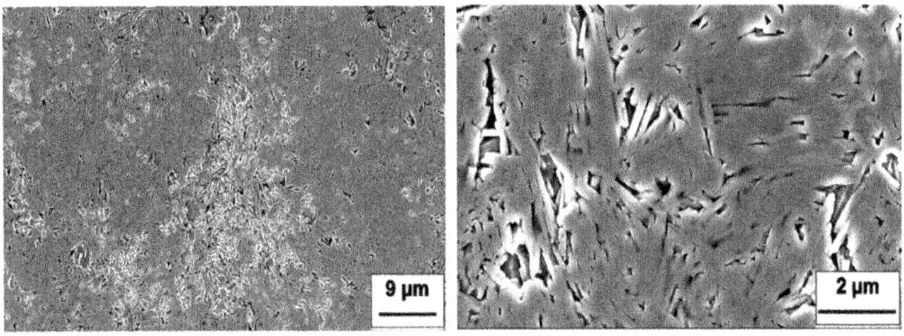

(a) (b)

Figure 3. Microstructures of TiB ceramic sintered by indirect current heating: (a) low and (b) high magnifications.

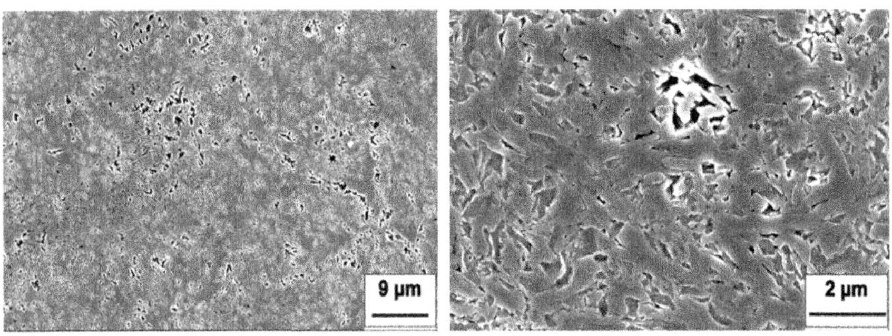

(a) (b)

Figure 4. Microstructures of TiB ceramic sintered by combined heating: (a) low and (b) high magnifications.

Figure 5 presents the X-ray diffraction analysis of TiB ceramic processed by the three approaches. It indicates that the amount of TiB and the residual beta titanium phase are nearly the same in all the process conditions. A difference was not expected; hence, the X-ray patterns confirm that the formation of TiB and the relative proportion of phases are independent of the process conditions.

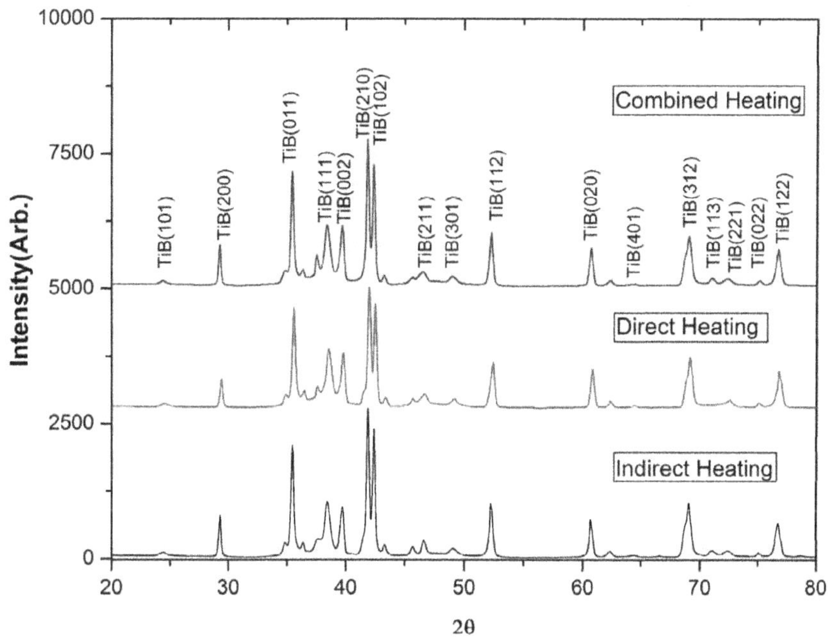

Figure 5: X-ray diffraction analysis

Figures 6, 7, and 8 illustrate the EDS mapping of the Ti, Fe, and Mo element distributions in the TiB ceramic under the three heating conditions. In all the sintered compacts, Ti and Mo appear to be uniformly distributed, but Fe appears to be non-uniformly distributed within the TiB whisker regions. However, as will be shown later, this is in contrast with the variation of hardness values, measured at random throughout the microstructure, which showed that the hardness was not largely affected by the Fe distribution.

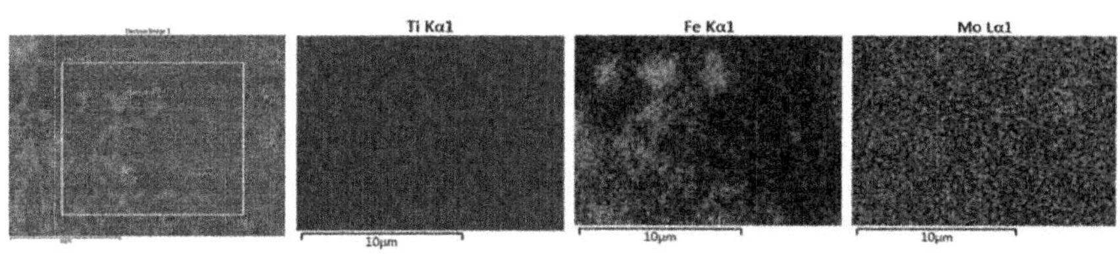

Figure 6: EDS analysis of direct heating specimen showing the distribution of elements Ti, Fe, and Mo; white rectangle defines area of analysis in remaining images.

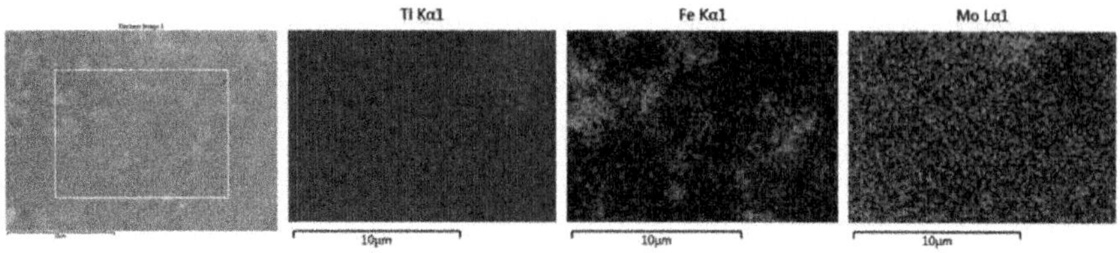

Figure 7: EDS analysis of indirect heating specimen showing the distribution of elements Ti, Fe, and Mo; white rectangle defines area of analysis in remaining images.

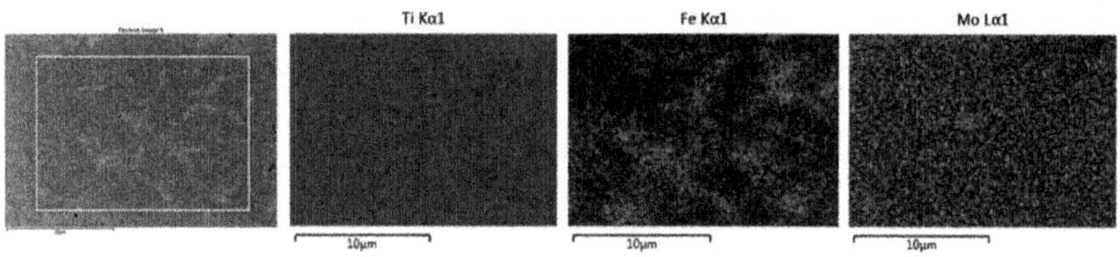

Figure 8: EDS analysis of combined heating specimen showing the distribution of elements Ti, Fe, and Mo; white rectangle defines area of analysis in remaining images.

III.3 Mechanical properties

The microhardness was sampled in the central region of each specimen (Vickers indenter, 1 kg, 15 s), with results given in Table 1. For each condition, ten measurements were made and the statistical variation between the measurements is found to be small, because the standard deviation of these measurements was less than 4%. This is somewhat surprising, because of the inhomogeneous distribution of Fe within TiB. It is possible that the hardness is largely determined by B-B covalent bonds in the orthorhombic lattice of TiB, because of large electron density accumulation, as was found[16] in a computational study. Future research will investigate this aspect in detail.

Table 1: Hardness of Nano Titanium Boride

Synthesis Process	**Vickers microhardness (kg/mm^2) @ 1 kgf**
Direct Heating	1,757 ± 62
Indirect Heating	1,788 ± 69
Combined Heating	1,766 ± 60

III.4. Comparison of Properties of Nano Titanium Boride with commercial Cerbec Silicon Nitride

Table 2 presents a comparison of mechanical properties of nano TiB that was synthesized by conventional hot press reaction sintering. Since the TiB made by the electric field activates sintering process is chemically and physically equivalent to that made by hot press, it is considered acceptable to use the properties of nano TiB, determined earlier, here to illustrate the advantages of this new material. The hardness levels (Table 1) of the TiB made by electric field activated reaction sintering is in the middle of the range of hardness values (1550-2000 kg/mm^2, Vickers hardness, measured at 1 kgf load) observed in various samples obtained in the hot press method.

Table 2. Comparison of mechanical properties of nano titanium boride with that of commercial silicon nitride, Cerbec NBD-200

Property	**Nano TiB**	**Silicon Nitride (Cerbec NBD-200)**
Microstructure	Nanostructured	Micron size grains
Density (g/cc)	4.5	3.16
Modulus (GPa)	370-425	320
Hardness (Hv)	1600-2000	1550
Flex Strength (MPa)	800-950	900
Fracture Toughness (MPa√m)	5-6	5.5
Thermal Expansion (X 10^{-6}/°C)	7.2 X 10^{-6}	2.9 X 10^{-6}
Thermal Conductivity (W/mK)	19.5	29
Electrical Resistivity (Ohm-cm)	~2x10^{-5}	~10^{14}
Hot Press Conditions (°C)	1350, 2 hrs, 12-15 MPa	1800, ~2 hrs
(Lub) Wear Volume Loss (mm^3) (Lub) Friction coeff.	0.0001 0.092	0.0003 0.086
(Dry) Wear Volume Loss (mm^3) (Dry) Friction coeff.	0.02487 0.563	0.06618 0.585

The other properties of nano titanium boride, synthesized by conventional hot pressing are given in Table 2, to illustrate the property benefits that can be achieved by the nanostructure of the ceramic. The details of property measurement methods are given elsewhere[17]. The properties are compared with that of a commercial version of silicon nitride, Cerbec, NBD-200[18]. The density of nano TiB is slightly higher than that of the Cerbec silicon nitride. The elastic properties, such as modulus and thermal expansion, are significantly higher than those of the silicon nitride. The elastic modulus is about 16-32% higher than that of Si_3N_4. The variability in the elastic modulus of TiB is not due to the material, but due to the estimation uncertainty—the experimentally extrapolated modulus, from measurements of elastic moduli of Ti-TiB composites, is about 370 GPa, whereas the DFT-based first principles calculations resulted in a value of 425 GPa. The preparation of single crystals of TiB, to measure the nine independent elastic constants, is very difficult, because of the peritectic reaction of formation of TiB. Regardless, it is clear that either of these estimates provides a significantly higher modulus value for the nano TiB relative to Si_3N_4. The significantly larger coefficient of thermal expansion for TiB (7.2 X 10^{-6}), compared to that of Si_3N_4 (2.9 X 10^{-6}) can be explained on the basis of the metallic character of TiB. The electrical resistivity is quite low and is of the order of that of metallic titanium (4.7 X 10^{-5} Ohm-cm) and hence some of the physical properties are close to that of titanium. The strength properties, including hardness, flexural strength, and fracture toughness are quite comparable and in some instances are higher than those of the silicon nitride. For instance, the average hardness of 1757 kg/mm^2 is significantly higher than that of Si_3N_4 (1550 kg/mm^2).

The nanostructured TiB shows a superior wear resistance, with respect to Si_3N_4. The tribological performance of TiB was evaluated using a ball-on-disk wear tester, both under lubricated and dry sliding wear conditions. The TiB balls met the specification of grade 25 and had a surface roughness of 40 nm Ra. The balls were run against a 75 mm diameter TiB disk (polished to a Ra of 89 nm) for a distance of 1 km. For comparison, tests using grade 5, TSN-03 Si_3N_4 balls were also run against the TiB disk. The coefficient of friction of the TiB ball is comparable to that of the Si_3N_4 ball (Table I). Wear volume losses from the balls were determined using a Zygo profilometer. The loss of material volume due to wear in TiB is about 33% of that in Si_3N_4 balls indicating that the present nanostructured TiB material is a good wear resistant material.

IV. CONCLUSIONS

A new ceramic material, nanostructured TiB, having high modulus, hardness, and strength is rapidly synthesized (30 min. @ 1350 °C with 10 MPa pressure) by electric field activated sintering. The bulk material is made of densely packed nanoscale TiB whiskers. X-ray diffraction and SEM analysis indicate that this material is almost 100% TiB and 100% dense. The nanostructured TiB ceramic is comparable and in some instances exceeds the specifications of Si_3N_4 in terms of mechanical and wear performance. The present research has shown that the TiB can be rapidly synthesized by electric sintering, at relatively lower temperatures and in relatively shorter times compared to that used in the production of Si_3N_4 ceramics. The TiB therefore may be attractive for some practical applications such as armor and wear resistant components.

ACKNOWLEDGEMENT:

The research was supported by the National Science Foundation, the program on Designing Materials to revolutionize our Engineering Future (DMREF), through the grant CMMI-1435758.

REFERENCES

[1] M. S. Yurlova, V. D. Demenyuk, L. Yu Lebedeva, D. V. Dudina, E. G. Grigoryev, and E. A. Olevsky. "Electric pulse consolidation: an alternative to spark plasma sintering." J. Mater. Sci., **49** [3] 952-985 (2014).

[2] S. Wei, Z.-H. Zhang, F.-C. Wang, X.-B. Shen, H.-N. Cai, S.-K. Lee, and L. Wang, "Effect of Ti content and sintering temperature on the microstructures and mechanical properties of TiB reinforced titanium composites synthesized by SPS process," Mater. Sci. Eng. A, **560** 249-255 (2013).

3 M. Selva Kumar, P. Chandrasekar, P. Chandramohan, and M. Mohanraj, "Characterization of titanium-titanium boride composites processed by powder metallurgy techniques," Mater. Charact., 73 [] 43-51, (2012).

[4] V. V. Patel, A. El-Desouky, J. E. Garay, and K. Morsi, "Pressure-less and current-activated pressure-assisted sintering of titanium dual matrix composites: Effect of reinforcement particle size," Mater. Sci. Eng. A, **507** 161-166, (2009).

[5] K. Yoshida, H. Mishina, S. Sasaki, M. Morita, and K. Mabuchi, "Mechanical properties of titanium cermets for joint prostheses," Mater. Trans., 47, [2] 418-425 (2006).

[6] H. Feng, Y. Zhou, D. Jia, Q. Meng, and J. Rao, "Growth Mechanism of In Situ TiB Whiskers in Spark Plasma Sintered TiB/Ti Metal Matrix Composites," Cryst. Grow. & Des., **6** [7] 1626-1630 (2006).

[7] H. Feng, Y. Zhou, D. Jia, and Q. Meng, "Microstructure and mechanical properties of in situ TiB reinforced titanium matrix composites based on Ti–FeMo–B prepared by spark plasma sintering," Comp. Sci. Tech., **64** [16] 2495-2500 (2004).
[8] D. Chakravarty and A. H. Chokshi, "Direct characterizing of densification mechanisms during spark plasma sintering," J. Am. Ceram. Soc., **97** [3] 765-771 (2014).
[9] X. Xu, T. Nishimura, N. Hirosaki, R. J. Xie, Y. Zhu, Y. Yamamoto, and H. Tanaka, "New strategies for preparing nanosized silicon nitride ceramics," J. Am. Ceram. Soc., **88** [4] 934-937 (2005).
[10] B. Basu, J. H. Lee, and D. Y. Kim, "Development of nanocrystalline wear-resistant Y-TZP ceramics," J. Am. Ceram. Soc., **87** [9] 1771-1774 (2004).
[11] K. Vanmeensel, A. Laptev, J. Hennicke, J. Vleugels, and O. Van der Biest, "Modelling of the temperature distribution during field assisted sintering," Acta Mater., 53, [16] 4379-4388 (2005).
[12] G. Molenat, "Temperature Control in Spark Plasma Sintering: An FEM Approach," J. Metall., **2010** (2010)
[13] T. Voisin, L. Durand, N. Karnatak, S. Le Gallet, M. Thomas, Y. Le Berre, J.-F. Castagné, and A. Couret, "Temperature control during Spark Plasma Sintering and application to up-scaling and complex shaping," J. Mater. Process. Tech., 213 [2] 269-278 (2013).
[14] K. Vanmeensel, A. Laptev, J. Hennicke, J. Vleugels, and O. Van der Biest, "Modelling of the temperature distribution during field assisted sintering," Acta Mater., 53, [16] 4379-4388 (2005).
[15] S. Madtha and K. S. Ravi Chandran, "Reactive-Sinter-Processing and Attractive Mechanical Properties of Bulk and Nanostructured Titanium Boride," J. Am. Ceram. Soc., 95 [1] 117-125 (2012).
[16] Panda, K. B., and KS Ravi Chandran. "First principles determination of elastic constants and chemical bonding of titanium boride (TiB) on the basis of density functional theory," Acta Mater., 54 [6] 1641-1657 (2006).
[17] S. Madtha, C. Lee, and K. S. Ravi Chandran. "Physical and mechanical properties of nanostructured titanium boride (TiB) ceramic," J. Am. Ceram. Soc., **91** [4] 1319-1321 (2008).
[18] Cerbec Ceramic Balls, Product Literature, CoorsTek, 10 Airport Park Road East Granby, CT 06026 USA

VERIFICATION OF EFFECTS OF ALTERNATIVE ELECTROMAGNETIC TREATMENT ON CONTROL OF BIOFILM AND SCALE FORMATION BY A NEW LABORATORY BIOFILM REACTOR

Hideyuki Kanematsu[1], Senshin Umeki[2], Nobumitsu Hirai[1], Yoko Miura[1], Noriyuki Wada[1], Takeshi Kougo[1], Kazuyuki Tohji[2], Hirokazu Otani[3], Kazuhiko Okita[4], Toshifumi Ono[4]

1: National Institute of Technology, Suzuka College, Suzuka, Mie, Japan

2: Graduate Schools of Environmental Studies, Tohoku University, Sendai, Miyagi, Japan

3: Technolabo. Co., LTD., Habu, Saitama, Japan

4. Tohoku Steel Co., Ltd., Murata-cho, Sibata-gun, Miyagi, Japan

ABSTRACT

Pipeworks in aqueous solution industry mostly develop scales and corrosion products at inner walls, which lead to serious materials deterioration and decrease its function such as heat efficiencies. The phenomena often involve biofouling and biofilm. However, there are not so many research activities to focus on this relation, even though biofouling is always followed by the scale formation. Authors have investigated the application of alternative electromagnetic field to control biofilm and biofouling successfully. However, the precise and detailed mechanism has been not clear yet. In this study, we focused on the pipeworks used for geothermal generation in hot spring resorts and their scale problems. The development of scale at the inner walls usually decreases the generating efficiency. A new desktop laboratory biofilm reactor has been specially built and to investigate the application effect of alternative electromagnetic field on the control of biofilm/biofouling for pipeworks of hot spring water.

INTRODUCTION

Japan is a volcanic country. It means that relatively shallow areas under the ground could have hot spring water. Hot spring water is usually heated to relatively high temperatures by magma movements and it contains many mineral components. Therefore, people believe it generally has a sort of healing power and many Japanese people love it. As for the healing power for human bodies, the heat would be the most important factor. However, it could be used not only for healing, but also for power generation. Actually, many trials have been done so far and put it into practical use to some extent. It is called "Geothermal Generation[(1)-(3)]".

Basically, the geothermal generation produces electricity by water vapor rotating steam turbine. Since it is one of the renewable energy, the negative production of carbon dioxide as global warming gas could be reduced, compared to thermal power generation. As for the geothermal resource, Japan has the third abundant geothermal resource. Therefore, the renewable energy should be used more positively. Under the condition where the safety and reliability of nuclear energy are wavering particularly in Japan, the expectation for geothermal generation has been very high.

The demerits for the geothermal generation are generally high initial cost (for development and exploration) and the vulnerability to natural disaster. In addition, it has a serious technical defect that is the scale problem in this paper. Fig.1 shows the schematic illustration for it[4]. The hot spring water exists deep under the ground. It has to be pumped up by pipeworks. However, the inner wall of pipework very often develops significant scales. The pipes diameter would be usually decreased and would lead to insufficient efficiency for geothermal generation.

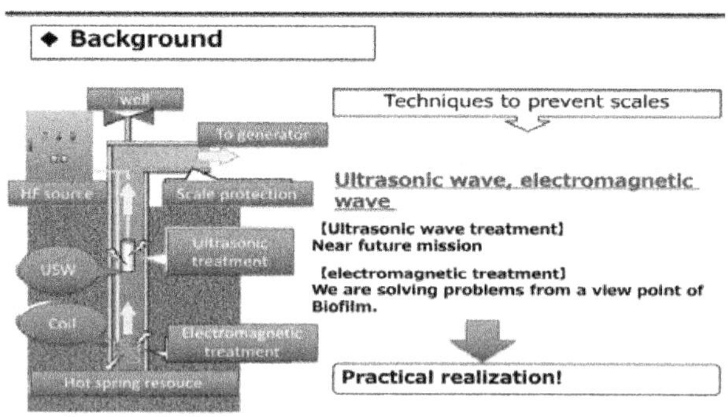

Fig.1 The background of this study – Geothermal Generation.

Previous studies by the authors have focused on the relation between scale formation and biofilm on pipes surfaces so far[5]. We have presumed that the scale would be formed through the biomineralization of biofilm to some extent[6], [7]. Therefore, the scale should be controlled with biofilm. In the past, we had successfully controlled biofilm by the application of alternating electromagnetic field[8]. In this paper, we investigated if it could control also the scale by using a new apparatus.

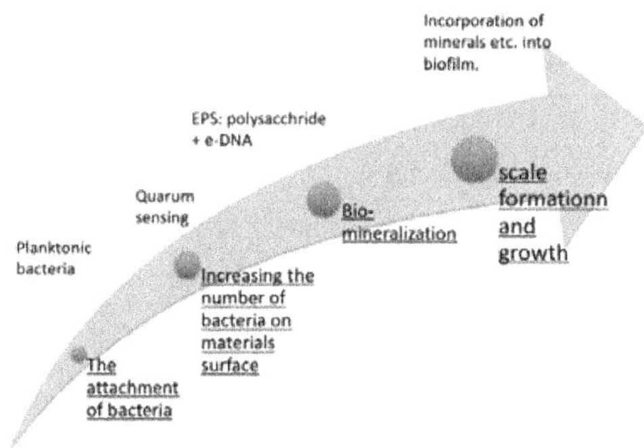

Fig.2 Biofilm and scale formations.

BIOFILM AND BIOMINERALIZATION

In every environment on our globe, bacteria exist everywhere. They usually flow everywhere on it. However, some bacteria such as (planktonic bacteria) exist minorly. Most of the bacteria in biofilm exist in an inhomogeneous film-like formed matter on materials surfaces. The formation has a close relation to bacterial activities. Planktonic bacteria lives usually in oligotrophic environments. They seek for nutrients (carbon compounds) to attach on materials surfaces. During repeating attachment and detachment process, the number of bacteria increases to a threshold value. Then bacteria excrete polysaccharide from their cells simultaneously (Quorum sensing phenomenon). Then biofilm forms and begins to grow. During the growth process, biofilm tends to incorporate many components in environments such as silicon, calcium, iron, zinc etc. As a result, complex combination of compounds would lead to the development of scales. Fig.2 shows schematic flow for the processes as a whole.

EXPERIMENTAL WORK

Three types of metallic tiny sheets were used as pipes specimens – pure iron, pure copper and pure aluminum whose purities were 99.9%. The specimen's dimensions were 10mm x 20mm (The thickness was 0.5mm). The specimens were polished by set in the vertical acrylic column (The diameter: 22 mm, the length: 450 mm), so that the specimen's surface was overlapped with the center plane of the column and so that the surface was parallel to the flow direction. Like our research activities in the past[(9),(10)], the 20 little hot spring water was provided by Naruko Hot Spring water. The components shown in Table (1) were kept in the bottom tank and the pump circulated water at the rate of 6l/min.

Table 1 Components of Uba-no-yu from Naruko hot sping water[r(11)]

Cation	Contents (mg/1kg)	Anion	Contents(mg/1kg)
Lithium	0.2	Fluorine	0.1
Sodium	268.1	Bromide	92.9
Potassium	24.4	Hydrogen sulfide	15.3
Ammonium	0.3	Thiosulfuric	7.8
Magnesium	24.2	Sulfate	189.5
Calcium	55.2	Hydrogen Phosphate	1.8
Strontium	0.4	Hydrogen Carbonate	673.1
Aluminum	<0.1	-	-
Manganese	0.9	-	-
Iron (II)	<0.1	-	-
Iron(III)	<0.1	-	-
Copper	<0.1	-	-
Zinc	<0.1	-	-

After the hot spring water flew out of the tank, it ran through the column from the lower part to the upper. Then it flew out from the pipe system and fell onto the intermediate plate where ambient air was blown down from a fan. In this way, the ambient germs were added into the hot spring water constantly. Finally, the hot spring water was returned to the tank. The process was repeated in a certain period and the biofilm formed on the specimen in an accelerated way. The temperature of hot spring water was kept at 30 degrees Celsius strictly. Such an artificial reactor on laboratory scale is generally called LBR (Laboratory Biofilm Reactor). Fig.3 shows the schematic illustration for the LBR used for this experiment. Fig.4 shows the enlarged part for LBR. Even though this type of LBR has been very often used by us so far, the high frequency unit was added for the 1st time in this study. It was made especially for this research project to provide high frequency to the specimens in the column through a solenoidal coil. The frequency of alternate current could be changed from 1kHz to 100 kHz (The maximum output current: 5 A). When the impedance would be changed, the inner automatic resonance circuit could automatically adjust itself to the resonance condition. By using the apparatus, we applied two special wavelengths (8kHz and 20kHz) to the specimens under a certain magnetic flux density (1mT: measured by Gaussmeter, Toyo Technica 5180), where we confirmed different phenomena on specimens surfaces as preliminary experiment in advance and their results were compared to those without applications of alternate currents. After the immersion and exposure for a couple of days in the LBR, specimens were taken out from the system and subjected to some evaluation tests. As the evaluation method, we have developed many kinds of techniques so far[12].

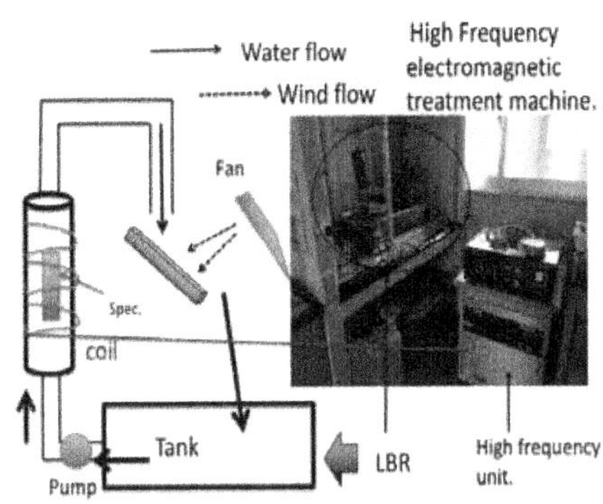

Fig.3. Experimental apparatus

Fig.4. The LBR

Since biofilm was very hard to fix correctly, being different from the bacteria themselves, plural methods should be used for the evaluation in some complementary styles. In this experiment, we picked up the two techniques. One of them was an optical microscope with 3D functions (Keyence Co. VW-9000). Fig.5 shows the appearance of the apparatus and an example of the 3D imaging. As usual, the specimen was put on the stage and was observed at the focal point. Then the stage was shifted slightly around the focal point and each image was taken and superimposed into a certain image. The computer software installed in the apparatus assigned colors to various points on the specimen's surface, depending on their heights. The sea-island structures observed from the vertical upper direction would demonstrate the existence of biofilm, since it generally has concavo-convex profiles in several micrometer orders.

The second technique was Raman Spectroscopy. Some researchers already confirmed the Raman spectra derived from biofilm in the past[13]-[16]. Therefore, we measured the specimens' surfaces by a Laser Raman spectroscopic apparatus (JASCO, NRS-3100) with its optical microscope as complementary approach. The specimens were irradiated by green laser (532nm) and the Raman Spectroscopic shits were measure particularly between 650 cm^{-1} and 1700 cm^{-1}. Fig. 6 shows the appearance of the apparatus.

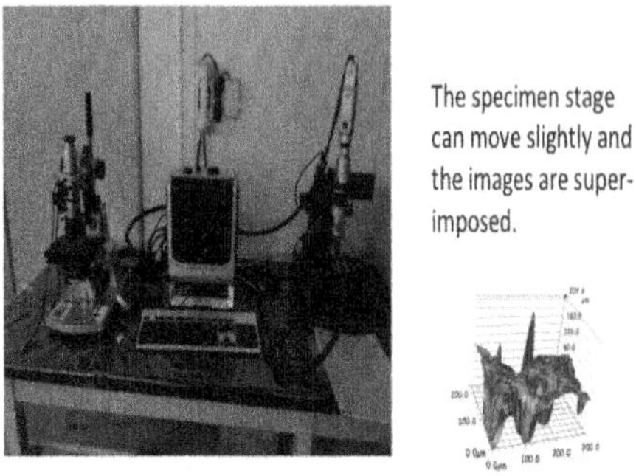

Fig.5. 3D imaging optical microscope.

Fig.6. Raman analysis.

RESULTS AND DISCUSSION

First, we had to fix the exposure time in the LBR. Fig.7 shows the observation results of iron specimens immersed in the LBR for one day and two days. The specimens were observed by the optical microscope with 3D imaging function. In these cases, any alternative currents were not applied. In both cases, the surfaces color tones were changed and some corrosion products were observed even by naked eyes. The 3D imaging pictures in both cases showed sea-island structures very clearly. These results mean that only one-day exposure (24 hours) would be a sufficient acceleration period for us to get the surface with biofilm or scales.

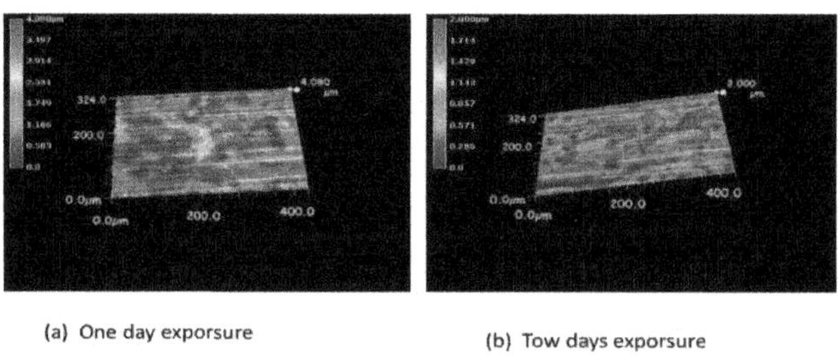

Fig.7 Iron specimens immersed in the LBR for one and two days.

Fig.8 shows the appearances for iron specimens immersed in the LBR for one day as well as the one before immersion as a reference. Compared to the specimen before the immersion, all specimens had the changes of color tones and scales to some extent. However, the observation results show clearly that the extent changed with the applied frequencies. When the frequency of 8kHz was applied to the specimen, the surface was kept clean pretty well. The color change was the least. On the other hand, the surface changed drastically and corrosion scales could be observed seriously at 20kHz. In the case, the surface was remarkably corroded. It was very hard to evaluate biofilm formation from the appearances. However, 3D imaging with the optical microscope could reveal the extent of biofilm formation and growth.

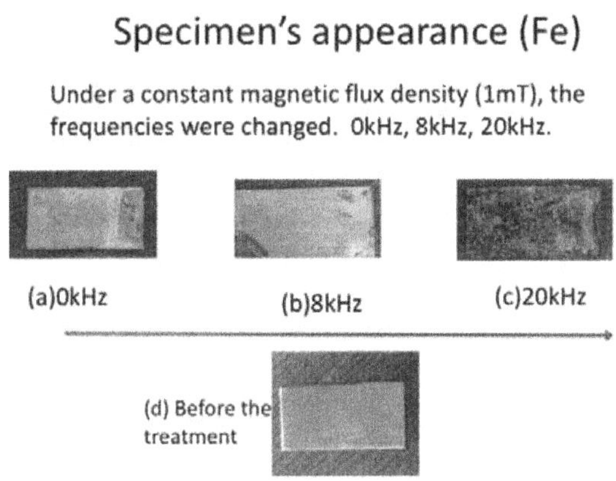

Fig.8 Iron specimens in the LBR for one day with and without the application of high frequency.

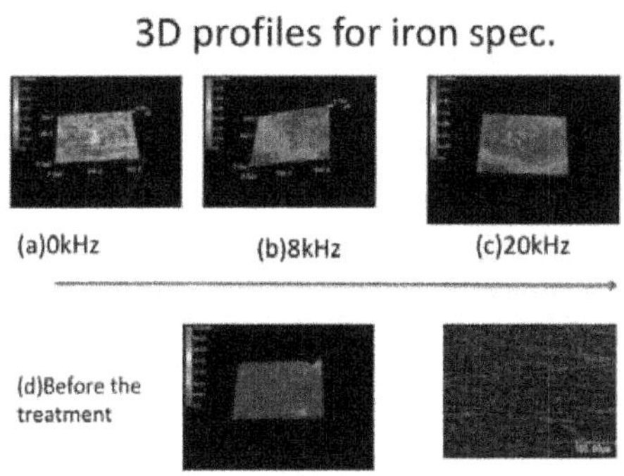

Fig.9. 3D imaging of iron specimens.

Fig.9 shows them concretely. At a glance, all of the specimens seemed to show biofilm formation except the specimen before the immersion where it showed only the monotone that indicates no biofilm formation. As for the specimens after immersion, the close examinations could tell the differences among them. The specimen without applied frequencies showed the sea-island structure as the most where the peak to bottom was approximately 4 mircometers. The specimen applied with the frequency of 8kHz showed only a gradient surface, where the color assigned by the computer of microscopy changed gradually. Even though the sea-island pattern was recognized to some extent (with only one micrometer amplitude) which was judged as the pattern corresponded to the specimen's slope on the observation stage of microscope as a whole. It showed that biofilm and scale were controlled pretty well. In the case of 20kHz, the surface pattern seemed to show only a gradient, even though the amplitude became larger, as compared to the 8kHz case. As shown above, the specimen's surface was corroded and corrosion scales formed significantly in that case.

Fig.10 shows the appearances observed by naked eyes for copper specimens. Since it was hard to recognize the changes by usual photos, the specimen's surface without the application of high frequency (0 kHz) seemed to be very similar to that in the case of 8kHz. However, when we observed the 3D images of both specimens (Fig.11), remarkable differences appeared. Before the immersion, the pattern corresponded just to the specimen's gradient. However, the 3D pattern of specimen without the application of high frequency showed a sea-island structure very clearly, while the pattern in the case of 8kHz showed a very clear surface, just like the chemical cleaned one. On the other hand, the specimen's surface in the case of 20 kHz, showed very rough profile as a whole and were different from any sea-island patterns. When we compared the result with the naked eye's observation, we could conclude that scales formed significantly in the case of 20kHz.

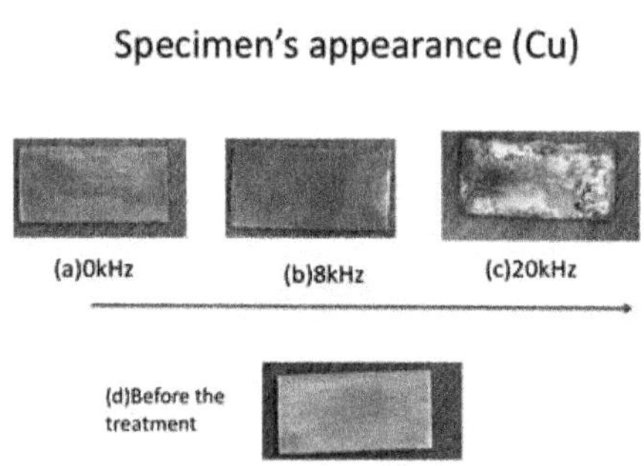

Fig.10 Copper specimens in the LBR for one day with and without the application of high frequency.

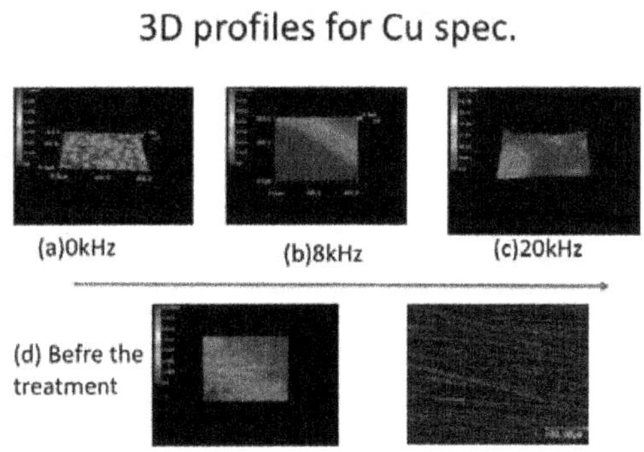

Fig.11 3D imaging for copper specimens.

Fig.12 showed the appearances of aluminum specimens, while Fig.13 showed its 3D imaging results. Again, for aluminum specimens, the visual photos were hard to tell the differences among them. However, the 3D imaging showed the differences clearly. The immersion in the LBR without the application of high frequency lead to the biofilm formation (0 kHz), while the specimen in the case of 8kHz showed no biofilm, but just a gradient of the specimen on the observation stage. In the case of 20kHz, the 3D pattern of the aluminum specimen showed the sea-island structure. However, we could observe while tiny compounds were scattered on the surface after the immersion. We presume a precipitation of calcium carbonate and it was reflected on the 3D image to some extent. At this point, it is difficult to make definitive statements if the sea-island structure would correspond only to biofilm formation. Probably, the mixture of biofilm and precipitated matters were reflected on the pattern. However, for the aluminum specimen that both the biofilm and scale were perfectly controlled in the case of 8kHz.

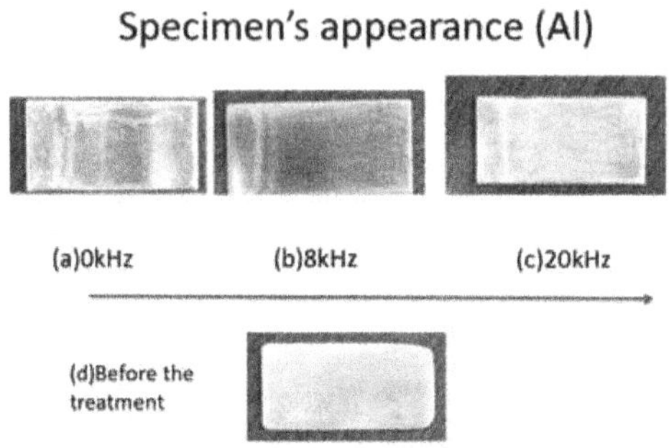

Fig.12 Aluminum specimens in the LBR for one day with and without the application of high frequency.

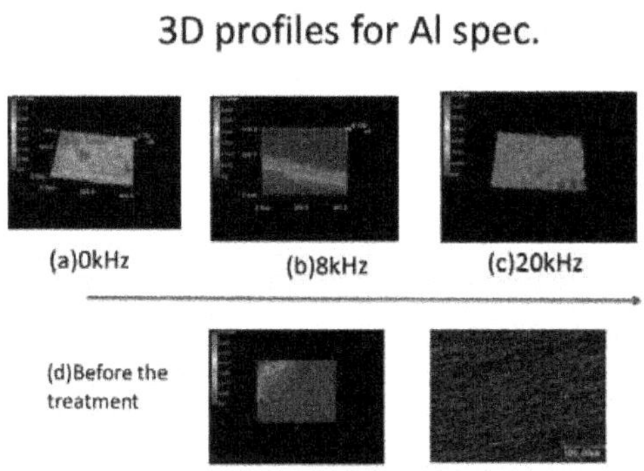

Fig.13. 3D imaging of Al specimens.

All of these results by the (conventional and 3D) microscopic and by naked observation show the following characteristics:

#1: Without the application of high frequency, the specimen tended to show biofilm significantly even in one day.

#2: When the frequency of 8kHz was applied, biofilm and scale were well controlled.

#3: When the frequency of 20kHz was applied, scales were seen for all specimens.

#4: The formation tendency was remarkable particularly for iron and copper. For aluminum, the scale was hard to form, even though biofilm could be observed.

Even though the 3D imaging might detect the concavo-convex surface profiles in several micrometer orders, they sometimes had the difficulty to distinguish between the biofilm and scales. As subsidiary measurement, we carried out Raman Spectroscopic analysis for all specimens.

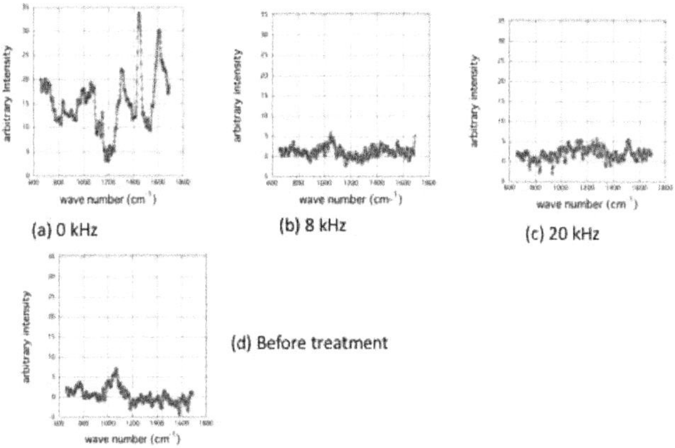

Fig.14. Raman shifts for iron specimens.

Fig. 14 shows the iron specimen's results on behalf of all kinds of specimens, since iron had the most remarkable formation tendency among all the tested specimens. Specimens without application of frequencies showed four typical peaks, even though the specimen before the immersion showed no remarkable peaks. They were observed around at 1120 cm^{-1}, at 1320 cm^{-1}, at 1440 cm^{-1} and at 1600 cm^{-1}. From some references in the past[(13)-(16)], we presume all of those peaks would correspond to the shifts derived from biofilm (1122 -1150 cm^{-1}: =C-C= lipid or C-N protein, 1320 cm^{-1}: C-H bend protein, 1366 cm^{-1}: C-H_2 deformation lipid, 1445 cm^{-1}: C-H_2 deformation lipid, 1540cm^{-1}: amide II, 1570cm^{-1}: amide II, 1616 cm^{-1}: amide I, for example.) However, the Raman shift in the case of 8kHz showed no remarkable peaks just like those without high frequency. The result was well coincident with the 3D observations by the optical microscope in the case of 8kHz. In the 20kHz case, the figure showed no remarkable peaks. It suggests that the peaks corresponding to biofilm components disappeared or hided due to the significant formation of scales. At any rate, all of these results for Raman shifts corresponded well to the results of microscopic observations.

From all of the results, we confirmed that the application of a specific frequency (8kHz) could control biofilm formation and the following formation of scale. In the past, some of the authors showed similar phenomena[(17),(18)]. However, the apparatus was not sufficient to fix the involved factors then, unfortunately, since the alternative current (corresponding directly to the density of the flux) was not regulated. For this experiment, we made sure that the applied current was kept constant during the immersion of specimens and the density of magnetic flux was actually measured and confirmed by a magnetograph (1mT). Therefore, we could fix in the series of experiments that the frequency of alternative current in a certain weak alternative electromagnetic field mainly played an important role to control biofilm followed by scales. We presume that hydroxide ion would adsorb to the materials surfaces selectively, when a certain frequency of alternative current would be applied. Therefore, the potential point in the zeta potential – pH curve would shift to the lower pH at the special frequency. One of the authors already observed the change of zeta potential under the alternative electromagnetic field for the specimens in hot spring(17),(18). If it would occur, the pH of aqueous environment in the vicinity of materials surfaces would increase to be alkaline locally. To confirm and generalize the relation and statement about the change and shift of zeta potentials, we need further experimentation as future research topic. However, at this point, many experimental results obtained so far suggest the phenomenon very strongly at least.

CONCLUSIONS

Three kinds of specimens, pure iron, pure copper and pure aluminum, were immersed in a specially designed Laboratory Biofilm Reactor (LBR) and the alternative electromagnetic field was applied to the system. All of the tests were carried out in 24 hours before taken out of the LBR's column and characterized using an optical microscopy with 3D imaging as well as using Raman spectroscopy. As a result, we could find that the formation of biofilm and the following scale formation could be controlled obviously at a certain frequency under a weak alternative electromagnetic field. This conclusion was confirmed by naked eyes, optical microscopic observation, 3D imaging and the Raman shifts. We presume that the selective adsorption and the following change of the zeta potential would be the main cause for the observed phenomenon.

ACKNOWLEDGMENT

The research project was carried out by the financial support of New Energy and Industrial Technology Development Organization (NEDO). The project was entitled Research and development of geothermal power generation technique/Innovative technology development for expanding introduction of geothermal power generation/"Research and development of physical removing scale technology for geothermal power generation with hot spring heat utilization". We greatly appreciate the organization for their constant financial support.

REFERENCES

(1) Watson, A., *Geothermal Engineering: Fundamentals and Applications.* 2013, New York: Springer. 336.

(2) Stober, I. and K. Bucher, *Geothermal Energy: From Theoretical Models to Exploration and Development.* 2013, Heidelberg: Springer. 291.

(3) DiPippo, R., *Geothermal Power Plants: Principles, Applicatoins, Case Studies and Environmental Impact, Thrid Edition.* 2012, Oxford, United Kingdom: Butterworth-Heinemann. 624.

(4) Umeki, S., *Research and development of geothermal power generation technique/Innovative technology development for expanding introduction of geothermal power generation/"Research and development of physical removing scale technology for geothermal power generation with hot spring heat utilization",* in *New Energy achievement debrief meeting.* 2015, NEDO: Yokohama, Japan. a poster.

(5) Kanematsu, H., *Microbiologically Influenced Corrosion in Marine Environment.* Rust Prevention & Control Japan, 2014. **58**(7): p. 265-271.

(6) Kanematsu, H., N. Hirai, Y. Miura, M. Tanaka, T. Kogo, and H. Itoh, *Various Metals from Water by Biofilm from an Ambient germs in a Reaction Container,* in *Materials Science and Technology.* 2013: Montreal, Quebec, Canada. p. 2154-2161.

(7) Ikegai, H., N. Hirai, M. Kobayashi, H. Toda, T. Moroboshi, S. Ikeda, K. Uesugi, D. Kuroda, and H. Kanematsu, *Biomineralization via microbiologically influenced corrosion induced by Pseudomonas aeruginosa biofilm.* CAMP-ISIJ, 2015. **28**: p. 468-470.

(8) Kanematsu, H., S. Sasaki, Y. Miura, T. Kogo, K. Sano, N. Wada, M. Yoshitake, and T. Tanaka, *Composite coating to control biofilm formation and effect of alternate electro-magnetic field.* Materials Technology, 2015. **30**(B1): p. 21-26.

(9) Kanematsu, H., T. Kougo, D. Kuroda, Y. Ogino, and Y. Yamamoto, *Biofilm Formation Derived from Ambient Air and the Characteristics of Apparatus.* Journal of Physics: Conference Series, 2013. **433**: p. 012031: 1-6.

(10) Kanematsu, H., T. Kogo, D. Kuroda, H. Itoh, and S. Kirihara, *Biofilm Formation and Evaluation for Spray Coated Metal Films on Laboratory Scale,* in *Thermal Spray 2013 - Innovative Coating Solutions for the Global Economy.* 2013, ASM International: Busan,

Republic of Korea. p. 520-525.

(11) Waters, G. *Chemical components of Uba-no-yu in Naruko hot spring water.* 2010; Available from: https://www.mizuhiroba.jp/waterdb/detail/o176.html.

(12) Kanematsu, H. and D.M. Barry, *Biofilm and Materials Science.* 2015, New York, the USA: Springer.

(13) Yuanqing Chao and Tong Zhang, *Surface-enhanced Raman scattering (SERS) revealing chemical variation during biofilm formation: from initial attachment to mature biofilm,* Anal Bioanal Chem. Sep. 2012, 404(5), p. 1465-1475.

(14) Esen Efeoglu, Mustafa Culha, *Surface-enhanced Raman scattering for Biofilm Characterization,* Spectroscopy, Oct. 2013,

(15) Yongqin Jiao, George D. Cody, Anna K. Harding, Paul Wilmes, Matthew Schrenk, Korin E. Wheeler, Jillian F. Banfield and Michael P. Thelen, Characterization of Extracellular Polymeric Substances from Acidophilic Microbial BiofilmsAppl Environ Microbiol, May. 2010, 76(9), p. 2916-2922.

(16) Jürgen Schmitt, David Nivens, David C. White and Hans-Curt Flemming, *CHANGES OF BIOFILM PROPERTIES IN RESPONSE TO SORBED SUBSTANCES – AN FTIR-ATR STUDY,* Wat. Sci. Tech. 1995, 32(8), p. 149-155

(17) Umeki, S., T. Kato, and N. Yoshikawa and S. Taniguchi, *Change in Zeta-Potential of Non-magnetic Colloid Particles by Weak AC Magnetic Field,* in *Proc. 3rd International Workshop on Water Dynamics, Sendai (2005).* 2005. p. 85-87.

(18) Umeki, S., T. Watanabe, H. Shimabukuro, T. Kato, N. Yoshikawa, S. Taniguchi, and K. Tohji, *Change in Zeta Potential by Alternating Electromagnetic Treatment as Scale Prevention Process.* Trans. Material Research Society of Japan, 2007. **32**(3): p. 611-614.

Microwave Assisted Sintering of Cold Iso-Statically Pressed Titanium 6-4 Powder Compacts

B.Y. Rock[1], M.A. Imam[2], T.F. Zarah[3]

[1]Naval Research Laboratory, Plasma Physics Division, Washington DC, USA
[2]George Washington University, School of Eng. and App. Sci. Washington DC, USA
[3]MATSYS, Inc., 45490 Ruritan Cir., Sterling, Virginia, USA

ABSTRACT

Cristal Metals Inc. has been producing powders of titanium and its alloys using its Armstrong Process®. The objective of this study is to determine the suitability of relatively large size compacts of the powder for consolidation to near-net shape by microwave energy. The results indicate that the use of microwave energy to sinter titanium powder produced from Cristal Metal powder is a cost-effective and energy-efficient method to achieve dense compacts of titanium. The in-depth heating permits rapid processing times that preserve the fine grain structure in the final product, resulting in excellent mechanical properties. The microwave sintering behavior of the compacts will be described and the characterization of the sintered specimens will be discussed.

INTRODUCTION

Titanium and its alloys offer excellent metallurgical and physical characteristics, but the high cost of component fabrication has restricted its use to mainly high performance aerospace and bio-medical applications [1]. The high cost arises from both the conversion of the ore to metal and from the metal component forming processes. A majority of the contribution to cost comes from the latter of these steps, with machining to shape a particularly cost intensive process, but the former step is also a significant contributor. The need to reduce the cost of components from both the reduction and forming ends has been the major driver of titanium research for many years.

Cristal Metals Inc. has developed a process for the production of titanium powders. This unique Armstrong Process® has the important advantages of being continuous, achieving full reaction efficiency, low reaction temperatures and direct alloy production. In particular, the process has demonstrated the ability to produce pre-alloyed Ti-6Al-4V powders. The development of this, as well as other methods for the direct production of economical, high quality titanium powders [1, 2, 3] presents an opportunity for alternative approaches to low cost titanium parts.

Powder metallurgical (PM) processing, particularly the conventional cold press and sinter technique can produce economical near-net-shape forms which require minimal post processing machining. For titanium alloy products, further economies can be obtained by the use of blended elemental powders, which are cheaper to produce than pre-alloyed powders. With the exception of fatigue strength, products produced by PM have been shown to be at least equivalent to ingot products. This failure is apparently from the residual porosity and may be mitigated by fuller densification.

An emerging technology with potential for meeting the goals of "better, faster, cheaper, and greener" is microwave sintering of powder compacts. Relative to conventional resistive elements, microwave radiation can be efficiently generated and transmitted to a green compact. Because the heating is direct and potentially volumetric, sintering cycles can be very rapid and the internal temperature distributions can be made uniform. In addition to economical considerations, the unique features of microwave heating lead one to expect improved mechanical properties and better overall performance advantages over conventional sintering, such as fine microstructure, higher strengths and densities, and greater uniformity.

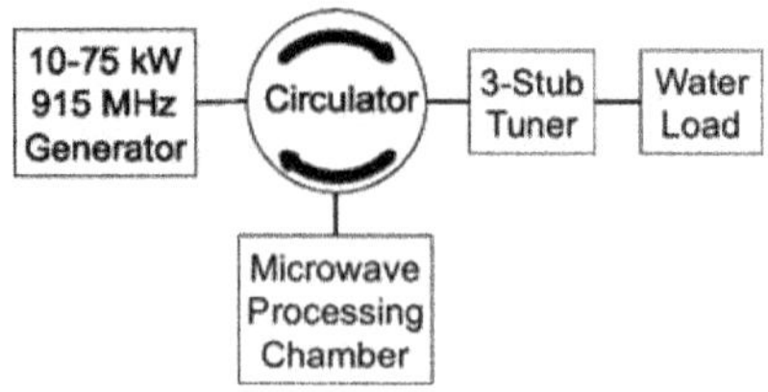

Figure 1: Schematic diagram of the microwave transmission system.

Titanium is known to be highly amenable to densification by diffusion bonding processes. This is because the particle surfaces in the powder self-clean starting between 500 and 700 °C, thus promoting inter-particle diffusion [4]. Since titanium is not absorptive of microwave energy at room temperature, it is necessary to arrange green compacts with microwave susceptors to make initial heating possible. This has proved to be an effective strategy for microwave sintering of relatively small titanium compacts [5, 6, 7]. However, the microwave absorption properties of titanium at elevated temperatures are unknown and it is not obvious that direct, volumetric heating takes place even in these regimes. It is thus in doubt whether larger titanium parts can be sintered to high uniform density by microwave energy even with susceptors. The purpose of this study is to investigate microwave sintering of relatively large titanium specimens compacted from Cristal Metals powders.

EXPERIMENTAL

Two different groups of Cristal Metals Ti-6Al-4V powder were used. One was pre-alloyed and the other, which we call ``blended" was pre-alloyed Ti+V powder blended with pre-alloyed Ti+Al powder in the correct amounts. Chemical analyses of pre-alloyed and blended powder are given in Table I. As can be seen in the table there was relatively high oxygen content in both types of powder.

The as-received powder was conditioned by blending for four hours in a V-blender and sieving to U.S. Mesh size 325. The resulting powder was then pressed in a Cold Isostatic Press (CIP). In order to study the microwave sintering of green bodies with initial densities of roughly 50%, 70% and 80%, specimens were prepared under CIP pressures of 30 ksi, 60 ksi and 90 ksi. The as pressed compacts were machined to be cylindrical in shape before sintering. The approximate dimensions of the cylinders were approximately 4 cm for diameter and 11.0 cm in length. The actual dimensions were dependent on CIP pressure and details of the packing.

Sintering was performed in a microwave furnace excited with a 915 MHz magnetron generator from Richardson Electronics. The microwave transmission system is shown schematically in Fig. 1. The microwave chamber was a cylindrical vacuum vessel with inner radius 19.6 cm and height 37.6 cm. The height of the chamber was adjustable with a movable floor on the chamber and it was set to give resonance with the TE_{11} cylindrical waveguide mode for the empty chamber without an aperture. The casket was initially placed on three alumina fiberboard rings to raise the approximate center of the titanium specimen one quarter of a guide wavelength from the bottom chamber wall. A test specimen that was processed

Table I: Weight percent of major elemental components in Crystal Metals powders used in this study.

	O	N	H	C	Fe	Na	Al	V
Pre-Alloyed	0.265	0.004	0.005	0.008	0.002	0.002	5.630	4.800
Blended	0.295	0.024	0.008	0.023	0.007	0.007	6.470	4.920

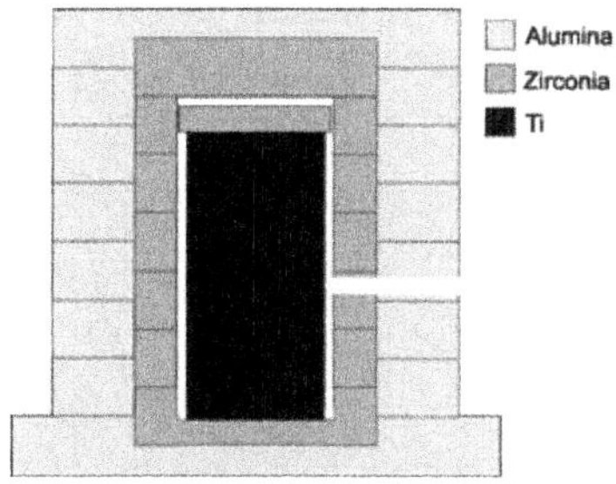

Figure 2: The microwave processing casket with susceptor and insulating materials labeled. The total size of the casket is approximately 8 inches tall and 6 inches in diameter.

this way showed deformation, differential densification at the bottom relative to the top, so all subsequent specimens were processed with their centers approximately one eighth of a guide wavelength from the bottom wall.

Microwaves were injected into the chamber via a centered 9.75 by 4.875 (i.e. WR975 sized) aperture in the top wall of the chamber. Analysis of the guided wave structures and the symmetries of the processing chamber and feeding transmission line indicate that the field excitation at the aperture is predominantly the TE_{11} cylindrical waveguide mode. Experience with microwave processing of large titanium specimens at 2.45 GHz revealed that when a tuner was used to impedance match the processing chamber, the standing wave established in the feeding waveguide led to field enhancement at the vacuum/air interface, often resulting in microwave induced glow discharge on the vacuum side of the interface. This experience, combined with the knowledge that the threshold breakdown and plasma maintenance field intensities decrease with frequency, lead us to abandon any attempt to impedance match the empty or loaded chamber with the exciting transmission line. Sintering without an impedance matching circuit element resulted in much reduced occurrence of microwave induced plasma, at the cost of significantly reduced energy efficiency in heating. In the present case, this was not a concern, but in an industrial process, the efficiency could be increased and occurrence of plasma minimized with an optimized waveguide to chamber transition and/or a better vacuum system.

The titanium compacts were packed in a suscepting and insulating ceramic casket. Yttria stabilized zirconia fiberboard was used as microwave susceptor. This material is also a good thermal conductor and helps to provide an isothermal sintering environment. To further promote isothermal conditions, alumina fiberboard was used as an additional thermal insulator. All of the materials were arranged in a cylindrically symmetric form. A schematic diagram of the processing casket is shown in Fig. 2.

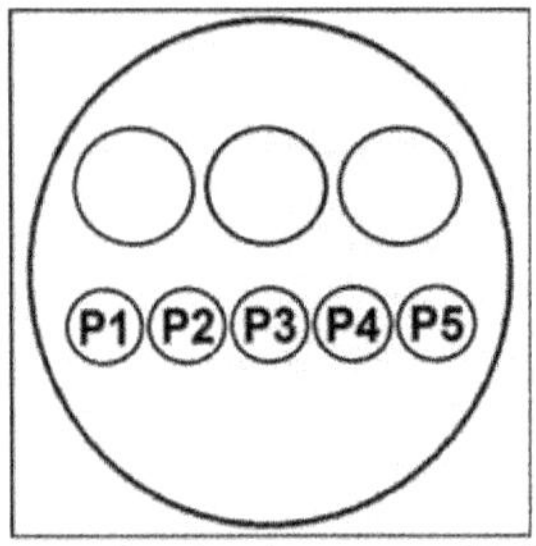

Figure 3: A schematic diagram showing the cores and indicating the labeling of the various density cores. Cored samples were cut from each of the sintered specimens for tensile, density and metallographic anlaysis. The cores were cut from the specimens with EDM.

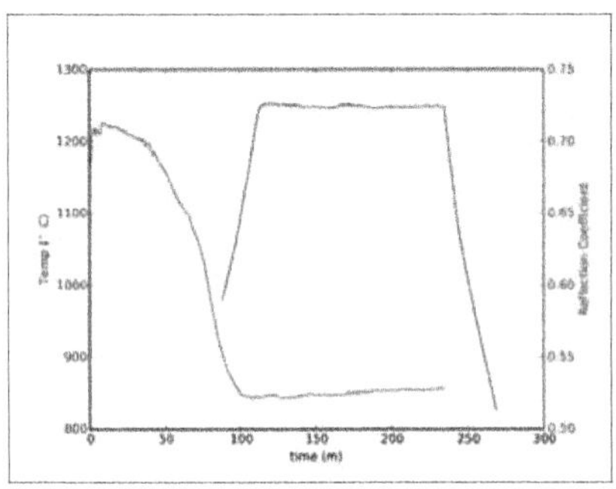

Figure 4: Temperature and reflection coefficient variation with time measured for a pre-alloyed powder compact CIP'd at 90 ksi and sintered for 120 minutes. The blended compacts showed a similar behavior as do compacts with different CIP pressures.

The temperature was measured with a two-color optical pyrometer with temperature range from 700-2000 °C. For some processing runs, a low temperature pyrometer 300-1000 °C was also used. A 6 mm hole was bored through the casket materials to provide an optical path for pyrometry. Processing was performed in an evacuated environment with base residual pressure of roughly 10 mtorr. Generally, titanium vacuum sintering should be performed at pressures well below 10 μtorr to avoid oxygen contamination, but this was not a concern here as the Oxygen content of the starting powders was already high (see Table I). Previous research on microwave sintering showed that small compacts could be sintered to high density at 1250 °C in 30 minutes [5]. For this study on relatively large specimens sintering was performed at 1250 °C for hold times of 30 minutes, 60 minutes and 120 minutes for each CIP compaction pressure. It was thought that longer hold times might be necessary for the temperature distribution to isotropize throughout the cylinder.

The temperature and heating rates of the specimens were determined by adjusting the incident microwave power. All runs began with approximately 200 W of microwave power, which was increased by 100 W roughly every 10 to 15 minutes. This was done until the specimens began to out-gas appreciably, at which point, the heating rate was limited to keep the furnace pressure below 200 mtorr.

After sintering, even though there was no noticable deformation of the cylinders, the specimens were machined on a lathe to square off their edges and make their surfaces straight and parallel. The total density of the cylinders were measured by an Archimedes principle. Cored cylinders were taken by EDM from several locations throughout the samples as shown in Fig. 3 for tensile, density and metallographic testing and analysis. The cylinders for density analysis were sliced into five pucks for further localization of the density measurements. Samples for tensile testing were also cut from the cored cylinders to have approximate dimensions 0.8 cm for diameter, 10.0 cm for length and with a reduced section approximately 4.5 cm in length. The gauge area was flat with dimension of 0.25 x 0.8 cm (cross sectional area of 0.2 cm^2).

RESULTS AND DISCUSSION

Previous published studies of microwave sintering of titanium of which we are aware all worked with relatively small compact sizes ($\lesssim$ 10 cm^3) and microwaves at 2.45 GHz [5, 7, 8, 6]. For the present study with larger ($\gtrsim$ 125 cm^3) specimens it was not obvious that even the longer wavelength 915 MHz microwaves would be effective for heating them, or that the heating would be uniform.

Figure 4 shows a typical heating curve for both the pre-alloyed and blended powder compacts. Because the pyrometer views through an rf shielding metallic mash, and only a small portion of the titanium specimen is optically visible, the pyrometer does not obtain sufficient infrared intensity to determine the temperature until the specimen is close to 1000 °C. Measurements with a different, low temperature pyrometer indicate that the specimens typically reach 300 °C in about 15 minutes and 400 °C in approximately one hour. The average heating rate from room temperature to 1250 °C was about 11°min. While the heating rate is more rapid than a typical vacuum furnace, it is a slow microwave heating pace. This was done intentionally to reduce the risk of producing a microwave discharge during the processing runs. With better pumping and a microwave impedance matching circuit element significantly faster heating rates should be possible.

In terms of energy efficiency, however, faster heating may not prove to be very effective for the initial stage of heating where, as shown in Fig. 4, almost 75% of the incident microwave power is reflected from the reaction chamber, and very little even of that which is absorbed is done so by the specimen, as opposed to the microwave susceptors. However, at higher temperatures considerably more of the microwave power is absorbed in the cavity. The reflection coefficient measured by the directional coupler is not the same a measurement of the absorption coefficient of the casket containing the titanium specimen, but it is a reasonable proxy for it, and it can be reasonably concluded that the heating of the casket becomes

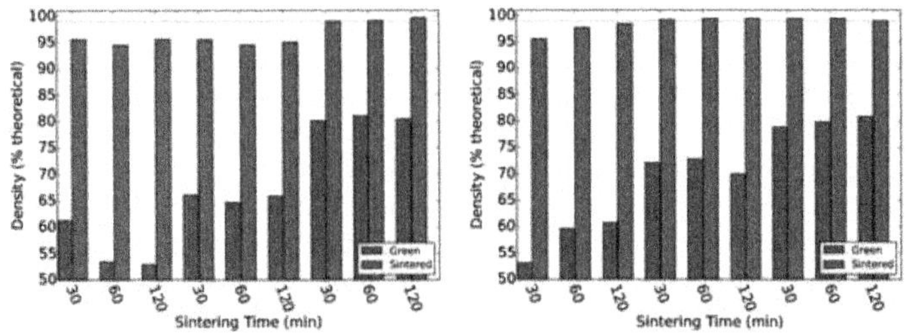

(a) (b)

Figure 5: The green and sintered density of specimens of pre-alloyed (a) and blended Ti64 powder compacts which were CIP'd at different pressures and sintered at 1250 °C for different times. In each chart, the first three groups of bars are for 30 ksi CIP, the second three for 60 ksi CIP and the last three for 90 ksi CIP. The dotted line indicates 99% density.

considerably more efficient at above about 600 °C. A future study will investigate the effects of heating rate on the metallurgical and mechanical properties of titanium specimens.

In order to achieve desired levels of mechanical properties such as strength, ductility, and fatigue strength, sintered materials should not only have a homogeneous chemical composition and microstructure, but also a relative density of greater than 98% of theoretical [5, 9, 10]. However, the relative densities of titanium alloys produced by the blended elemental approach normally do not exceed 95% [11]. Sintered titanium samples have been reported to obtain a higher density when sintered using TiH_2 powder, but with a 43 ksi CIP, even small specimens require 4 hours of conventional sintering to surpass 95% theoretical density and fall short of the 98% threshold [12]. Sintered materials can be subjected to hot isostatic pressing or other hot deformation process [13, 14] to further increase their density. However, this increases the number of production steps, increases the cost of parts, and coarsens the grains. All of these effects negate the advantages of the PM approach.

The relative densities, as measured by Archimedes' method, for the green and sintered state of the

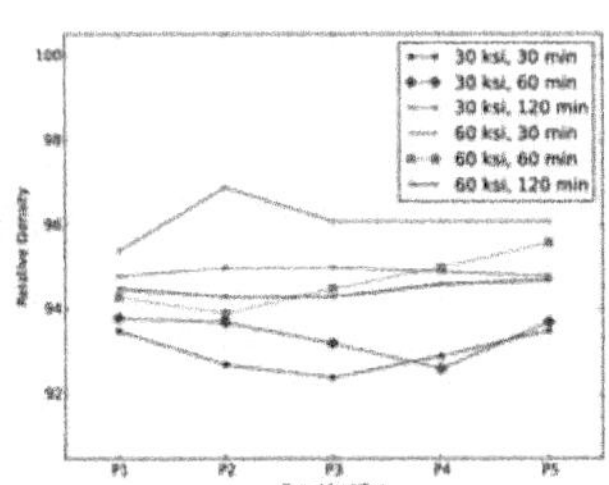

Figure 6: Density variations as measured from cores taken from different locations in the cylinders sintered from the pre-alloyed powder. The core identifiers are the same as in Fig. 3.

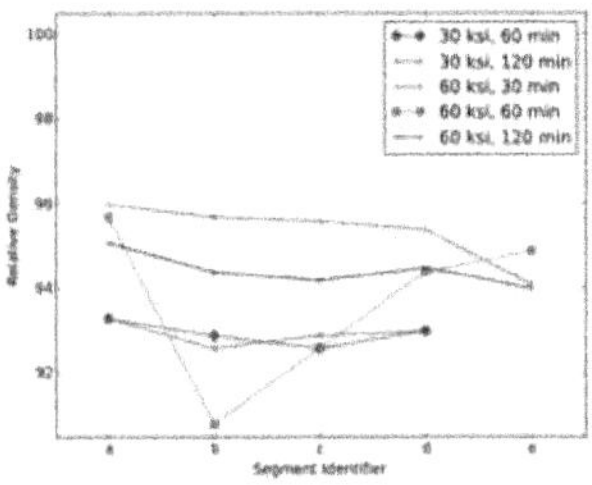

Figure 7: Density variations of pucks sectioned from a density core. Each density core was sliced into five pucks. The pucks were labeled a, b, c, d, e (from top to bottom). The measurements shown here are for P1 cores from pre-alloyed powder specimens.

specimens prepared for the present study are shown in Fig. 5a for pre-alloyed powder, and Fig. 5b for the blended powder. It can be seen that the sintered density is greater than 95% for all of the specimens except one, which had an anomalously low green density. Furthermore, the densities of all of the specimens was uniform throughout the volume. Results of the density measurements of the denist cores from the pre-alloyed powder specimens are shown in Fig. 6. The local densities of the cores is shown in Fig. 7. The variations in these figures are also representative of the blended powder specimens.

It is interesting that the soak time had little effect on the final density for all of the specimens, and that the green density (CIP pressure) also had little effect on the final density above about 60% relative green density. Associated with the densification came shrinkage of the linear dimensions of the green compact by between 10% and 20% depending on the initial and final densities. This shrinkage was observed to be uniform and isotropic.

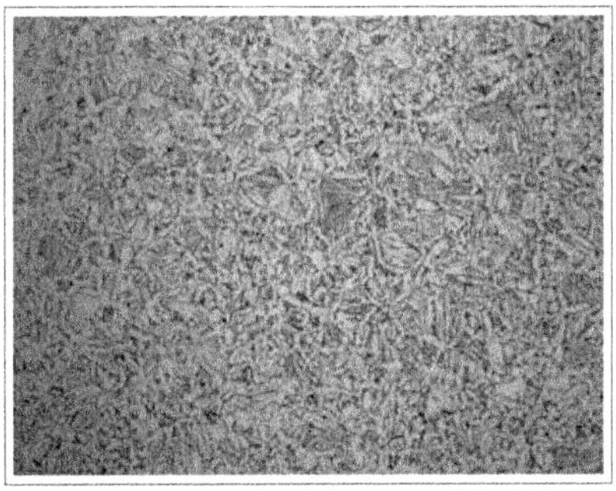

Figure 8: A typical optical micrograph of the sintered Ti-6Al-4V showing α and β phases. The specimen shown here was from a blended powder specimen that was CIP'd at 30 ksi and sintered for 60 minutes.

Table II: Preliminary results for the tensile strength (ksi) for specimens of pre-alloyed and blended powders pressed at various CIP pressures and sintered at 1250 °C for various times. Data that are not yet available are labeled N.A.

	Soak (min)	30 ksi	60 ksi	90 ksi
Pre-Alloyed	30	110.6	135.2	N.A.
	60	124.4	124.6	N.A.
	120	127.3	129.8	N.A.
Blended	30	114.8	126.1	108.6
	60	122.3	106.3	N.A.
	120	121.2	122.3	111.8

Tensile test results of an average ultimate tensile strength is shown in Table II for sintered specimens from the pre-alloyed and blended powder compacts. The ductility of the samples was less than 3% because of relatively high oxygen content. Higher green density (CIP pressure) for blended powder showed decrease in ultimate tensile strength. It is possible that there may be larger grain size compared to lower CIP pressure. This hypothesis is supported by the lack of further densification with both increasing the sintering time for these specimens, and in increasing the CIP pressure from 60 ksi. This is may not be the case with the pre-alloyed samples but the data at 90 ksi CIP pressure is not yet available. Further investigation and more data are needed to clarify the results.

For metallographic analysis, the sample was polished to mirror finish. An optical micrograph of the sample is shown in Fig. 8. The micrograph shows a microwave sintered Ti-6Al-4V specimen showing alpha and beta phases. Some etch pits or porosity is also evident in the micrograph.

CONCLUSION

We have successfully sintered the Cristal Metals Ti-6Al-4V powder compacts to high density with microwave energy. In contrast to previous reports, the specimens processed here were relatively large. Measurements of the density at various locations throughout the sintered specimens show that the density of the specimens were uniform volumetrically. Both pre-alloyed and blended elemental Cristal Metals, inc. powders sintered to high density regardless of sintering time or green density. However, the final density was most sensitive to the green density, with green densities above 70% required for sintered density above 99%. Tensile tests of the cored samples demonstrate the the material is well sintered and structurally sound.

It is evident that microwave sintering is a potentially low-cost, energy-efficient process for processing powder metallic materials to near-net shape parts. We believe the economics of microwave processing will be superior to any conventional process. The process can be also be used for composites, laminates, direct alloying, and functionally gradient materials.

References

[1] M. Imam, F. Froes, and K. L. Housley, ``Titanium and titanium alloys," in Kirk-Othmer Encyclopedia of Chemical Technology, 2010, pp. 1--41.

[2] D. Van Vuuren, M. Imam, S. Oosthuizen, B. Rock, T. Zahrah, and H. Chikwanda, ``Characterization of csir-ti powder and its consolidation by microwave sintering," in Ti 2015: The 13th World Conference on Titanium, San Diego, Ca, in press.

[3] M. Imam, B. Rock, and T. Zahrah, ``Consolidation of cristal metals powder of titanium and its alloys by microwave energy to near-net shape," in Ti 2015: The 13th World Conference on Titanium, San Diego, Ca, in press.

[4] M. Qian, ``Cold compaction and sintering of titanium and its alloys for near-net-shape or preform fabrication," Int. J. Powder Matall., vol. 46, no. 5, pp. 29--44, 2010.
[5] M. Imam, J. Feng, B. Rock, and A. Fliflet, ``Processing of titanium and its alloys by microwave energy," Adv. Mat. Res., vol. 1019, pp. 11--18, 2014.
[6] S. Luo, C. Guan, Y. Yang, G. Schaffer, and M. Qian, ``Microwave heating, isothermal sintering, and mechanical properties of powder metallurgy titanium and titanium alloys," Metall. Mater. Trans. A, vol. 44, pp. 1842--1851, 2013.
[7] S. Luo, M. Yan, G. Schaffer, and M. Qian, ``Sintering of titanium in vacuum by microwave radiation," Metall. Mater. Trans. A, vol. 42, pp. 2466--2474, 2011.
[8] S. Luo, Y. Yang, G. Schaffer, and M. Qian, ``Novel fabrication of titanium by pure microwave radiation of titanium hydride powder," Scripta Materialia, vol. 69, pp. 69--72, 2013.
[9] V. Moxson, P. Sjoblom, and M. Trzcinski, ``Ti-6al-4v properties achieved via extra low chlorine titanium powder," in Non-Ferrous Materials Advances in Powder Metallurgy and Particulate Material, vol. 6, 1992, pp. 125--140.
[10] V. Moxson, O. Senkov, and F. Froes, ``Production and application of low cost titanium powder products," Int. J. Powder Metall., vol. 34, no. 5, pp. 45--53, 1998.
[11] S. Abkowitz and D. Rowell, ``Superior fatigue properties for blended elemental p/m ti-6al-4v," JOM, vol. 38, no. 8, pp. 36--39, 1986.
[12] J. Oh, K. Heo, W. Kim, G. Choi, and J. Lim, ``Sintering properties of ti-6al-4v alloys prepared using ti/tih_2 powders," Mater T. JIM, vol. 54, no. 1, pp. 119--121, 2013.
[13] S. Abkowitz, S. M. Abkowitz, P. F. Weihrauch, and M. G. Wells, ``low cost p/m manufacture of titanium alloy components for fatigue critical applications," in P/M in Aerospace, Defense and Demanding Applications, San Diego, CA, 1993, pp. 241--250.
[14] F. Froes, D. Eylon, and H. Bomberger, ``Powder metallurgy of titanium alloys a review," in Titanium Technology: Present Status and Future Trends, 1985, pp. 49--59.

MICROWAVE HEATING OF ENSEMBLES OF SPHERICAL METAL PARTICLES SURROUNDED BY INSULATING LAYERS

K. I. Rybakov[1,2], V. E. Semenov[1]
[1] Institute of Applied Physics, Russian Academy of Sciences, Nizhny Novgorod, Russia
[2] Lobachevsky State University of Nizhny Novgorod, Nizhny Novgorod, Russia

ABSTRACT

Microwave sintering of metal powder compacts is one of the key directions of research in the field of microwave processing of materials. It has been demonstrated both theoretically and experimentally that a significant factor in the microwave heating of conductive powder compacts is the presence of insulating (e.g., oxide) layers on the surface of the individual particles. These layers facilitate partial penetration of microwaves into the powder compacts and thereby enhance the efficiency of microwave absorption. In this work, the effective complex dielectric permittivity and magnetic permeability of an ensemble of spherical metal particles surrounded by insulating layers are calculated within the effective medium approximation. For this purpose, we account for the non-uniform distribution of both electric and magnetic field in individual particles and perform averaging over the ensemble. The developed model makes it possible to simulate microwave heating of metal powder compacts.

INTRODUCTION

Starting from the pioneering work by Roy *et al.*, [1] there has been significant interest in the microwave sintering of powder metals, metal–ceramic composites and other particulate conductive materials.[2,3] There have been a vast number of experimental efforts to determine optimum conditions for efficient microwave heating of these materials. In particular, it has been argued that depending on the properties, the materials are heated in a different manner at the points of electric and magnetic field maxima in standing-wave applicators.[4] There have also been theoretical works and numerical calculations aimed on the explanation of the microwave heating behavior of conductive powder materials.[5–8]

It has been accepted that the description of microwave absorption in particulate materials can rely on their effective dielectric and/or magnetic properties. Many models have been developed to determine the effective properties based on the properties of the individual particles and their distribution in the material. In this paper a new model is suggested. It is intended to provide a more accurate description of the effective dielectric properties of particulate materials with high conductivity, such as powder metals.

THEORY

One of the efficient methods to describe the properties of inhomogeneous media is the so called effective medium approximation (EMA), originally proposed by Bruggeman.[9] The EMA implies that the inclusions of each phase are embedded in the matrix with the effective properties that have to be found. For example, one can easily obtain the EMA equations describing the dielectric permittivity of a system comprising a mixture of two or more phases. Particles of each phase can be viewed as spherical inclusions in the matrix. The electric field in each of such spherical inclusions is obtained within the quasi-electrostatic approach. The electric field inside the spherical inclusion and outside of it obeys the equations

$$\nabla \times \mathbf{E} = 0, \quad \nabla \cdot \mathbf{E} = 0. \tag{1}$$

When a uniform external electric field $\mathbf{E}_0$ is imposed on the system, the field inside the spherical inclusion is also uniform and can be written as

$$\mathbf{E}_{(i)} = A\mathbf{E}_0, \tag{2}$$

whereas the field outside is

$$\mathbf{E}_{(e)} = \frac{B}{r^3}\left[3\mathbf{n}(\mathbf{n}\cdot\mathbf{E}_0) - \mathbf{E}_0\right] + \mathbf{E}_0, \tag{3}$$

where A and B are arbitrary constants which are determined from the conditions at the boundary between the spherical inclusion and the effective medium surrounding it. These boundary conditions implying continuity of the tangential component of the electric field $\mathbf{E}_\tau$ and the normal component of the electric induction $D_n = \varepsilon E_n$ give a system of two equations with respect to A and B. As a result, the electric field within the inclusions is obtained as

$$\mathbf{E}_k = \frac{3\varepsilon_{\text{eff}}}{2\varepsilon_{\text{eff}} + \varepsilon_k}\mathbf{E}_0, \tag{4}$$

where ε_k is the dielectric permittivity of the kth phase, ε_{eff} is the (sought-for) effective dielectric permittivity of the composite medium.

In particular, when describing the effective dielectric properties of porous materials, such as powder compacts, one of the phases involved in the EMA is the void space with $\varepsilon = 1$. The electric field inside the spherical pores (voids) in the "effective medium" is

$$\mathbf{E}_{(v)} = \frac{3\varepsilon_{\text{eff}}}{2\varepsilon_{\text{eff}} + 1}\mathbf{E}_0. \tag{5}$$

It should be remembered that the quasi-electrostatic approximation is applicable provided that the particle size is small compared to the electromagnetic wavelength within the material:

$$k_0 a\left|\sqrt{\varepsilon_k}\right| << 1, \tag{6}$$

where $k_0 = \omega/c$ is the vacuum wavenumber, ω is angular frequency of microwaves, c is the velocity of light, and a is the particle radius.

The effective dielectric permittivity is determined from the relationship

$$\langle\mathbf{D}\rangle = \varepsilon_{\text{eff}}\mathbf{E}_0, \tag{7}$$

where $\mathbf{D}$ is the electric induction, and the angular brackets denote averaging over the volume. Obviously,

$$\langle\mathbf{D}\rangle = \sum_k C_k\varepsilon_k\mathbf{E}_k, \tag{8}$$

where C_k is the relative volumetric fraction of the kth phase in the composite, $\sum_k C_k = 1$.

Inserting (8) and (4) into (7), we obtain a kth-degree equation for ε_{eff}:

$$\sum_k C_k \frac{\varepsilon_k - \varepsilon_{\text{eff}}}{2\varepsilon_{\text{eff}} + \varepsilon_k} = 0. \tag{9}$$

One of the advantages of the EMA is that $\varepsilon_{\text{eff}} \to \varepsilon_k$ when $C_k \to 1$ for each of the phases. For this reason the EMA is widely used in the modeling of the sintering processes where concentrations of the phases, including the void phase, may vary in a broad range. It should be noted that in the intermediate range between these limiting cases the EMA cannot be justified rigorously and in fact remains just an interpolation formula. However, in many cases the EMA-based calculations demonstrate reasonable agreement with the experimental data.

If the EMA is applied to the description of a compacted metal powder, the imaginary part of the dielectric permittivity (or conductivity) of the composite exhibits a behavior that is similar to percolation transition. When the volumetric fraction of metal is about 0.3, the conductivity grows sharply by many orders of magnitude. This reflects the establishment of connectivity between individual metal particles. In the microwave heating perspective, this means that any metal powder compact with the porosity less than 70 % should reflect microwaves almost completely and therefore it cannot be microwave heated. However, this is refuted by the experimental results that often demonstrate that metal powder compacts can be microwave heated and sintered to full density.

We suggested an explanation of this behavior based on the insulating effect of thin oxide layers on the surface of the conductive powder particles. We developed a modification of the EMA that accounted for the presence of the insulating "shells" on the spherical particles.[5] In this modified EMA, the electric field within the particle is calculated within a quasi-electrostatic approach for a spherical conductive particle with the complex dielectric permittivity $\varepsilon_c = \varepsilon'_c + i\sigma/\varepsilon_0\omega$ (where ε'_c is the real part of the dielectric permittivity, σ is the effective conductivity of the material, and ε_0 is the electric constant) surrounded by a spherical insulating layer with the (real) dielectric permittivity ε_i (Fig. 1). This quasi-electrostatic problem is solved in a similar way as the problem of the sphere without shell discussed above, but it requires four arbitrary constants to be determined from two pairs of boundary conditions at the boundaries between the conductive particle and the insulating shell and between the shell and the outer effective medium. To account for non-zero porosity of the powder compacts, this system includes spherical voids as a second type of inclusions. As a result of volume averaging according to Eq. (7), the following quadratic equation is obtained for ε_{eff}:[10]

$$2\beta\varepsilon_{\text{eff}}^2 + [\beta(3C_s - 2) + \alpha\varepsilon_i(1 - 3C_s)]\varepsilon_{\text{eff}} - \alpha\varepsilon_i = 0, \tag{10}$$

where C_s is the volume fraction of solid (conductive + insulating) in the compact, $\alpha = 2\varepsilon_i + \varepsilon_c + 2(\varepsilon_c - \varepsilon_i)/\xi^3$, $\beta = 2\varepsilon_i + \varepsilon_c + (\varepsilon_i - \varepsilon_c)/\xi^3$, $\xi = b/a$, a is the radius of the conductive particle without shell, and b is the outer radius of the shell.

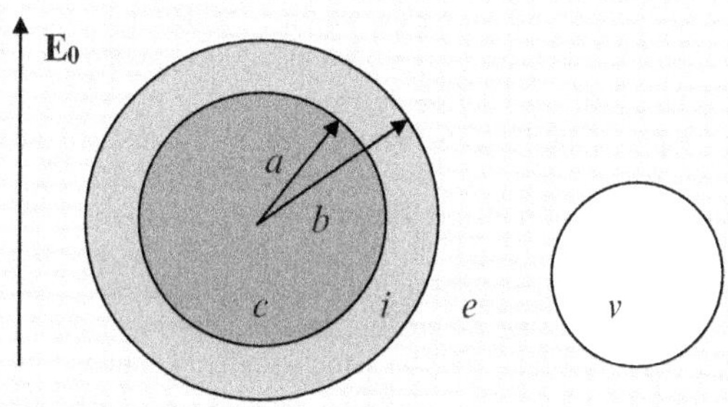

Figure 1. A conductive particle (c) surrounded by an insulating shell (i) and a spherical void (v) in the effective medium (e).

By applying this approach to such conductive powder systems as silicon / silicon oxide it has been shown that the presence of even very thin shells can drastically change the effective dielectric permittivity, and, in particular, its imaginary part $\varepsilon''_{\text{eff}}$ that characterizes microwave absorption. The obtained results were in reasonable agreement with experiment.[5]

It has been widely discussed in the recent literature that microwave absorption in metal powders relies on the magnetic-type losses in the material. The alternating magnetic field in the electromagnetic wave incident on a particle gives rise to the rotational electric field within the particles, which in turn drives eddy currents that generate heat. The properties of materials containing non-ferromagnetic metal particles can be described by the complex magnetic permeability that depends on the volume fraction of particles, their size, conductivity of metal, and frequency. The effective complex magnetic permeability of a material containing a small volume fraction, C_m, of non-ferromagnetic metal particles can be obtained as

$$\mu_{\text{eff}} = 1 + \chi \approx 1 + C_m p_m, \tag{11}$$

where χ is magnetic susceptibility and p_m is the magnetic polarizability of particles. To obtain p_m, it is necessary to solve the wave equation for the magnetic field in the metal particles:

$$\nabla^2 \mathbf{H} + k^2 \mathbf{H} = 0, \tag{12}$$

where $k = k_0 \sqrt{\varepsilon_c} \cong (1+i)/\delta$ is the wavenumber in the metal, and

$$\delta = c\sqrt{2\varepsilon_0/(\sigma\omega)} \tag{13}$$

is the skin depth. The calculations described in the textbook by Landau and Lifshitz[12] give the following expression for the magnetic polarizability (per unit volume) of a metal particle:

$$p_m = -\frac{3}{2}\left(1 - \frac{3}{k^2 a^2} + \frac{3}{ka}\operatorname{ctg} ka\right), \tag{14}$$

where a is particle radius. Maximum absorption (determined by the imaginary part of the magnetic polarizability, p''_m) is observed when the particle size is on the order of skin depth (13). We tested this approach by comparing the calculated and measured absorption in metal-ceramic composites and found reasonable, although not exact, agreement.[11]

When a plane electromagnetic wave is incident on a particle, both the electric field and the magnetic field contribute to the energy absorption and hence heating of the particle. It can be shown that the contribution of the magnetic-type effects into absorption becomes prevailing over the contribution of the dielectric losses calculated within the quasielectrostatic approximation when the conductivity of metal and hence the value of $|\varepsilon_c|$ is so high that

$$k_0 a |\varepsilon_c| >> 1. \tag{15}$$

Note that there can be a domain of conductivity values in which both inequalities (6) and (15) are satisfied and therefore full microwave absorption can be characterized by $\varepsilon''_{\text{eff}}$ and μ''_{eff} obtained according to Eqs. (10) and (11), respectively.

However, once the electromagnetic wavelength within the material (or the skin depth (13)) becomes comparable or smaller than the metal particle size, the quasi-electrostatic approximation applicability criterion (6) is violated. When calculating the effective dielectric properties of the materials containing such particles, one should also take into account the non-uniform structure of the *electric* field in the particles. In the rest of this paper we report the results of such calculations and compare them with the previous approaches. The electric field structure in the metal particles with insulating shells is obtained by solving the wave equation using the method described by Landau and Lifshitz. The result is then used to construct the EMA.

Let us consider a spherical conductive particle (*c*) surrounded by an insulating layer (*i*). In the domains (*c*) and (*i*) shown in Fig. 1 the electric field **E** should obey the equations

$$\nabla^2 \mathbf{E} + k^2 \mathbf{E} = 0, \quad \nabla \cdot \mathbf{E} = 0. \tag{16}$$

The field satisfying these equations can be presented in the form

$$\mathbf{E} = \nabla \times \nabla \times (f \cdot \mathbf{E}_0), \tag{17}$$

where $\mathbf{E}_0$ is the external electric field imposed on the system and $f(r)$ is a spherically symmetric solution of a scalar equation

$$\nabla^2 f + k^2 f = 0. \tag{18}$$

For the domain (*c*) the solution that remains finite at $r = 0$ is

$$f_c = A\frac{\sin k_c r}{r}, \tag{19}$$

and for the domain (*i*) the solution is

$$f_i = C\frac{\sin k_i r}{r} + D\frac{\cos k_i r}{r}, \tag{20}$$

where $k_c = k_0\sqrt{\varepsilon_c}$, $k_i = k_0\sqrt{\varepsilon_i}$, *A*, *C* and *D* are three arbitrary constants.

Performing the vector differentiation in Eq. (17), one can obtain the following expression for the electric field in each of the domains (*c*) and (*i*):

$$\mathbf{E}_{(c,i)} = \left(\frac{f'}{r} + k^2 f\right)\mathbf{E}_0 - \left(\frac{3f'}{r} + k^2 f\right)\mathbf{n}(\mathbf{E}_0 \cdot \mathbf{n}), \tag{21}$$

where **n** is a unit normal vector in the radial direction (the second derivative of *f* cancels out due to Eq. (18)).

In the external domain (*e*) the electric field, as previously, obeys Eq. (1) and it can be presented in the form (3), which provides the fourth arbitrary constant, *B*.

The four constants *A*, *B*, *C* and *D* are determined from the boundary conditions at the boundaries $c-i$ and $i-e$.

As discussed above, the effective medium approximation implies that the inclusions of each phase are embedded in the matrix with the effective dielectric permittivity $\varepsilon_{\mathrm{eff}}$ that has to be found. For the description of metal powder compacts of various density/porosity, the phases involved are the metal particles with the insulating layers described above and the void space.

The effective dielectric permittivity is determined, as previously, from the relationship (7). The volume average on the lhs of Eq. (7) is obtained as

$$\langle\mathbf{D}\rangle = \varepsilon_c C_s \frac{a^3}{b^3}\frac{1}{V_c}\int_{(c)}\mathbf{E}_{(c)}dV + \varepsilon_i C_s\left(1-\frac{a^3}{b^3}\right)\frac{1}{V_i}\int_{(i)}\mathbf{E}_{(i)}dV + (1-C_s)\mathbf{E}_{(v)} = \varepsilon_{\mathrm{eff}}\mathbf{E}_0, \tag{22}$$

where C_s is the overall concentration of solid material (metal + oxide) in the powder compact, $V_c = 4\pi a^3/3$ is the volume of the conductive particle without the insulating layer, $V_i = 4\pi(b^3 - a^3)/3$ is the volume of the insulating layer, $dV = 2\pi r^2 dr \sin\vartheta d\vartheta$, r and ϑ are spherical coordinates counted from the center of the particle ($\vartheta = 0$ corresponds to the direction of the external field $\mathbf{E}_0$).

As a result, Eq. (22) becomes a cubic equation with respect to $\varepsilon_{\mathrm{eff}}$:

$$\begin{aligned}&2w_1w_2\varepsilon_{\mathrm{eff}}^3 + \left[6C_s\varepsilon_i w_2 q_7 - 6C_s\varepsilon_i w_1 q_8 + (3C_s-2)w_1w_2 + 2(w_1w_4 + w_2w_3)\right]\varepsilon_{\mathrm{eff}}^2 + \\ &+\left[6C_s\varepsilon_i w_4 q_7 + 3C_s\varepsilon_i w_2 q_7 - 6C_s\varepsilon_i w_3 q_8 - 3C_s\varepsilon_i w_1 q_8 + (3C_s-2)(w_1w_4 + w_2w_3) + 2w_3w_4\right]\varepsilon_{\mathrm{eff}} + \\ &+3C_s\varepsilon_i w_4 q_7 - 3C_s\varepsilon_i w_3 q_8 + (3C_s-2)w_3w_4 = 0,\end{aligned} \tag{23}$$

where

$$w_1 = q_9 + \frac{\varepsilon_i q_2 q_4 - \varepsilon_c q_1 q_5}{\varepsilon_i q_3 q_4 + \varepsilon_c q_1 q_6} q_{10}, \quad w_2 = \frac{\varepsilon_i q_3 q_4 + \varepsilon_c q_1 q_6}{\varepsilon_i q_2 q_4 - \varepsilon_c q_1 q_5} q_9 + q_{10},$$

$$w_3 = \varepsilon_i\left(\frac{\varepsilon_i q_2 q_4 - \varepsilon_c q_1 q_5}{\varepsilon_i q_3 q_4 + \varepsilon_c q_1 q_6} q_8 - q_7\right), \quad w_4 = \varepsilon_i\left(q_8 - \frac{\varepsilon_i q_3 q_4 + \varepsilon_c q_1 q_6}{\varepsilon_i q_2 q_4 - \varepsilon_c q_1 q_5} q_7\right),$$

$$q_1 = \frac{k_c\cos k_c a}{a} - \frac{\sin k_c a}{a^2}, \quad q_2 = \frac{k_i\cos k_i a}{a} - \frac{\sin k_i a}{a^2}, \quad q_3 = \frac{k_i\sin k_i a}{a} + \frac{\cos k_i a}{a^2},$$

$$q_4 = \frac{k_c^2\sin k_c a}{a} + \frac{k_c\cos k_c a}{a^2} - \frac{\sin k_c a}{a^3}, \quad q_5 = \frac{k_i^2\sin k_i a}{a} + \frac{k_i\cos k_i a}{a^2} - \frac{\sin k_i a}{a^3},$$

$$q_6 = \frac{k_i^2\cos k_i a}{a} - \frac{k_i\sin k_i a}{a^2} - \frac{\cos k_i a}{a^3}, \quad q_7 = \frac{k_i\cos k_i b}{b^2} - \frac{\sin k_i b}{b^3}, \quad q_8 = \frac{k_i\sin k_i b}{b^2} + \frac{\cos k_i b}{b^3},$$

$$q_9 = \frac{k_i^2\sin k_i b}{b} + \frac{k_i\cos k_i b}{b^2} - \frac{\sin k_i b}{b^3}, \quad q_{10} = \frac{k_i^2\cos k_i b}{b} - \frac{k_i\sin k_i b}{b^2} - \frac{\cos k_i b}{b^3}.$$

RESULTS AND DISCUSSION

The results of calculations demonstrate that the account of the wave structure of the electromagnetic field in the metal particles result in significant differences in the effective dielectric properties of metal powder compacts compared to those obtained within the previous models. While the real part of the obtained effective dielectric permittivity, ε'_{eff}, is in most cases very close to the value given by the "quasi-static" EMA with insulating shells (Eq. (10)), the imaginary part, ε''_{eff}, that characterizes absorption is in many cases much higher.

The effective dielectric permittivity depends on many parameters, such as the relative volumetric concentration of solid in the compact, metal conductivity, microwave frequency, size of powder particles, the thickness of the insulating layer, etc.

Figure 2 shows the dependence of ε_{eff} on the relative volumetric concentration of solid in the compact. Both the real and imaginary part of ε_{eff} display percolation-like behaviour, which is better demonstrated using the log scale (Fig. 2, c and d). While the values of the real part, ε'_{eff}, are virtually the same in the wave EMA and quasi-static EMA, the values of the imaginary part are significantly different.

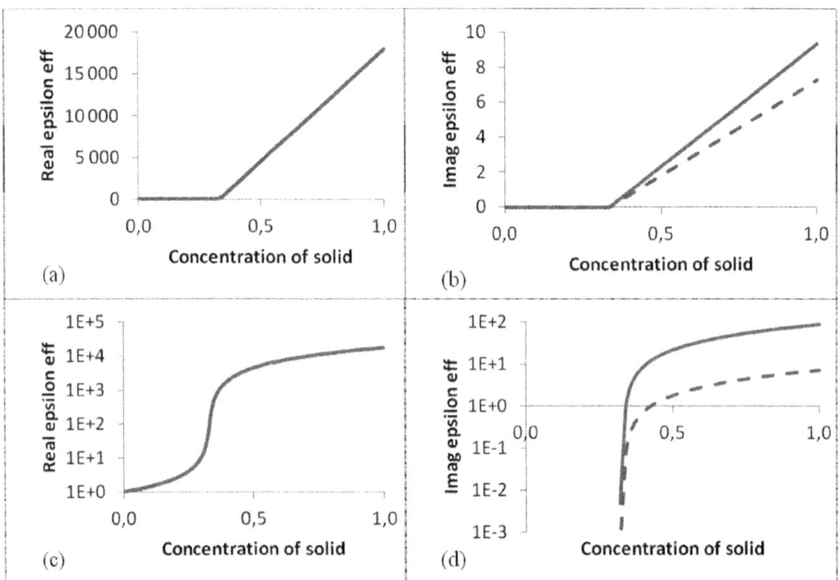

Figure 2. Effective dielectric permittivity ε_{eff} of copper powder with insulating oxide layers on particles vs. relative volumetric concentration of solid: (a), (b) powder particle radius 1 μm, oxide layer thickness 1 nm, (c), (d) powder particle radius 10 μm, oxide layer thickness 10 nm. Frequency 24 GHz. Solid line corresponds to the wave EMA approach, and dashed line – to the quasi-static EMA; for the real part of ε_{eff} (figs. (a) and (c)) both appoaches give the same result. Note log scale in figs. (c) and (d).

Figure 3 shows the dependence of ε''_{eff} on the conductivity of metal, σ, for two frequencies, 2.45 GHz (regular microwaves) and 24 GHz (millimeter waves). The conductivity range chosen for the plots embraces most metals and alloys, from manganin and constantan ($\sigma \sim 2 \cdot 10^6$ S/m) to copper and silver ($\sigma \sim 6 \cdot 10^7$ S/m). It can be seen that the wave EMA coincides with the δ quasi-static EMA in the domain of lower conductivities and at the lower frequency – i.e., where the skin depth δ (13) is much higher than the particle size. Away from this domain, ε''_{eff} scales as $\sigma^{-1/2}$ for the wave EMA and σ^{-1} for the quasi-static EMA (Fig. 3, b).

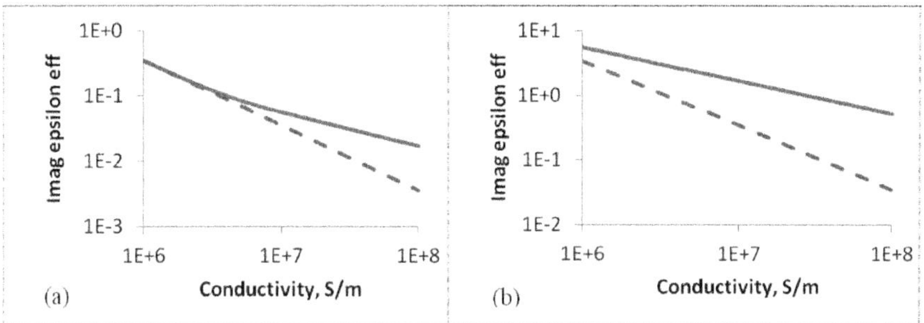

Figure 3. Imaginary part of the effective dielectric permittivity of compacted metal powder with insulating oxide layers vs. conductivity of metal: (a) at a frequency of 2.45 GHz, (b) at 24 GHz. Powder particle radius 10 µm, oxide layer thickness 100 nm, relative volumetric concentration of solid $C_s = 1$. Solid line corresponds to the wave EMA approach, and dashed line – to the quasi-static EMA.

It is also instructive to compare the imaginary part of the effective dielectric permittivity with the imaginary part of the effective magnetic permeability, μ''_{eff} (Eqs. (11), (14)) of the metal powder compact with insulating oxide layers. Figure 4 shows the results of such a comparison vs. metal conductivity for two frequencies and two powder particle sizes.

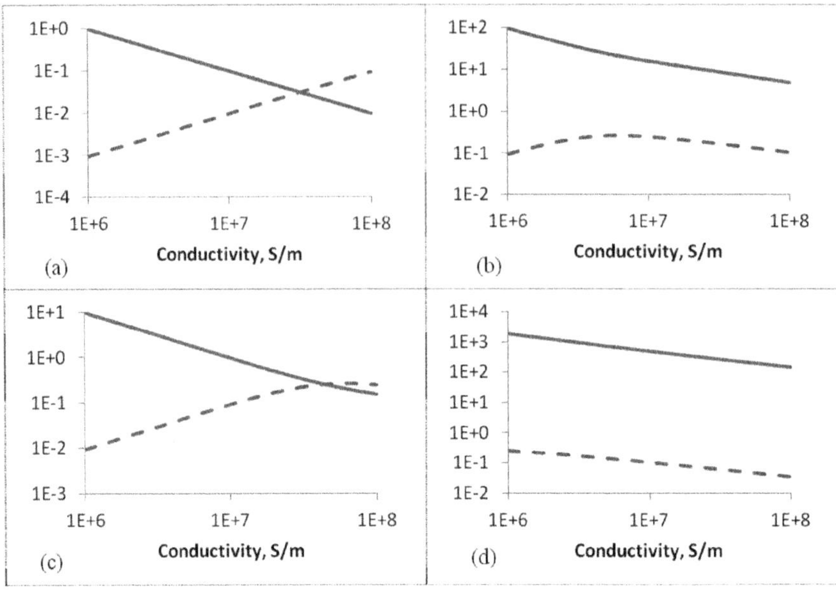

Figure 4. Imaginary part of the effective dielectric permittivity (solid line) and imaginary part of the effective magnetic permeability (dashed line) of compacted metal powder with insulating oxide layers vs. conductivity of metal: (a) powder particle radius 1 μm, frequency 2.45 GHz; (b) powder particle radius 10 μm, frequency 2.45 GHz; (c) powder particle radius 1 μm, frequency 24 GHz; (d) powder particle radius 10 μm, frequency 24 GHz. Oxide layer thickness 3 nm, relative volumetric concentration of solid $C_s = 0.5$.

As mentioned above, $\mu''_{\rm eff}$ exhibits a maximum when the skin depth δ (13) is on the order of metal particle size. Such maxima are seen in Fig. 4 b and c. The comparison shows that in many powder compacts $\varepsilon''_{\rm eff}$ can well be much higher than $\mu''_{\rm eff}$. In fact, in the wave EMA model $\varepsilon''_{\rm eff}$ grows with decreasing the insulating shell thickness. This can be explained by the fact that the electric field strength increases in the thin insulating layers which effectively act as capacitances between the metal particles.

It should be remembered that the microwave absorption is not solely determined by the values of $\varepsilon''_{\rm eff}$ and $\mu''_{\rm eff}$. The absorbed microwave power w (per unit volume) is proportional to them but also to the squares of the electric and magnetic field amplitudes:

$$w = \frac{\omega}{2}\left(\varepsilon_0 \varepsilon''_{\rm eff} \mathbf{E}^2 + \mu_0 \mu''_{\rm eff} \mathbf{H}^2\right). \tag{24}$$

Yet, the distribution of the electric and magnetic fields **E** and **H** in the object undergoing microwave heating depends in turn on its effective dielectric and magnetic properties. Therefore, to assess microwave absorption, the obtained values of $\varepsilon''_{\rm eff}$ and $\mu''_{\rm eff}$ should be first used to solve the macroscopic electrodynamic problem for the object positioned in the microwave applicator.

CONCLUSIONS

A model for the calculation of effective dielectric properties of powder metal compacts has been developed. The model is based on the effective medium approximation and accounts for the wave structure of the electromagnetic field within the metal particles. The model is applicable when the wavelength in the material of particles is comparable or smaller than their size, which is characteristic of most metals. The parameters in the model are the particle radius, thickness of the insulating oxide layers on the particles, conductivity of metal, dielectric permittivity of the oxide, relative concentration of solid in the compact (or porosity), and the microwave frequency. The calculations demonstrate that the nonuniform nature of the electric field results in highly significant values of the imaginary part of the effective dielectric permittivity, which cannot be neglected when assessing the feasibility of microwave heating of conductive particulate materials.

ACKNOWLEDGEMENT

This research was supported in part by Russian Science Foundation, grant # 15-12-00046.

REFERENCES

[1] R. Roy, D. Agrawal, J. Cheng, S. Gedevanishvili, Full sintering of powdered-metal bodies in a microwave field, *Nature*, **399**, 668–70 (1999).
[2] M. Gupta, W. L. E. Wong, *Microwaves and Metals*, Wiley, Singapore, 2007. – 228 pp.
[3] N. Yoshikawa, Fundamentals and Applications of Microwave Heating of Metals, *J. Microwave Power and Electromagn. Energy*, **44**, 4–13 (2010).
[4] J. Cheng, R. Roy, D. Agrawal, Experimental proof of major role of magnetic field losses in microwave heating of metal and metallic composites, *J. Mater. Sci. Lett.*, **20**, 1561–3 (2001).
[5] K. I. Rybakov, V. E. Semenov, S. V. Egorov, A. G. Eremeev, I. V. Plotnikov, Yu.V. Bykov, Microwave heating of conductive powder materials, *J. Appl. Phys.*, **99**, 023506 (2006).
[6] P. Mishra, G. Sethi, A. Upadhyaya, Modeling of Microwave Heating of Particulate Metals, *Metallurgical and Materials Transactions B* **37**, 839–45 (2006).
[7] V. D. Buchelnikov, D. V. Louzguine-Luzgin, G. Xie, S. Li, N. Yoshikawa, M. Sato, A. P. Anzulevich, I. V. Bychkov, A. Inoue, Heating of metallic powders by microwaves: Experiment and theory, *J. Appl. Phys.*, **104**, 113505 (2008).
[8] M. Ignatenko, M. Tanaka, M. Sato, Absorption of microwave energy by a spherical nonmagnetic metal particle, *Jpn. J. Appl. Phys.*, **48**, 067001 (2009).
[9] D. A. G. Bruggeman, Berechnung verschiedener physikalischer Konstanten von heterogenen Substanzen, I. Dielektriziätskonstanten und Leitfähigkeiten der Mischkörper aus isotropen Substanzen, *Ann. Phys-Berlin, Series 5*, **24**, 636–79 (1935).
[10] K. I. Rybakov, E. A. Olevsky, E. V. Krikun, Microwave sintering – fundamentals and modeling, *J. Amer. Ceram. Soc.*, **96**, 1003–20 (2013).
[11] S. V. Egorov, A. G. Eremeev, I. V. Plotnikov, K. I. Rybakov, V. V. Holoptsev, Yu. V. Bykov, Absorption of microwaves in metal-ceramic powder materials, *Radiophysics and Quantum Electronics*, **53**, 354–62 (2010).
[12] L. D. Landau, E. M. Lifshits, L. P. Pitaevskii, *Electrodynamics of Continuous Media*, Pergamon, New York, 1960. § 45, p. 193.

SINTERING OF OXIDE CERAMICS UNDER RAPID MICROWAVE HEATING

Yu. V. Bykov[1], S. V. Egorov[1], A. G. Eremeev[1], V. V. Kholoptsev[1], I. V. Plotnikov[1], K. I. Rybakov[1,2], A. A. Sorokin[1]
[1] Institute of Applied Physics, Russian Academy of Sciences, Nizhny Novgorod, Russia
[2] Lobachevsky State University of Nizhny Novgorod, Nizhny Novgorod, Russia

ABSTRACT

Rapid microwave sintering of Y_2O_3 nanopowder samples has been investigated. Using a 24 GHz / 6 kW gyrotron system for microwave processing of materials, the compacted powder samples were heated at rates in the range 10...200 °C/min to maximum temperatures 1400...1700 °C with zero hold time. At higher heating rates a sharp increase in the material's effective conductivity has been observed, which is similar to the flash sintering effect. Maximum densities of the sintered Y_2O_3 samples, above 98 %, have been obtained with a heating rate of 100 °C/min and maximum temperatures 1600...1700 °C. The duration of the high-temperature stage of sintering was about 8 min and the microwave power absorbed in the samples, per unit volume, was about 40 W/cm^3.

INTRODUCTION

Recently a significant interest has been drawn to the investigation of the flash sintering effect that occurs when heating is accompanied with the application of a dc or low-frequency ac voltage to the sample.[1-7] In a number of experiments it has been demonstrated that the onset of flash sintering correlates with a sharp increase in the electric conductivity of the samples undergoing heating.[8] The change in the conductivity manifests itself by an increase of the electric current between the electrodes applied to the sample, the voltage being kept constant. For the flash sintering effect to occur it is necessary to have a significant power input produced by the electric current (usually higher than 10 W/cm^3) that would provide additional Joule heating of the sample to several hundred degrees above the furnace temperature.

We have demonstrated recently that the flash sintering effect is also present under microwave heating of compacted ceramic materials.[9,10] Dense, optically transparent yttria-based ceramics have been obtained under heating at a rate of 100 °C/min to 1500 °C with zero hold time. It should be noted that both the dc/ac and microwave flash sintering effects occur under similar condition of heat generation and removal. In both cases the heating is due to the action of internal heat sources: electric current in the former case and volumetric absorption of the microwave energy in the latter. Also in both cases the heat is removed from the sample surface by convective and – at higher temperatures – radiative heat flows. Based on the similarity of the thermal conditions, it is reasonable to assume that the mechanisms of the flash sintering, governed by thermally activated mass transport, are also similar.

As opposed to the electric current-assisted sintering, in the case of microwave sintering the changes in the conductivity of the material cannot be easily detected by voltage and current measurements (unless electrodes are applied purposely for conductivity measurement using low currents with negligible contribution into heating). However, the changes in the effective high-frequency conductivity of material can be characterized in the microwave systems capable of precise automatic control of the heating process. Microwave heating occurs due to the absorption of the electromagnetic energy. The absorbed power, per unit volume, can be expressed as

$$p_{abs} = \sigma_{eff} E^2, \tag{1}$$

where σ_{eff} is the effective conductivity and E is the electric field strength in the material. Provided that the variations in the dielectric permittivity of the material and in the heat removal conditions during sintering are insignificant, an increase in the effective conductivity would result in a deviation of the heating rate. If the process control system uses a fast enough feedback loop, such a deviation from the prescribed heating schedule results in an automatic reduction of the microwave power. Therefore, sudden changes in the microwave power can be indicative of the changes in the effective conductivity of the material in the course of the sintering process. This is, in fact, a modification of the method of microwave differential thermal analysis (MDTA) that is widely used to detect chemical reactions or phase transformations in the materials that result in changes in the dielectric losses during heating.[11,12]

In this work the method based on the analysis of the records of the automatically regulated microwave power is used for the characterization of rapid sintering of Y_2O_3 and TiO_2 samples compacted from pure nanosized powders. For Y_2O_3 samples the parameters of the sintering regimes – the maximum temperature, the heating rate, and the absorbed power per unit volume – have been determined that result in obtaining ceramics with a density above 98 % of the theoretical value. In the case of rapid microwave heating of TiO_2 samples, possible reasons of the absence of the flash sintering effect are discussed.

EXPERIMENTAL

Green Y_2O_3 samples were prepared from Y_2O_3 nanopowder produced at the Institute of Chemistry of High-Purity Substances, Russian Academy of Sciences, Nizhny Novgorod, Russia by the method of self-propagating high-temperature synthesis (SHS).[13] As precursors, coarse Y_2O_3 powder and high-purity nitric and acetic acids were used. The coarse powder was dissolved in a mixture of acetic and nitric acids taken in the molar ratio 2:1. The solution was evaporated at a temperature of 110 °C, and yttrium acetonitrate was extracted as a white crystalline mass. It was divided into portions of about 2 g, which were put into quartz crucibles and introduced into a furnace preheated to 700 °C. Yttrium acetonitrate ignited, and after the end of burning a solid foam consisting of ultrafine and nanosized Y_2O_3 particles was obtained. For full oxidation of organic products yttrium oxide was annealed at a temperature of 1100 °C. The BET particle size of the so obtained powder was 0.13 μm. According to SEM, after annealing the powder was agglomerated with the average agglomerate size about 15 μm (Fig. 1). The impurity content was determined by atomic emission spectrometry with inductively coupled plasma (ICP–AES) using the instrument iCAP 6300. The main impurities were Si (50ppm), Ca (15ppm), Fe (11ppm), and Zn (3 ppm).

Figure 1. SEM image of the agglomerated Y_2O_3 powder after annealing at 1100 °C.

For sintering the samples were compacted by uniaxial pressing at 200–300 MPa. The compacted disks, 13 mm in diameter and 1.5 – 2.5 mm in thickness, had a density of 42 ± 1 % of the theoretical value. The value of 5.01 g/cm^3 was used as the theoretical density for Y_2O_3.

The samples were heated in the applicator of a gyrotron system capable of delivering up to 6 kW of power at 24 GHz with a feedback computer control of the power.[14] The samples were placed in the center of a quartz crucible of 110 mm in diameter and 110 mm in height and thermally insulated by coarse Y_2O_3 powder. The temperature of the samples (T) was measured by a B-type thermocouple with the tip embedded in the shallow hole drilled to the center of the sample where the temperature should be maximal during the volumetric microwave heating. Two additional thermocouples were positioned within the Y_2O_3 powder filling at distances 16 mm (T_1) and 35 mm (T_2) from the sample edge. During the microwave heating processes, the temperature was always lower away from the sample ($T > T_1 > T_2$), which is a confirmation that the heating occurs predominantly due to the absorption of microwave power (1) in the sample.

At the initial stage of heating, the samples were heated in air at a rate of 10 °C/min up to an intermediate temperature, which was 800 °C, and held at this temperature for one hour to remove the absorbed water and residues of the organics. At and below this temperature no densification of samples was detected. Then, if the vacuum sintering had been planned, the applicator was evacuated to 1 – 10 Pa and the samples were heated to the preset target temperature with a preset heating rate in the range 10 … 200 °C/min. When the target sintering temperature was reached, the microwave power automatically switched off and the sample cooled down together with the thermal insulation.

To measure the dc conductivity of the samples during heating platinum electrodes were imposed on the samples. A stabilized dc voltage of 1 V was applied to the electrodes, and the current flowing through the sample was measured to an accuracy of ±0.1 μA.

The density of the sintered samples was measured by Archimedes weighing in distilled water with the accuracy of ± 0.01 g/cm^3. The phase composition of samples was identified by X-ray diffraction (Ultima IV, Rigaku, Japan). The microstructure analysis was performed using scanning electron microscopy (JEOL JSM-6390 LV, Japan), both on as-sintered and mirror polished and thermally etched surfaces. The average grain size was obtained by averaging over 250 – 300 grains using a mean linear intercept method.[15]

RESULTS AND DISCUSSION

At the first stage of experimental studies, a series of sintering tests was undertaken on Y_2O_3 samples with the heating rates of 10, 25, 50, 100, and 200 °C/min in the temperature range from 800 to T_{max} = 1400 °C. Shown in Fig. 2 are the dependencies of the relative densities of the obtained sintered samples and the maximum microwave power input in the applicator on the heating rate. It can be seen that in this series, in which the maximum sintering temperature was 1400 °C, the density of the samples decreases gradually as the heating rate increases. The decrease in the density of the sintered samples is due to the reduction of the effective sintering time with an increase in the heating rate. This observation is typical of many sintering studies which use different heating rates. A significant decrease in the final density is especially noticeable at the maximum heating rate, 200 °C/min. As seen from Fig. 2, the change in the heating rate from 10 to 200 °C/min requires an increase in the input microwave power from 400 to 2000 W. However, based on the sintering results it can be assumed that the corresponding increase in the electromagnetic field strength does not lead to an increase in the final density of samples.

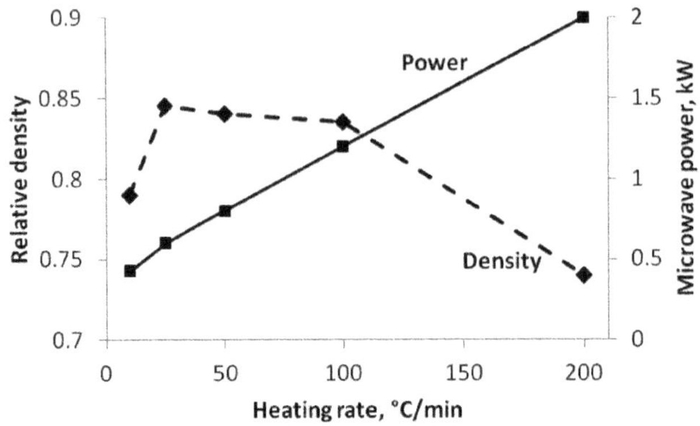

Figure 2. Relative density of the sintered Y_2O_3 samples (dashed line) and microwave power input to the applicator (solid line) vs. heating rate at the high-temperature stage of sintering. The maximum sintering temperature 1400 °C.

Shown in Fig. 3 are the time dependencies of the temperature of the sample and the microwave power input to the applicator for the sample heated at a rate of 25 °C/min to T_{max} = 1400 °C. It can be seen that at the final stage of heating (starting at $T \approx$ 1050 °C) the microwave power required to sustain the prescribed heating rate virtually does not increase in spite of the growth of the heat losses from the sample, which in vacuum have mostly radiative nature and hence are proportional to T^4. This fact evidences in favor of strong dependence of the effective high-frequency conductivity, σ_{eff}, of Y_2O_3 on temperature.

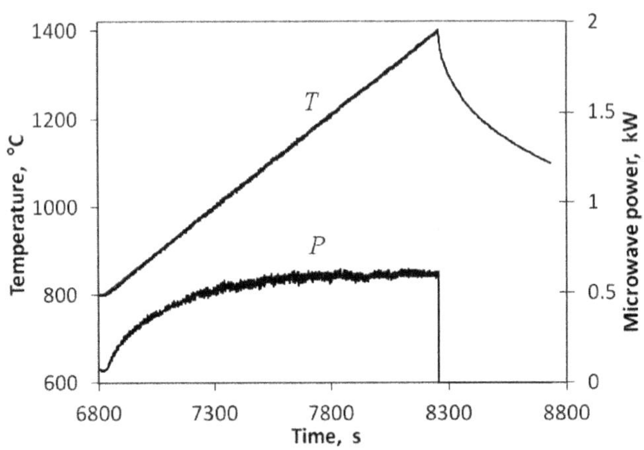

Figure 3. The temperature of the sample (T) and the microwave power (P) input to the applicator for the sample vs. time during the high-temperature sintering stage of the process with the heating rate 25 °C/min and the maximum sintering temperature T_{max} = 1400 °C.

At the next stage of experimentation, sintering processes with higher maximum temperatures, T_{max} = 1500 ... 1700 °C, were accomplished. Shown in Fig. 4 are the time dependencies of temperature and microwave power for the sample microwave heated at the same rate of 25 °C/min to a higher maximum temperature, T_{max} = 1500 °C. It can be seen that at higher temperatures the microwave power that sustains the prescribed heating rate not only does not increase but even decreases smoothly by about 15 % towards the end of the heating stage.

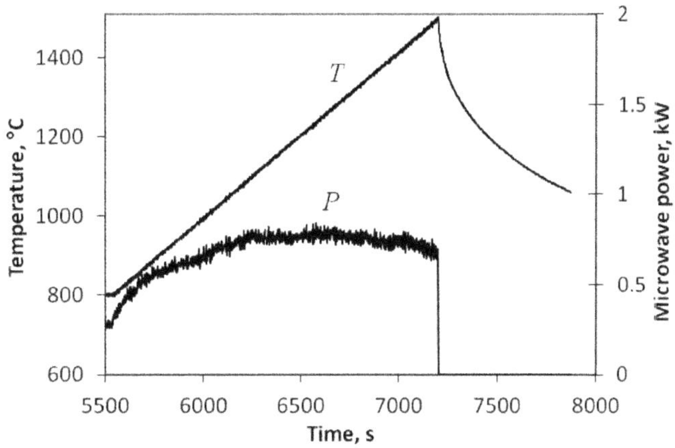

Figure 4. Same as in Fig. 3 but the maximum sintering temperature T_{max} is 1500 °C.

The experiments with higher heating rates revealed that the power decreases more noticeably with increasing the heating rate. In the process carried out at a rate of 50 °C/min the power decreases by 30 % from its maximum value. With further increase in the heating rate the nature of the power evolution changes from gradual decrease to sharp drop. Shown in Fig. 5 are the time dependencies of temperature and power for the process carried out at a rate of 100 °C/min. At $T \approx$ 1300 °C the input microwave power drops by half, from 1.4 kW to 0.7 kW, and remains virtually constant during the subsequent increase of the sample temperature.

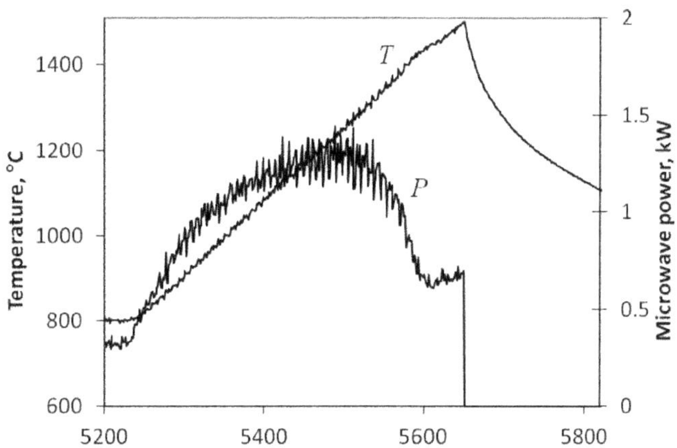

Figure 5. Same as in Fig. 3, but the heating rate is 100 °C/min and the maximum sintering temperature T_{max} is 1500 °C.

The dependency of the final density of the sintered samples on the heating rate in the processes with the target temperature 1500 °C is shown in Fig. 6. At low heating rates the density decreases with the increase in the heating rate, which is explained by the reduction of the effective sintering time and, as mentioned above, is typical of non-isothermal sintering. However, with a increase of the heating rate to 50 and 100 °C/min the final density increases significantly in spite of further reduction in the sintering time.

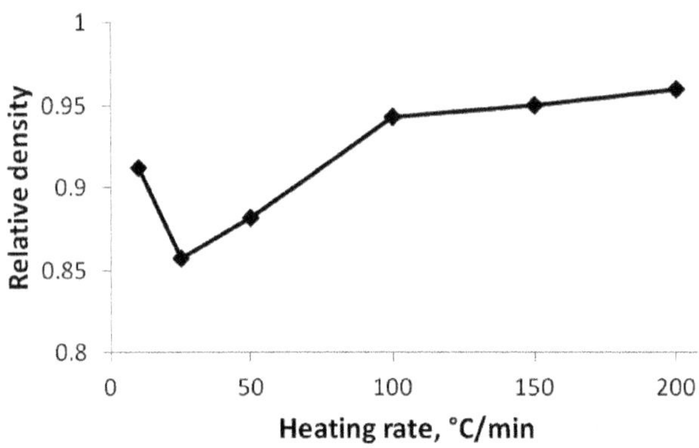

Figure 6. Relative density of the sintered Y_2O_3 samples vs. heating rate at the high-temperature stage of sintering. The maximum sintering temperature 1500 °C.

Along with the density, the behavior of the microwave power at higher heating rates is different as discussed above. The decrease in the microwave power required to sustain the constant-rate heating process can only be explained by a sharp increase in the microwave absorption coefficient, i.e. in the effective high-frequency conductivity of the material of the sample, σ_{eff}. It follows from the results of the sintering experiments carried out at heating rates in the range 50 – 200 °C/min to a maximum temperature of 1500 °C that the temperature at which the conductivity exhibits the sharp increase ($T = 1325 \pm 25$ °C) does not depend on the heating rate, although the power varies by a factor of 4. This suggests that the microwave electromagnetic field does not affect the conductivity. Another evidence in favor of the absence of the electromagnetic field influence on the conductivity are the results of the dc conductivity measurements during microwave heating of the sintered samples (with a relative density of 98 %). The activation energy of conductivity, derived from the measurements during microwave heating at a rate of 5 °C/min to 1400 °C, was 1.96 ± 0.15 eV, which coincides with the 1.94 eV value that is characteristic of high-purity polycrystalline Y_2O_3.[16] It should be noted that during these measurements the voltage was 1 V and the current through the samples did not exceed 60 µA, making the contribution of Joule heating negligible compared to the absorbed microwave power.

In the similar rapid sintering processes carried out to even higher maximum temperatures, 1600 and 1700 °C, with zero hold time, samples with densities above 98 % were obtained in spite of the severe agglomeration of the Y_2O_3 powder. The highest final densities, 98.3 and 98.1 %, were obtained in the processes carried out at a heating rate of 100 °C/min to 1600 and 1700 °C, respectively. The densities obtained in these rapid processes occurred to be higher than those obtained by isothermal sintering: the sample heated at a rate of 10 °C/min to 1600 °C with a hold time of 15 min had a density of 97.6 %.

Figure 7 shows the time dependencies of temperature and power in a process with a preset heating rate of 100 °C/min and the maximum temperature of 1700 °C. It can be seen that the power drop that occurs due to the increase in the effective conductivity of the sample as discussed above is accompanied by a sharp increase in the temperature which is not immediately compensated by the automatic control system. In the temperature range 1325 – 1435 °C the actual rate of temperature increase approaches 200 °C/min which is twice the value of the preset heating rate. Using the values of the heating and cooling rates immediately before and after the time instant when the maximum temperature is reached, the specific absorbed power can be determined from the energy balance equations.[10] For example, for the case of heating at a rate of 100 °C/min to the maximum temperature of 1700 °C the specific absorbed power is 42 W/cm^3.

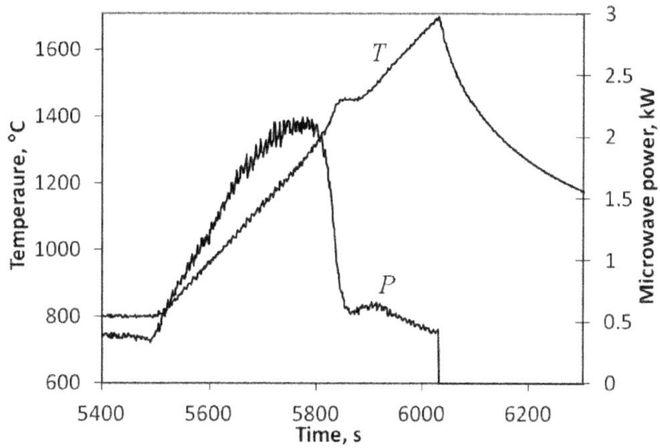

Figure 7. Same as in Fig. 3, but the heating rate is 100 °C/min and the maximum sintering temperature T_{max} is 1700 °C.

Shown in Fig. 8 are SEM images of the surfaces of the as-sintered samples sintered to densities above 98 %. Pores of both intergranular and intragranular type are present in the sample. The sample sintered in the process with T_{max} = 1600 °C (density 98.3 %) exhibits a bimodal grain size distribution with the characteristic sizes 1 – 3 μm and 5 – 9 μm. In the sample sintered in the process with T_{max} = 1700 °C the average grain size is much higher (~ 15 μm) although its density is slightly lower. It can be seen that many larger grains have an internal substructure with characteristic dimensions on the order of 1 – 3 μm. The elements of this substructure are not divided by any pronounced boundaries and probably result from some kind of fusing of micron-sized grains.

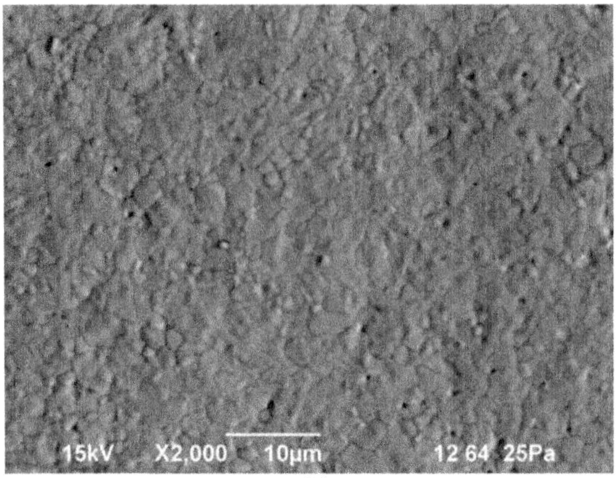

(a)

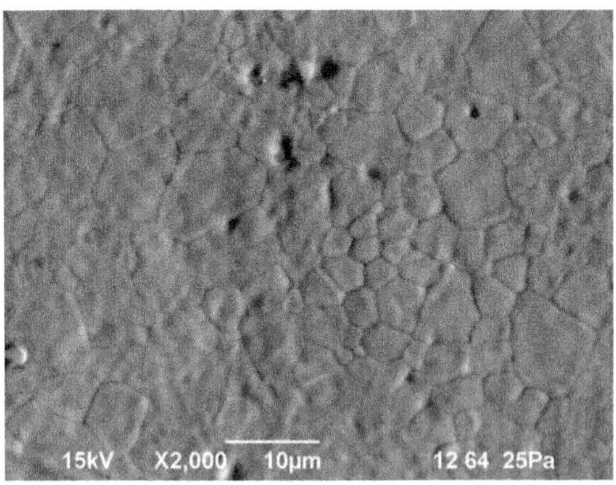

(b)

Figure 8. SEM images of the surfaces of the Y_2O_3 samples obtained in rapid microwave sintering processes with a heating rate of 100 °C/min: (a) maximum sintering temperature T_{max} = 1600 °C, (b) T_{max} = 1700 °C.

Figure 9 shows the results of an XRD study of the samples microwave sintered to T_{max} = 1700 °C at two heating rates, 100 °C/min (b) and 150 °C/min (c). It can be seen that the latter XRD spectrum exhibits a wide plateau around the (4,0,0) peak, which might be an evidence in favor of the existence of an amorphized or liquid phase in this sample.

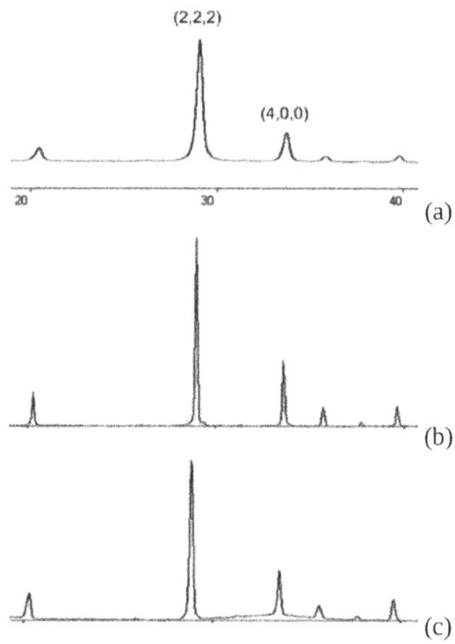

Figure 9. XRD spectra of (a) the initial Y_2O_3 powder; (b) grinded surface of a microwave sintered sample (heating rate 100 °C/min, maximum temperature T_{max} = 1700 °C); and (c) as-sintered sample (150 °C/min to T_{max} = 1700 °C).

CONCLUSIONS

Using a 24 GHz / 6 kW gyrotron system for microwave processing of materials, the compacted Y_2O_3 nanopowder powder samples were heated in the vacuum at rates in the range 10...200 °C/min to maximum temperatures 1400...1700 °C with zero hold time. At higher heating rates a sharp increase in the material's effective conductivity has been observed, which is similar to the flash sintering effect. Rapid densification of the material occurs simultaneously with the increase in the conductivity. Maximum densities of the sintered Y_2O_3 samples, above 98 %, have been obtained with a heating rate of 100 °C/min and maximum temperatures 1600...1700 °C. The duration of the high-temperature stage of sintering was about 8 min and the microwave power absorbed in the samples, per unit volume, was about 40 W/cm^3. The obtained density values are higher than those obtained by a slower-rate (10 °C/min) sintering process with a 15 min isothermal hold at the maximum temperature. From the practical viewpoint, it is important that the onset of rapid densification is easily identified without dilatometry by a sharp decrease in the level of microwave power required to sustain the preset rate of heating of the sample.

ACKNOWLEDGEMENT

This research was supported in part by Russian Foundation for Basic Research, grant # 13-03-01154.

REFERENCES

[1] D. Yang, H. Conrad, Enhanced Sintering Rate of Zirconia (3Y-TZP) by Application of a Small AC Electric Field, *Scripta Mater.*, **63**, 328–31 (2010).

[2] M. Cologna, A.L.G. Prette, R. Raj, Flash-Sintering of Cubic Yttria-Stabilized Zirconia at 750 °C for Possible Use in SOFC Manufacturing, *J. Am. Ceram. Soc.*, **94**, 316–9 (2011).

[3] A.L.G. Prette, M. Cologna, V. Sglavo, and R. Raj, Flash-Sintering of Co_2MnO_4 Spinel for Solid Oxide Fuel Cell Applications, *J. Power Sources*, **196**, 2061–5 (2011).

[4] M. Cologna, J.S.C. Francis, R. Raj, Field Assisted and Flash Sintering of Alumina and its Relationship to Conductivity and MgO-doping, *J. Eur. Ceram. Soc.*, **31**, 2827–37 (2011).

[5] A. Karakuscu, M. Cologna, D. Yarotski, J. Won, J. S. C. Francis, R. Raj, B. P. Uberuaga, Defect Structure of Flash-Sintered Strontium Titanate, *J. Am. Ceram. Soc.*, **95**, 2531–6 (2012).

[6] H. Yoshida, Y. Sakka, T. Yamamoto, J.-M. Lebrun, and R. Raj, Densification Behavior and Microstructural Development in Undoped Yttria Prepared by Flash-Sintering, *J. Eur. Ceram. Soc.*, **34**, 991–1000 (2011).

[7] E. Zapata-Solvas, S. Bonilla, P. R. Wilshaw, R. I. Todd, Preliminary Investigation of Flash Sintering of SiC, *J. Eur. Ceram. Soc.*, **33**, 2811–6 (2013).

[8] R. Raj, Joule Heating during Flash-Sintering, *J. Eur. Ceram. Soc.*, **32**, 2293–301 (2012).

[9] K.I. Rybakov, Yu.V. Bykov, A.G. Eremeev, S.V. Egorov, V.V. Kholoptsev, A.A. Sorokin, V.E. Semenov, Microwave ultra-rapid sintering of oxide ceramics, in *Processing and properties of advanced ceramics and composites VII*, edited by M. M. Mahmoud, A. S. Bhalla, N. P. Bansal, J. P. Singh, R. Castro, N. J. Manjooran, G. Pickrell, S. Johnson, G. Brennecka, G. Singh, D. Zhu (*Ceramic Transactions*, Vol. 252). Wiley, 2015. P. 57-66.

[10] Yu.V. Bykov, S.V. Egorov, A.G. Eremeev, V.V. Kholoptsev, K.I. Rybakov, A.A. Sorokin, Flash microwave sintering of transparent Yb:$(LaY)_2O_3$ ceramics, *J. Am. Ceram. Soc.*, 2015, published online 26.08.2015. DOI: 10.1111/jace.13809.

[11] E. Karmazsin, Use of low- and high-power microwave energy for thermal analysis, *Thermochim. Acta*, **110**, 289–95 (1987).

[12] G.M.B. Parkes, G. Bond, P.A. Barnes, E.L. Charsley, Development of a new instrument for performing microwave thermal analysis, Rev. Sci. Instrum., **71**, 1968–75 (2000).

[13] S.S. Balabanov, E.M. Gavrishchuk, A.M. Kut'in, D.A. Permin, Self-propagating high-temperature synthesis of Y_2O_3 powders from $Y(NO_3)_{3x}(CH_3COO)_{3(1-x)}\cdot nH_2O$, *Inorg. Mater.*, **47**, 484–488 (2011).

[14] Yu. Bykov, A. Eremeev, M. Glyavin, V. Kholoptsev, A. Luchinin, I. Plotnikov, G. Denisov, A. Bogdashev, G. Kalynova, V. Semenov, N. Zharova, 24–84-GHz Gyrotron Systems for Technological Microwave Applications, *IEEE Trans. Plasma Sci.*, **32**, 67 – 72 (2004).

[15] M.I. Mendelson, Average grain size in polycrystalline ceramics, *J. Am. Ceram. Soc.*, **52**, 443–6 (1969).

[16] N.M. Tallan and R.W. Vest, Electrical properties and defect structure of Y_2O_3, *J. Am. Ceram. Soc.*, **49**, 401–4 (1966).

ROLES OF ELECTROMAGNETICALLY-ENHANCED FREE ENERGY ON NON-THERMAL MICROWAVE EFFECTS IN MATERIALS PROCESSING – A REVIEW AND DISCUSSION

Boon Wong, Ph.D.
Torrance, California, USA

ABSTRACT

A key requirement for the occurrence of non-thermal microwave effects in dielectrics is the use of coherent, polarized, and resonant (CPR) microwave irradiation. These effects in materials processing are envisioned to be a result of an isothermal uniform energy transfer (work) process sequence between a microwave-irradiated dielectric and its surroundings. The sequential work process starts with an application of CPR microwave irradiation on the dielectric at a given temperature, producing a high resonant electromagnetic (EM) work input, which then directly contributes to the electromagnetically-enhanced free energy (EMEFE) of the irradiated system.

This report first identifies the sources of EMEFE, followed by an analytical discussion of its roles on the occurrence of non-thermal microwave effects in materials processing. As per the analysis, EMEFE acts as (1) the driving force for isothermal reaction feasibility, and (2) the motivating force for process-kinetics enhancement. Proposed rationale also suggests that isothermal sintering-kinetics enhancement, occurring in any "wave-field-homogeneous"[Appendix] dielectric powdered system under uniform CPR microwaves, is caused by increased mobility of vacancies. Increased vacancy mobility, which usually improves the operating mass transport mechanism and enhances solid flux within the powdered system, may subsequently speed up densification – changing the process rate from a slow, high-activation to a fast, low-activation kinetics.

In addition, a plausible correlation between micro-EMEFE change and ponderomotive action in microwave processing of some ionic solids is also presented (Appendix). Nevertheless, a comprehensive microscopic characterization of the driving force, EMEFE change, and its significance on sintering-kinetics enhancement under CPR microwaves must be critically explored.

INTRODUCTION

While debates on the reality of non-thermal microwave effects in materials processing have been ongoing and widely documented in the past twenty-five years,[1] a key requirement for the occurrence of non-thermal microwave effects in a dielectric, however, has been clearly identified – use of coherent, polarized, and resonant (CPR) microwave irradiation.[2],[3]* Recently, this author proposed an understanding on non-thermal microwave effects as a result of a well-controlled, isothermal orderly-energy-transfer (work) process sequence between a microwave-irradiated dielectric system and its surroundings.[4]-[6] This sequential work process, which is best realized and presented using classical (non-quantum) physics and chemistry, is summarized as follows:

The electromagnetic (EM) work input to a dielectric system from incident CPR microwaves is believed to be tremendously augmented and maximized. This resonant EM work (energy) input, with an estimate approaching typical bond energy of solid dielectrics,[5],[6] is many orders of magnitude greater than the EM work done by non-CPR irradiation. As a result, a drastic gain in free energy of the CPR-irradiated system due to the augmented (resonant) EM work input may enormously enhance the isothermal reactivity (reaction feasibility and kinetics at a given temperature) of the system, thereby allowing occurrence of various non-thermal microwave effects during materials processing.

This report identifies and quantifies the phenomenological driving and motivating (stimulating) forces under which all the observed non-thermal microwave effects in materials processing are governed. Enhancements of process feasibility and kinetics under CPR microwave irradiation are also characterized from a classical, non-quantum perspective.

BACKGROUND – RELEVANT THERMODYNAMICS AND ELECTRODYNAMICS

According to the 1st and 2nd laws of thermodynamics, the fundamental expression accounting for the change in effective internal energy of a stationary dielectric system under an uniform EM (microwave-) field is:[4],[7]

$$dU \leq TdS - \Pi dV + \sum(\mu dN) + \sum(\gamma dA) + V\mathbf{E}\bullet d\mathbf{P} + V\mu_o\mathbf{H}\bullet d\mathbf{M} \tag{1}$$

where TdS is the entropic energy, ΠdV is the expansion work, $\sum(\mu dN)$ is the chemical work, $\sum(\gamma dA)$ is the surface work, $V\mathbf{E}\bullet d\mathbf{P} + V\mu_o\mathbf{H}\bullet d\mathbf{M}$ is the total EM work attributed to polarization and magnetization, respectively within the volume, V, of the dielectric body under uniform EM field or microwave irradiation, and $\mu_o \approx 1.26 \times 10^{-6}$ H/m, is the permeability of vacuum.

The equation in Exp. (1) applies to processes that are reversible, whereas the inequality is for irreversible (spontaneous) changes.

Thermodynamic Potential Function Involving EM Fields

Let us define an appropriate thermodynamic potential – a Gibbs-type (free energy) function, Φ, for scenarios involving EM fields as:

$$\Phi \equiv U - TS + \Pi V - \sum(\gamma A) - V\mathbf{E}\bullet\mathbf{P} - V\mu_o\mathbf{H}\bullet\mathbf{M} \tag{2}$$

Via Legendre differential transformation – differentiating Eqn. (2), then substituting Exp. (1) for dU into the subsequent differential equation, we obtain:

$$d\Phi \leq -SdT + Vd\Pi - \sum(Ad\gamma) - V\mathbf{P}\bullet d\mathbf{E} - (\mathbf{E}\bullet\mathbf{P})dV - V\mu_o\mathbf{M}\bullet d\mathbf{H} - \mu_o(\mathbf{M}\bullet\mathbf{H})dV + \sum(\mu dN) \tag{3}$$

In scenarios of condensed systems ($dV \approx 0$) at constant temperature ($dT = 0$), pressure ($d\Pi = 0$), specific surface free energy ($d\gamma = 0$), and composition ($dN = 0$), Exp. (3) becomes:

$$d\Phi]_{T,\Pi,\gamma,N} \leq - V\mathbf{P}\bullet d\mathbf{E} - V\mu_o\mathbf{M}\bullet d\mathbf{H} \tag{4}$$

Rearranging Exp. (4), we obtain:

$$d\Phi]_{T,\Pi,\gamma,N} + V\mathbf{P}\bullet d\mathbf{E} + V\mu_o\mathbf{M}\bullet d\mathbf{H} \leq 0$$

or we may simply write:

$$(dG^o + V\mathbf{P}\bullet d\mathbf{E} + V\mu_o\mathbf{M}\bullet d\mathbf{H}) \leq 0 \qquad (4a)$$

where $dG^o = d\Phi]_{T,\Pi,\gamma,N}$ is the differential change of the traditional Gibbs free energy of the system under zero EM field at constant temperature, pressure (1 atm), specific surface free energy, and composition. For condensed dielectric matter, volume change of the system becomes negligible. As a result, $dG^o = d\Phi]_{T,\Pi,\gamma,N} \approx dA^o$,[4] i.e., the differential changes of the traditional Gibbs and Helmholtz free energies of a condensed system under zero EM field are approximately identical at constant temperature, specific surface free energy, and composition.

The equation in Exp. (4a) applies to processes that are reversible (at equilibrium), whereas the inequality is for irreversible (spontaneous) changes. The **E** and **H** in the expression are the macroscopic electric field found in the dielectric system and the magnetizing field, respectively under the uniform EM field, **P** and **M** are the polarization and magnetization vectors in the dielectric system, respectively, and V is the volume of the system.

Simplifying the mathematical notation, we further assume the scenario of an isotropic dielectric body under uniform EM field. Hence replacing the scalar (dot) products $V\mathbf{P}\bullet d\mathbf{E}$ and $V\mu_o\mathbf{M}\bullet d\mathbf{H}$ by $VPdE \equiv \delta W_P$ and $V\mu_oMdH \equiv \delta W_M$, respectively in Exp. (4a), we obtain:

$$(dG^o + VPdE + V\mu_oMdH) \leq 0, \text{ or}$$

$$(dG^o + \delta W_P + \delta W_M) \leq 0 \qquad (4a')$$

where P and M are the components of the polarization and magnetization vectors in the directions of the E and H fields, respectively, δW_P and δW_M are the infinitesimal amounts of polarization and magnetization work input (as differential free energy gains) to the system, respectively.

Now, the sum of the differential changes of all free energies, i.e., $(dG^o + \delta W_P + \delta W_M)$ stated in Exp. (4a'), may adequately provide criteria determining equilibrium (= 0) and spontaneity (< 0) of isothermal processes in isotropic, homogeneous, condensed dielectric systems under uniform EM fields.

EM Energy Input to Dielectrics under Uniform Microwave Irradiation

Microwaves, like all other electromagnetic waves (EMWs), contain electromagnetic energy that is stored in the electric and magnetic fields of the waves. When a beam of microwaves interacts with a dielectric medium, energy transfers may take place between them. On the one hand, some of the electromagnetic energy of the waves may be dissipated, converting to thermal energy, ε_{Th}, due to dielectric and magnetic losses during the polarization and magnetization cycles, respectively. As a result, temperature of the dielectric rises – a phenomenon called thermal microwave effect or dielectric heating. On the other hand, other portions of the electromagnetic energy of the waves could do (non-thermal) work, W_{EM}, on the material resulting from polarization and magnetization, which may then directly increase the (isothermal) free energy of the dielectric, however, without causing a change in temperature – a phenomenon called non-thermal microwave effect (which is the focused subject of this paper). A characterization of EM energy transfers to dielectrics under microwave irradiation is briefly summarized below:

When one-dimensional, uniform plane-microwaves interact with an isotropic and homogeneous dielectric solid, the total electromagnetic energy transfer, ε_{EM}, to the dielectric body may be evaluated using the Poynting's theorem in classical electrodynamics:[8]

$$\varepsilon_{EM} = \varepsilon_E + \varepsilon_M = \int_V\int(EdP)dV + \int_V\int(\mu_o HdM)dV \quad (5)$$

where ε_{EM}, ε_E, and ε_M are the (total) electromagnetic, electric, and magnetic energies loaded into the dielectric body with a volume, V, respectively, E and H are the magnitudes of the electric and magnetizing fields of the uniform incident waves, respectively, dP and dM are the corresponding differential changes of the magnitudes of the polarization and magnetization in the dielectric system, respectively.

If the dielectric under consideration is linear, isotropic, and homogeneous (LIH) in nature, the magnitudes of its polarization, P, and magnetization, M, can then be mathematically expressed by Eqn. (6a) and Eqn. (6b):[9]

$$P = \varepsilon_o(\tilde{\varepsilon}_r - 1)E \quad (6a)$$

$$M = (\tilde{\mu}_r - 1)H \quad (6b)$$

where $\tilde{\varepsilon}_r$ and $\tilde{\mu}_r$, are the complex relative permittivity and relative permeability of the dielectric, respectively, $\varepsilon_o \approx 8.85 \times 10^{-12}$ F/m, is the permittivity of vacuum.

Mathematically, complex functions, $\tilde{\varepsilon}_r$ and $\tilde{\mu}_r$ of the dielectric may be expressed as:[10]

$$\tilde{\varepsilon}_r = (\varepsilon_r + j\varepsilon'') \quad (7a)$$

$$\tilde{\mu}_r = (\mu_r + j\mu'') \quad (7b)$$

where ε_r in Eqn. (7a) and μ_r in Eqn. (7b) are, respectively, the real parts of the complex relative permittivity, $\tilde{\varepsilon}_r$, and permeability, $\tilde{\mu}_r$; ε'' in Eqn. (7a) and μ'' in Eqn. (7b) are the dielectric loss factor and magnetic loss factor of the dielectric – the imaginary parts of $\tilde{\varepsilon}_r$ and $\tilde{\mu}_r$, respectively.

Applying Eqn. (7a) and Eqn. (7b), we substitute Eqn. (6a) for P, Eqn. (6b) for M into Eqn. (5). Upon subsequent integration, we then obtain the total electromagnetic energy transfer, ε_{EM}, to the dielectric from the incident microwaves as follows:

$$\varepsilon_{EM} = \varepsilon_E + \varepsilon_M \approx (1/2)V[\varepsilon_o(\varepsilon_r - 1 + j\varepsilon'')E^2 + \mu_o(\mu_r - 1 + j\mu'')H^2]$$
$$= (1/2)V[\varepsilon_o(\varepsilon_r - 1)E^2 + \mu_o(\mu_r - 1)H^2] + j[(1/2)V(\varepsilon_o\varepsilon''E^2 + \mu_o\mu''H^2)], \text{ or}$$

$$\varepsilon_{EM} = W_{EM} + j\varepsilon_{Th} \quad (8)$$

where $W_{EM} = W_P + W_M = (1/2)V[\varepsilon_o(\varepsilon_r - 1)E^2 + \mu_o(\mu_r - 1)H^2]$, is the non-thermal EM work done, and $\varepsilon_{Th} = {}^D\varepsilon_{Th} + {}^M\varepsilon_{Th} = (1/2)V(\varepsilon_o\varepsilon''E^2 + \mu_o\mu''H^2)$, is the total thermal dissipation (loss).

Consequently, the time-averaged (non-thermal) electromagnetic work input into the dielectric under sinusoidal microwaves may be expressed as:[11]

$$\langle W_{EM}\rangle = \langle W_P\rangle + \langle W_M\rangle = (1/4)V[\varepsilon_o(\varepsilon_r - 1)E_o^2 + \mu_o(\mu_r - 1)H_o^2] \quad (9a)$$

The time-averaged electromagnetic energy dissipated resulting in dielectric heating under sinusoidal microwaves may also be evaluated:

$$<\varepsilon_{Th}> = <^{D}\varepsilon_{Th}> + <^{M}\varepsilon_{Th}> = (1/4)V(\varepsilon_o\varepsilon''E_o^2 + \mu_o\mu''H_o^2) \quad (9b)$$

where $<W_P>$, $<W_M>$, $<^{D}\varepsilon_{Th}>$, and $<^{M}\varepsilon_{Th}>$ are the time-averaged polarization work, magnetization work, dielectric energy loss, and magnetic energy loss, respectively, E_o and H_o are the amplitudes of the electric and magnetizing fields of the incident sinusoidal waves, respectively.

In scenarios of irradiated non-magnetic dielectrics, $<W_M> = 0$ in Eqn. (9a); hence the EM work input simply is:

$$<W_{EM}> \approx <W_P> = (1/4)V[\varepsilon_o(\varepsilon_r - 1)E_o^2] \quad (10a)$$

The polarization work input, $<W_P>$, in Eqn. (10a) is also the time-averaged gain in free energy (sum of the potential and kinetic energies) by the oscillating dipoles inside the non-magnetic dielectric during each cycle of irradiation.

Since $<^{M}\varepsilon_{Th}> = 0$ during dielectric heating of non-magnetic systems, time-averaged EM energy dissipated (as thermal energy) of the incident microwaves in the dielectric is also simply:

$$<\varepsilon_{Th}> \approx <^{D}\varepsilon_{Th}> = (1/4)V(\varepsilon_o\varepsilon''E_o^2) \quad (10b)$$

Here are notes on the applications of Eqn. (10a) and Eqn. (10b) for microwave processing of non-magnetic dielectrics:

(1) Both equations hold if, and only if, the incoming EMW-field throughout the entire irradiated volume, V, of the dielectric is uniform. This scenario could be achieved when the irradiated dielectric system under study possesses a (geometric) size compatible with the effective penetration depth of the incident wave-field. The scenario also presumes absence of the structural irregularities in the irradiated system that may cause local field amplification or diminution.

(2) Only the EM energy input to non-magnetic dielectrics from incident microwaves in form of work, $<W_{EM}>$, expressed by Eqn. (10a), may contribute to the isothermal free energy gain of the system, thereby creating non-thermal microwave effects during processing. EM energy thermally dissipated, $<\varepsilon_{Th}>$, initially input from the microwave-field expressed by Eqn. (10b), will only participate to dielectric heating, i.e., raising the temperature of the system.

(3) E_o in both equations is the amplitude of the oscillating electric field E inside the dielectric per Maxwell's equations.[12]

(4) Both ε_r and ε'' in the equations are dielectric properties. Their magnitudes, however, are very strong functions of frequency resonance between the vibrating dipoles and oscillating electric field of the applied microwave irradiation.

Resonant Polarization Work under CPR Microwave Irradiation

From the discussion in last section, energy transfer via work to a non-magnetic dielectric system under uniform, coherent, and polarized microwave irradiation can be approximated by the time-averaged polarization work, $<W_P>$, whose magnitude is dominantly dictated by the real-part of the relative permittivity, ε_r, at a given (uniform) microwave-field amplitude, E_o as already quantified by Eqn. (10a). On the other hand, ε_r is a function of the field-frequency. The abrupt increment of ε_r at a resonant condition reaching a tremendously large (maximum) magnitude, ε_r^{max}, is well predicted by Lorentz's classical harmonic oscillator model[10b],[13]. Although not usually expected to occur at frequencies $< 10^{13}$ rad/s, this phenomenon has been experimentally observed in some solid-state dielectrics, e.g., GaAs, at microwave frequencies.[14a] At resonance, $\varepsilon_r^{max} >>> \underline{\varepsilon}_r$. ε_r^{max} under CPR microwave irradiation may become many orders (up to 10 to 11 orders) larger than its non-resonant counterpart, $\underline{\varepsilon}_r$.[5],[6]

The "unexpected" occurrence of ionic/covalent dielectric resonance at microwave frequencies is now comprehended as follows: under harmonic EM waves, the effective resonant frequency of an engineering ceramic dielectric is usually lower than the natural frequency of its handbook-perfect, single-crystal counterpart (typically > 10^{13} rad/s). In many occasions, the effective resonant frequencies of ionic and covalent dielectrics may even drop to a magnitude within the frequency range of RF/microwaves.[14] These tremendous decreases of resonant frequencies of ceramic dielectrics under oscillating EM fields are mainly due to: (1) the contribution of a strong Lorentz (local) electric field to the individual dipole oscillations inside the densely-packed solid media,[13b] and/or (2) bond weakening of individual dipoles located at or near to the microstructural irregularities of the ceramic dielectrics, e.g., vacancies, dislocations, grain boundaries, and internal surfaces (pores).[14b]

We may now compare resonant work, $<\hat{W}_P>$, with non-resonant work, $<W_P>$ using Exp. (11) shown below:

$$<\hat{W}_P> \approx (1/4)[V\varepsilon_o(\varepsilon_r^{max} - 1)E_o^2] >>> <W_P> \approx (1/4)[(V\varepsilon_o(\underline{\varepsilon}_r - 1)E_o^2] \quad (11)$$

Although the resonant polarization work input, $<\hat{W}_P>$, per Exp. (11) may vary with the incident microwave intensity (E_o^2), its magnitude on the other hand could simply be made tremendous via the huge ε_r^{max} contribution itself at resonance. For example: even under a relatively low E_o ($\leq$ 40 kV/m) of any CPR irradiation, the resulting $<\hat{W}_P>$ with ε_r^{max} per Exp. (11) could be comparable to typical bond energy in most ceramics .[5] Therefore, $<\hat{W}_P>$ may subsequently provide an enormous free energy gain for the system in favor to process feasibility and kinetics, regardless of the low E_o of the microwaves commonly applied in materials processing.

ELECTROMAGNETICALLY-ENHANCED FREE ENERGY (EMEFE) IN MICROWAVE PROCESSING OF DIELECTRICS

Let us now identify the sum of the (differential) changes of isothermal free energies at constant specific surface free energy and composition, initially formulated in Exp. (4a'), as the isothermal change in electromagnetically-enhanced free energy, dF, of a condensed dielectric system under uniform microwave irradiation: $dF \equiv (dG^o + \delta W_P + \delta W_M)$. We further assume that the dielectric is a non-magnetic material, i.e., $\delta W_M = 0$. According to Exp. (4a'), spontaneity (feasibility) of any isothermal process under conditions considered involving EM or microwave-field is always dictated by the reduction of EMEFE:

$$dF = dG^o + \delta W_P < 0 \quad (12)$$

The total change in EMEFE of an isothermal feasible process from its initial-reactant to final-product states according to Exp. (12) is:

$$\Delta F = {}^fF - {}^iF = ({}^fG^o + {}^fW_P) - ({}^iG^o + {}^iW_P) < 0, \text{ or}$$

$$\Delta F = ({}^fG^o - {}^iG^o) + ({}^fW_P - {}^iW_P) \approx \Delta G^o + (<{}^fW_P> - <{}^iW_P>) < 0 \quad (13)$$

where the superscripts i and f denote the energy quantities at their initial-reactant and final-product states of the process, respectively, $\Delta G^o = ({}^fG^o - {}^iG^o)$ is the process free energy change at a given temperature under zero EM (microwave-) field, $<W_P>$ is the free energy gain in the system through the time-averaged polarization work input.

Assuming that the dielectric properties of the reactants and products of the process are LIH but, their compositions and/or structures are different, we may then approximate the polarization work under sinusoidal irradiation at the initial and final states, respectively as follows:

$$ {}^{i}W_P = (1/2)[V\varepsilon_o({}^{i}\varepsilon_r - 1){}^{i}E^2] \approx (1/4)[V\varepsilon_o({}^{i}\underline{\varepsilon}_r - 1){}^{i}E_o^2] = <{}^{i}W_P> \quad (13a)$$

$$ {}^{f}W_P = (1/2)[V\varepsilon_o({}^{f}\varepsilon_r - 1){}^{f}E^2] \approx (1/4)[(V\varepsilon_o({}^{f}\underline{\varepsilon}_r - 1){}^{f}E_o^2] = <{}^{f}W_P> \quad (13b)$$

For scenarios under which uniform CPR microwaves are applied, the change in EMEFE of the process, ΔF, in a non-magnetic dielectric system,** first stated in Exp. (13), may further be expressed as:

$$\Delta F \approx [\Delta G^o + (<{}^{f}\hat{W}_P> - <{}^{i}\hat{W}_P>)] < 0 \quad (14)$$

where $<{}^{i}\hat{W}_P> = (1/4)[V\varepsilon_o({}^{i}\varepsilon_r^{max} - 1){}^{i}E_o^2]$ and $<{}^{f}\hat{W}_P> = (1/4)[V\varepsilon_o({}^{f}\varepsilon_r^{max} - 1){}^{f}E_o^2]$ – functions of both resonant relative permittivity, ε_r^{max}, and irradiation intensity (power), E_o^2, are the time-averaged free energy gains in the system through resonant polarization work input at the initial-reactant and final-product states during the microwave-enhanced process, respectively. Also note that the resonant $<\hat{W}_P>$ given in Exp. (14) could be orders of magnitude greater than the non-resonant $<W_P>$ given in Exp. (13), as initially indicated by Exp. (11).

FEASIBILITY AND KINETICS OF MICROWAVE-ENHANCED PROCESSES

From the point of view of thermodynamics, isothermal feasibility of a microwave-enhanced process is governed by the change in EMEFE of a system during the process. Exp. (14) demonstrates that the reaction feasibility of an isothermal interaction between a non-magnetic dielectric and its surroundings under resonant microwaves is dictated by ΔF of the process, which in turn is highly controlled by the contribution from the net resonant polarization work input, $(<{}^{f}\hat{W}_P> - <{}^{i}\hat{W}_P>)$, to the system. Consequently, lowering EMEFE of a system during any microwave-enhanced process, i.e., $\Delta F < 0$, as illustrated in Fig. 1, is considered to be the driving force (DF) for isothermal feasibility of the process.

Yet, ΔF itself is usually quite "silent" about the kinetics of any process. The rate and mechanism of a process and the way by which its rate can be modified, on the other hand, may be understood and quantified by classical kinetics.

Process (Reaction) Kinetics -- The Basics

In manufacturing, many materials processes that we encounter are chemical reaction-controlled and interfacial diffusion-controlled types. Their process kinetics may be formulated using classical kinetic theory and Boltzmann's molecular energy distribution law.[6],[13a],[15]

Accordingly, the net-forward rate, $\Re$, of any of these processes is:

$$\Re \sim \exp[-Æ/(RT)]\{1 - \exp[\Delta F/(RT)]\} \quad (15)$$

where ~ is a mathematical symbol denoting "proportional to", R is the gas constant, T is the

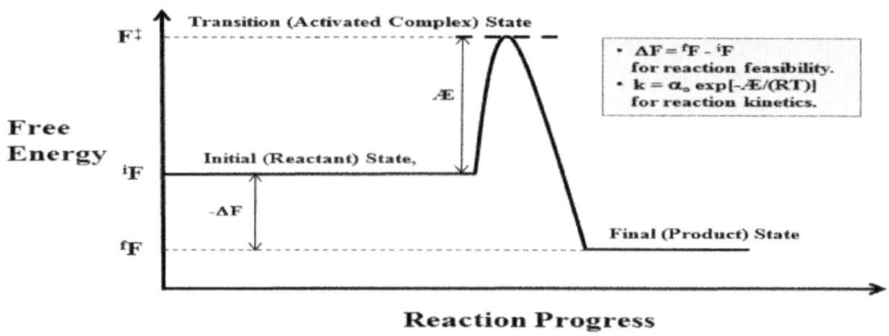

Figure 1. Process (reaction) feasibility and kinetics.

thermodynamic temperature, Æ and ΔF are the activation free energy required for the rate-controlling mechanism of the process and the total change in free energy (driving force) for the process going from the initial-reactant state to the final-product state, respectively, as shown in Fig. 1. As long as the process is thermodynamically feasible, $\Delta F < 0$, the process rate, $\mathfrak{R}$, in Exp. (15) is always positive and finite.

In many (chemical) reaction-controlled processes, $|\Delta F| >> RT$; $\{1 - \exp[\Delta F/(RT)]\}$ in Exp. (15) approaches unity. Under this circumstance, the isothermal reaction-controlled-process rate is mainly dictated by the activation free energy, Æ, as shown in Exp. (15a):

$$^{R}\mathfrak{R} \sim \exp[-Æ/(RT)] \tag{15a}$$

Exp. (15a) is actually the statement of the well-known Arrhenius-Eyring relationship and a key outcome of the transition-state theory.[15a],[15b],[16] Arrhenius-Eyring relationship, originally formulated for chemical reaction kinetics, states that for a given reactant composition (concentration), the forward rate of a reaction is governed by a (specific) rate constant, k, i.e., $^{R}\mathfrak{R} \sim k$ where k is dictated by the rate-controlling mechanism (path) and is an exponential function of temperature. Therefore,

$$^{R}\mathfrak{R} \sim k = \alpha_o \exp[-Æ/(RT)] \tag{16}$$

where α_o is the (thermal) frequency factor – $\alpha_o \approx 10^{13}$ Hz for the temperature range under consideration, R is the gas constant and T is the thermodynamic temperature.

On the other hand, in solid-state (interfacial) diffusion-controlled processes, sintering included, $|\Delta F| << RT$. The term $\{1 - \exp[\Delta F/(RT)]\} \approx 1 - [1 + \Delta F/(RT)] = -\Delta F/(RT)$ in Exp. (15). Hence the diffusion-controlled-process rate becomes:

$$^{D}\mathfrak{R} \sim \exp[-Æ/(RT)][-\Delta F/(RT)] \tag{15b}$$

or via Exp. (16):

$$^{D}\mathfrak{R} \sim k[-\Delta F/(RT)] = \{\alpha_o \exp[-Æ/(RT)]\}[-\Delta F/(RT)] \tag{17}$$

Under this circumstance, the rate of an isothermal diffusion-controlled process, $^{D}\mathfrak{R}$, per Exp. (17) is still dominantly governed by the rate constant, k, (or the mobility[Appendix]) due to its exponential dependence on Æ of the rate-controlling mechanism. Yet, the driving force, ΔF, e.g., a total surface energy reduction or a ponderomotive force induced by a local wave-field on interface-charged species in an ionic ceramic during sintering,[17],[Appendix] which linearly governs the rate, may also affect the process kinetics, however to a much less degree as expected.

Contribution of Resonant EM Work on Activation Free Energy and Rate Constant

Activation free energy, Æ, as per the classical theory of kinetics, is a free-energy barrier for the process to overcome in order to proceed in time. Therefore from this point of view, reactants can convert into products in time during the process if, and only if, the properly aligned reacting molecules acquire sufficient energy from an external field, e.g., an EM (microwave-) field, to overcome the free-energy barrier, Æ, for the reaction. Under purely thermal (zero-microwave-field) conditions, the traditional form of Arrhenius-Eyring relationship holds:

$$k^{o} = \alpha_{o}\exp[-Æ^{o}/(RT)] \quad (18a)$$

where k^{o} and $Æ^{o}$ are the rate constant and activation free energy under purely thermal (zero-microwave-field) conditions, respectively.

On the other hand, under CPR microwave irradiation within the same temperature range, a modified rate constant, k^{EM}, holds:***

$$k^{EM} = \alpha_{o}\exp[-(Æ^{o} - \langle {}^{i}\hat{W}_{P}\rangle)/(RT)] = \alpha_{o}\exp[-Æ^{\S}/(RT)] \quad (18b)$$

where $Æ^{\S} = (Æ^{o} - \langle {}^{i}\hat{W}_{P}\rangle)$, is the (effective) activation free energy under both thermal and CPR-irradiation conditions, $\langle {}^{i}\hat{W}_{P}\rangle$ is the time-averaged, resonant polarization work input to the initial-state reactants under irradiation.[6],**** Comparing Eqn. (18b) to Eqn. (18a), the significant reduction of the activation free energy from $Æ^{o}$ to $Æ^{\S}$ due to the contribution of $\langle {}^{i}\hat{W}_{P}\rangle$ under resonant irradiation may result in $k^{EM} >> k^{o}$. Consequently, the rate of every thermodynamically feasible ($\Delta F < 0$) process at a given temperature could be highly enhanced under CPR irradiation.

ROLES OF EMEFE ON NON-THERMAL MICROWAVE EFFECTS UNDER CPR IRRADIATION

Significance of non-thermal microwave effects in sintering has been recently reviewed and demonstrated.[17] The occurrence of these effects, on the other hand, is believed to be highly dictated by the EMEFE of any system during processing.[4]-[6] The roles of EMEFE on two important non-thermal microwave effects in materials processing are discussed below:

EMEFE – The Driving Force for Reaction-Feasibility Promotion

In the reaction shown in Fig. 2, the reactants at the initial system state are assumed to be highly polarizable thereby resulting in a tremendous resonant $\langle {}^{i}\hat{W}_{P}\rangle$ under CPR irradiation. Conversely, the products at the final system state are assumed to be relatively non-polarizable, thus producing negligible $\langle {}^{f}\hat{W}_{P}\rangle$ under irradiation. The incredible $\langle {}^{i}\hat{W}_{P}\rangle$ input radically increases the EMEFE of the initial-reactant state from ${}^{i}G^{o}$ to ${}^{i}F$ thereby assuring a favorable EMEFE change (driving force) for the forward reaction, i.e., $\Delta F < 0$ as per Exp. (14). Consequently, tremendous resonant $\langle {}^{i}\hat{W}_{P}\rangle$ input to reactants at the initial state of the process under

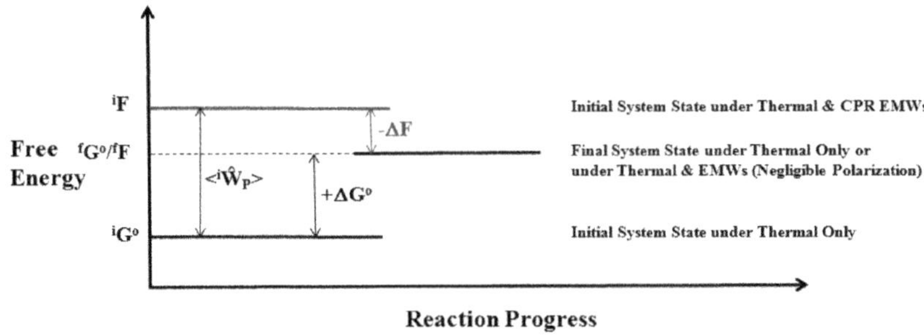

Purely Thermal	**Thermal + EM**
$\Delta G^o > 0$, reaction not feasible	**$\Delta F < 0$, reaction feasible**

Figure 2. Promotion of reaction feasibility under CPR microwave irradiation. The final-state free energy remains essentially unchanged during the process (${}^{f}G^o \approx {}^{f}F$) due to the insignificant polarizability of the products ($<{}^{f}\hat{W}_P>$ is negligible) under the irradiation.

CPR microwaves helps attain the forward-process feasibility. In many occasions, $<{}^{i}\hat{W}_P>$ may even promote a thermally non-feasible reaction ($\Delta G^o > 0$) to an energetically favorable one under irradiation ($\Delta F < 0$), producing a non-thermal microwave effect as illustrated in Fig. 2.

This type of non-thermal microwave effect is frequently observed and has been applied to technology of solid-state phase transformations in the industry: the production of thermally-unstable high-temperature products at relatively low temperatures. A classic example is the formation of the high-temperature hexagonal $BaTiO_3$ phase under microwave irradiation at as low as 300 °C.[2a],[18] This microwave effect may also dictate results in crystalline-to-amorphous phase transformations at temperatures below melting under well-controlled irradiation.

EMEFE – The Motivating Force for Process-Kinetics Enhancement

Fig. 3 depicts a thermodynamically feasible process even under a purely thermal condition ($\Delta G^o < 0$). Nevertheless, its relatively high activation free energy, $Æ^o$, which produces a very low rate constant k^o as per the Arrhenius-Eyring relationship – Eqn. (18a), results in a very slow process under this purely thermal condition.

Although the scenario illustrated in Fig. 3 assumes both reactants (at the initial state) and products (at the final state) are equally polarizable,***** i.e., $<{}^{i}\hat{W}_P> = <{}^{f}\hat{W}_P>$, a significant promotion of the initial free energy of the system due to the high $<{}^{i}\hat{W}_P>$ input under CPR irradiation**** may remarkably reduce the free energy difference between the transition and initial states thereby considerably lowering the activation free energy of the forward process from $Æ^o = (F^{\ddagger} - {}^{i}G^o)$ to $Æ^{\S} = (F^{\ddagger} - {}^{i}F)$. As a result, $k^{EM} >> k^o$ as predicted – Eqn. (18b) for k^{EM} and Eqn. (18a) for k^o. Under this circumstance, the isothermal kinetics of the CPR-irradiated process may become astounding.

There are two common kinds of process-kinetics motivations attributed to the above microwave effect documented in the literature:

(1) Motivation of (chemical) reaction-controlled process kinetics. Example: the rate enhancement on reduction reaction of FeO to γFe under irradiation.[3] The reduction rate enhancement in this example is attributed to the tremendous contribution of the resonant EM work input to the free energy of the reactants, $<{}^{i}\hat{W}_P>$, which then remarkably reduces the activation free energy

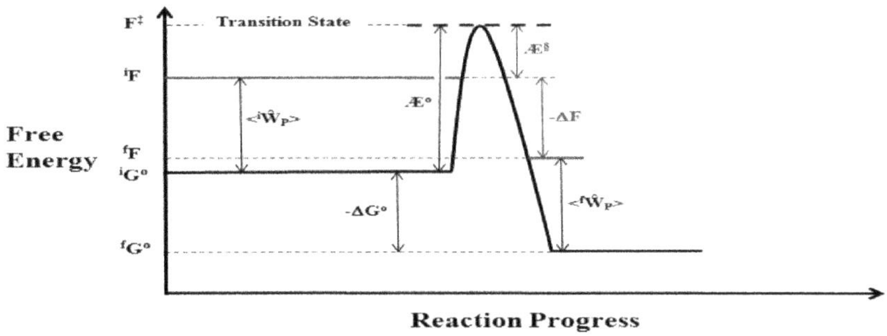

Purely Thermal		Thermal + EM
$\Delta G^o < 0$, feasible		$\Delta F < 0$, feasible
	$-\Delta G^o = -\Delta F$, since $\langle {}^i\hat{W}_P \rangle = \langle {}^f\hat{W}_P \rangle$	
$Æ^o$	$>$	$Æ^§$
k^o	$<<$	k^{EM}
Slow rate		Rapid rate

Figure 3. Process-kinetics enhancement under CPR microwave irradiation. This scenario illustrates that the kinetics of both the irradiated and the purely thermal processes take place under identical driving forces ($-\Delta F = -\Delta G^o$).

(Æ) from a high $Æ^o$ to a low $Æ^§$ (Fig. 3). As a result, the rate constant $k^{EM} >> k^o$; process kinetics becomes extraordinary. Reduction of activation free energy in this enhanced kinetics is envisioned to result from improving geometric alignment and stimulating collision energy of the involved reactant molecules by irradiation, augmenting activation entropy while reducing activation enthalpy.[4],[6] Hence the observed irradiation-enhanced FeO reduction reaction is the result of a mechanism change of the rate-controlling step of the reaction.
(2) Motivation of (interfacial) diffusion-controlled process kinetics. Examples include the irradiation-enhanced sintering-related processes, e.g., the densification and grain growth rate augmentations in alumina and other engineering ceramics.[19]-[23]

During CPR-wave sintering of a "wave-field-homogeneous"[Appendix] dielectric powder-compact, the tremendous contribution of the resonant EM/polarization work, $\langle {}^i\hat{W}_P \rangle$, again significantly lowers the activation free energy (Æ) of a purely thermal process from $Æ^o$ to $Æ^§$, while maintaining the same sintering driving force (total surface free energy reduction) as that for the process under purely thermal condition, $\Delta F = \Delta G^o < 0$ (Fig. 3). Therefore, the drastic process-rate enhancement in this scenario, according to Exp. (17) and Exp. (18b), is solely caused by a modification of the rate-controlling mass transport mechanism under uniform microwave irradiation. From an atomistic point of view,[24] one may envision this irradiation-induced change of the rate-controlling mass transport mechanism in sintering as follows: Microscopically, atomic (ionic) vacancy diffusion is widely accepted as the dominant rate-controlling mass transport mechanism in most ceramic sintering processes.[25] During sintering under CPR irradiation, tremendous instantaneous $\langle {}^i\hat{W}_P \rangle$ input may "open up" the atomically dense-packed, periodic bulk structure of individual grains in the powder-compact, extensively turning them into an atomically looser-packed, relatively random "grain-boundary-like" structure. Energetically, one may then visualize the kinetics enhancement as follows: the resonant work, $\langle {}^i\hat{W}_P \rangle$, significantly reduces the activation free energy of the purely-thermal process, $Æ^o$, via reductions of both free energies of vacancy formation and migration, hence causing a considerable increase in vacancy mobility for the mass transport in the irradiated solid. As a result, the use of CPR irradiation may improve the rate-controlling mass transport mechanism in sintering, i.e., changing it from a slow, high-Æ vacancy diffusion process through a closely-packed, vacancy-poor lattice structure (under purely-thermal condition) to a fast, low-Æ vacancy transport process via a relatively open, vacancy-rich "grain-boundary-like" structure (under irradiation). Subsequently, astonishing enhancement of the sintering rate may occur.

CONCLUSIONS

Non-thermal microwave effects are real physical phenomena. A key requirement for their occurrence in irradiated dielectric materials during processing is the application of CPR irradiation of microwaves. These effects are envisioned to be a result of a well-controlled, isothermal work-process sequence between an irradiated dielectric system and its surroundings. The sequential work process starts with an application of CPR microwave irradiation on the system at a given temperature, producing an incredibly high EM work input, which then directly contributes to the EMEFE of the irradiated dielectric system.

EMEFE plays at least two critical roles on the occurrence of non-thermal microwave effects: (1) its change in a process, ΔF, as expressed by Exp. (14) and illustrated in Fig. 2, is the driving force determining reaction feasibility, and (2) its magnitude at initial-reactant state of a process, ${}^{i}F = {}^{i}G^{o} + \langle {}^{i}\hat{W}_{P} \rangle$, governing $Æ^{\S}$ and thus k^{EM} as expressed by Exp. (18b) and demonstrated in Fig. 3, is the motivating force for process-kinetics enhancement. Under CPR irradiation, tremendous gain in initial-state EMEFE of an irradiated system via remarkable resonant $\langle {}^{i}\hat{W}_{P} \rangle$ input may dominantly control the isothermal reactivity of the system, thereby accounting for non-thermal microwave effects such as promotion of process feasibility.

In addition, a tremendous increase of the initial-state EMEFE of an irradiated system via a high resonant $\langle {}^{i}\hat{W}_{P} \rangle$ input may also cause a significant reduction of activation free energy (Æ) for mobility (rate constant), which then enhances the isothermal process kinetics. Because resonant $\langle {}^{i}\hat{W}_{P} \rangle$ input, as per Exp. (14), is a function of both resonant relative permittivity and irradiation intensity, in addition to the impact from incredibly large resonant relative permittivity, increasing power of the incident CPR-waves may also raise the initial-state EMEFE of the system, thereby lowering Æ for mobility (rate constant), and eventually enhancing kinetics of the irradiated process, as experimentally observed and reported in the literature.[17],[21]

According to current rationale, enhanced sintering kinetics of a "wave-field-homogeneous"[Appendix] ceramic powder-compact under uniform CPR microwaves are understood to result from significantly improved process-rate-controlling vacancy mobility in the irradiated body, modifying the operating mass transport mechanism from a slow, high-activation to a fast, low-activation path.

A plausible correlation between micro-EMEFE change and ponderomotive action in microwave processing of ionic ceramics is presented in the Appendix of this paper. For the near future, however, a thorough microscopic characterization of the driving force, EMEFE change, and its significance on microwave-sintering-kinetics enhancement must be critically investigated.

FOOTNOTES

*Coherent (C): incident waves are monochromatic and single-phase. Polarized (P): the directions of E-field oscillations of the microwaves are uniformly perpendicular to the axis of propagation. Resonant (R): the field frequency of the applied microwave irradiation is within the effective resonant frequency range of the oscillating dipoles in the irradiated dielectric. CPR irradiation may be found, for example, in a single-mode resonant cavity of standing microwaves whose E and H fields are polarized and independently localize at specific zones in the cavity.

**Following a similar procedure, Exp. (4a') may also be applied to obtain an expression analogous to Exp. (14) for the change of EMEFE that dictates process spontaneity in a magnetic system. However, magnetic resonance under CPR microwaves, typically governed by the resonance of unpaired d- or f-orbital electron spins in the magnetic material, is by and large considered to be a non-classical, quantum phenomenon. Therefore, characterization of the maximum relative permeability, μ_r^{max}, at the resonant frequency as well as subsequent evaluation of the significance of resonant magnetization work, $<\hat{W}_M>$, to EMEFE in a magnetic solid under CPR irradiation are beyond the scope of current discussion.

***In Eqn. (18b), α_o remains as the thermally-induced molecular vibratory frequency factor ($\approx 10^{13}$ Hz). The applied microwave frequencies during microwave processing are usually orders of magnitude lower, i.e., $< 10^{11}$ Hz. Therefore, molecular (ionic) collision frequency during processing under microwave irradiation is expected to be the same as that under purely thermal condition, presumably both processes are conducted at the same temperature.

****Reminder: $<^i\hat{W}_P>$ is a function of both resonant relative permittivity ($^i\varepsilon_r^{max}$) and applied microwave intensity ($^iE_o^2$) at the initial-reactant state of the irradiated system – Exp. (14).

*****During microwave sintering of a simple oxide, e.g., alumina, both the composition and crystallographic structure of the dielectric are expected to remain unchanged. Hence the polarizabilities of the initial-reactant and final-product bodies are equal during the entire process.

APPENDIX: MICRO-EMEFE CHANGE ON MICROWAVE SINTERING KINETICS – ITS PLAUSIBLE CORRELATION WITH MICROWAVE-INDUCED PONDEROMOTIVE FORCES IN SOLID-STATE IONIC PLASMAS

In any (interfacial) diffusion-controlled process such as sintering, the rate per classical kinetic theory presented in the main text is:

$$^{D}\mathfrak{R} \sim \{\alpha_o \exp[-\text{Æ}/(RT)]\}[-\Delta F/(RT)] \qquad (17)$$

where ~ is a mathematical symbol denoting "proportional to", $\{\alpha_o \exp[-\text{Æ}/(RT)]\}$is a kinetic factor and $[-\Delta F/(RT)]$ is a thermodynamic factor.

Assuming that molar EMEFE change over a vacancy jump distance, a, in x-direction during vacancy diffusion in microwave sintering is: $(\Delta F/n) = (a/n)(dF/dx)$, where n is the number of moles. The thermodynamic factor in Exp. (17) then becomes:

$$[-\Delta F/(RT)] = -[a/(RT)](dF/dx) \qquad (A1)$$

On the other hand, according to thermodynamics discussed earlier in this report, EMEFE change during isothermal sintering of a non-magnetic, isotropic, linear dielectric powder-compact under sinusoidal CPR microwaves may be expressed as follows:

$$dF = -[\Sigma(\gamma dA) + VEdP] = -[\Sigma(\gamma dA) + <V\varepsilon_o(\varepsilon_r^{max} - 1)EdE>] \qquad (A2)$$

where $\Sigma(\gamma dA)$ is the differential change in surface energy and $<V\varepsilon_o(\varepsilon_r^{max} - 1)EdE>$ denotes the time-average of the change in resonant polarization (EM) energy.

Substituting Exp. (A2) for dF into Exp. (A1), followed by presuming $\varepsilon_r^{max} >>> 1$ at a resonant condition, we then obtain:

$$[-\Delta F/(RT)] = -[a/(RT)](dF/dx) = [a/(RT)]\{\Sigma[\gamma(dA/dx)] + <0.5V\varepsilon_o\varepsilon_r^{max}(dE^2/dx)>\} \qquad (A3)$$

Hence the mass flux, J_{mass}, resulting from the vacancy-transport-process rate is:

$$J_{mass} = -J_{vacancy} = -Cv \approx -C(a^D\Re) \sim -Ca\{\alpha_o\exp[-Æ/(RT)]\}[-\Delta F/(RT)] \qquad (A4)$$

where C is the local vacancy concentration at the unit area of consideration and v is the average velocity of diffusing vacancies.

Substituting Exp. (A3) for $[-\Delta F/(RT)]$ into Exp. (A4), followed by replacing the resonant permittivity $\varepsilon_o\varepsilon_r^{max}$ with ε^{max}, we obtain:

$$J_{mass} \sim -C\{(RT)^{-1}(a^2\alpha_o)\exp[-Æ/(RT)]\}\{\Sigma[\gamma(dA/dx)] + <0.5V\varepsilon^{max}(dE^2/dx)>\} \qquad (A5)$$

Exp. (A5) suggests that besides vacancy concentration, the mass flux in any microwave-enhanced sintering process is governed by (1) the mobility, $M \sim (RT)^{-1}(a^2\alpha_o)\exp[-Æ/(RT)]$ and (2) the driving force, $DF = \Sigma[\gamma(dA/dx)] + <0.5V\varepsilon^{max}(dE^2/dx)> = \delta W_S/dx + <\delta\hat{W}_P/dx>$. Hence DF for microwave-enhanced sintering is a combination of two components: the surface energy gradient and the time-averaged resonant polarization energy gradient, respectively.

For the scenario when all dipoles anywhere in a "wave-field-homogeneous" dielectric powder-compact are able to perform uniform polarization during microwave sintering, $<\delta\hat{W}_P/dx>$ within the compact is basically non-existent. Therefore, DF in Exp. (A5) is simply the surface energy gradient, $\delta W_S/dx$, which is the same as that of conventional sintering. This scenario has been assumed and applied in discussions throughout the main text of this paper. On the other hand, in some microwave-enhanced sintering processes of ceramics, local amplification/intensification of the electric field of the incident waves within microstructure[26],[27] may cause the highest resonant $<\hat{W}_P>$ input, hence the highest EMEFE gain to occur in the vicinity of the relatively defective neck-surfaces between grains in the powder-compact. Under this circumstance, the local non-uniform wave-fields in the microstructure may create micro-EMEFE changes (differences) or time-averaged resonant polarization energy gradients between the neck-surfaces and elsewhere along their neighboring grain boundaries thereby providing an additional component of the driving force for the mass flux in sintering as stated in Exp. (A5). Accordingly, the augmented DF could then increase the mass flux (vacancy flow in opposite direction) within the powder-compact, speeding up the reduction of the free surfaces (pores) surrounded by grains. As a result, densification may significantly be enhanced (see attached illustration below).

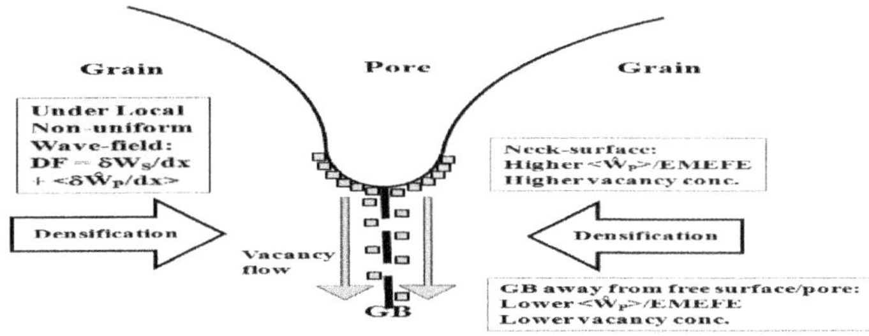

One may also observe that the micro-EMEFE-derived component of the driving force in microwave-enhanced sintering, $<\delta\hat{W}_P/dx> = <0.5\varepsilon^{max}(dE^2/dx)(V)>$, has a similarity with the (time-averaged) microwave-induced ponderomotive force acting on mobile ions to enhance solid-state ionic diffusion, i.e., PMF ~ $<0.5\varepsilon(dE^2/dx)>$, as proposed by Booske et al..[27] Nonetheless, a critical study is required to microscopically characterize and evaluate this additional DF component in comparison to the conventional one contributed from the surface energy reduction. Moreover, the significance of the combined DF, contributed from both sources, on subsequent sintering-kinetics enhancement under irradiation must also be properly identified and quantified.

REFERENCES

[1] Examples of documents are:
[1a] *Proceedings of Symposium on Microwaves: Theory and Application in Materials Processing*, D. E. Clark, F. D. Gac, and W. H. Sutton (eds.), *Ceramic Transactions*, Vol. 21, ACerS, Westerville, OH (1991).
[1b] *Proceedings of the 2nd Global Congress on Microwave Energy Applications*, R. L. Schulz and D. C. Folz (eds.), 48HrBooks, USA (2013).
[2] *Conference Proceedings, MS&T'09*, Pittsburgh, PA (2009).
[2a] R. Roy and M. L. Rao, The Birth of a New Field of Materials Science: Resonant Polarized Radiation Interactions with Matter, 563-72.
[2b] J. Brooks, Einstein's Hidden Variable: Part B – The Resonance Factor, 585-96.
[2c] B. C. Towe, Induced Ultra-high Frequency Ultrasonic Vibration as the Driving Force for Reported Sub-thermal Microwave Effects on Materials, 597-607.
[2d] K. I. Rybakov et al., Effects of Intensity and Polarization of Microwave Field in High-temperature Processing of Nanostructured Materials, 632-43.
[2e] M. L. Rao et al., Dramatic Structuring of Liquid Water using Polarized Microwave, Radiofrequency Radiation, and Crystal-induced Epitaxy, 655-66.
[3] M. Sato et al., The Microwave Technologies and Activities in Japan, presented at MS&T'11, Columbus, OH (2011).
[4] B. Wong, Roles of Electromagnetic Fields on Materials Processing and Performance – A Thermodynamic and Kinetic Perspective, presented at MS&T'11, Columbus, OH (2011), also in *Ceramic Transactions*, ACerS, Vol. 234, 151-63, John Wiley & Sons, New Jersey (2012).
[5] B. Wong, Non-thermal Microwave/RF Effects on Materials Processing – A Phenomenological Perspective, presented at 2GCMEA, Long Beach, CA (2012), also in *Proceedings of the 2nd Global Congress on Microwave Energy Applications*, R. L. Schulz and D. C. Folz (eds.), 204-18, 48HrBooks, USA (2013).

[6] B. Wong, Understanding Non-thermal Microwave Effects in Materials Processing – A Classical Non-quantum Approach, presented at MS&T'13, Montreal, Quebec (2013), also in *Ceramic Transactions,* ACerS, Vol. 249, 329-38, John Wiley & Sons, New Jersey (2014).
[7] See texts/articles such as:
[7a] E. A. Guggenheim, *Thermodynamics,* 4th edition, North-Holland, Amsterdam (1959).
[7b] C. J. Adkins, *Equilibrium Thermodynamics,* 3rd edition, Cambridge University Press, Cambridge (1983).
[7c] J. M. Honig, *Thermodynamics,* 3rd edition, Academic Press, Amsterdam (2007).
[7d] J. M. Honig and L. L. Van Zandt, Thermodynamic Characterization of Electromagnetic Fields, *J. Franklin Inst.* Vol. 323, No. 3, 297-343 (1987).
[7e] R. P. Lungu, Thermodynamic of Electric and Magnetic Systems, *Trend in Electromagnetism – From Fundamentals to Applications,* V. Barsan (ed.), InTech Europe, Rijeka, Croatia (2012).
[8] Refer to texts:
[8a] I. S. Grant and W. R. Phillips, *Electromagnetism,* 2nd edition, John Wiley & Sons, Chichester, England (1990).
[8b] J. F. Nye, *Physical Properties of Crystals,* Clarendon Press, Oxford, UK (1957).
[8c] D. J. Griffiths, *Introduction to Electrodynamics,* 3rd edition, Pearson Prentice-Hall, New Jersey (1999).
[9] S. O. Kasap, *Principles of Electronic Materials and Devices,* 3rd edition, McGraw-Hill, New York (2006).
[10] See texts:
[10a] A. von Hippel, *Dielectrics and Waves,* Artech House, Boston (1954).
[10b] N. Braithwaite, *Electromagnetism – Electromagnetic Waves,* Book 3, The Open University, Latimer Trend and Company, United Kingdom (2006).
[11] R. E. Collin, *Foundations for Microwave Engineering,* McGraw-Hill, New York (1966).
[12] C. Kittel, *Introduction to Solid State Physics,* Ch.12, 3rd edition, John Wiley & Sons, New York, NY (1966).
[13] See texts:
[13a] R. P. Feynman, R. B. Leighton, and M. Sands, *The Feynman Lectures on Physics,* Addison-Wesley, Reading, MA (1963-64).
[13b] A. J. Dekker, *Solid State Physics,* Prentice-Hall, Englewood Cliffs, NJ (1957).
[13c] A. J. Dekker, *Electrical Engineering Materials,* Prentice-Hall, Englewood Cliffs, NJ (1959).
[13d] G. Bekefi and A. H. Barrett, *Electromagnetic Vibrations, Waves, and Radiation,* The MIT Press, Cambridge, MA (1977).
[14] Resonance of solid-state dielectrics under microwaves may be found in:
[14a] R. D. Larrabee and W. A. Hicinbothem, Jr., Resonance Behavior of the Dielectric Constant of GaAs at Microwave Frequencies, *Applied Physics Letters,* 10, [12], 334-36 (1967).
[14b] J. H. Booske, R. F. Cooper, and I. Dobson, Mechanisms for Nonthermal Effects on Ionic Mobility during Microwave Processing of Crystalline Solids, *J. Mater. Res.,* 7, 495-501 (1992).
[15] Refer to chapters in:
[15a] G. M. Barrow, *Physical Chemistry,* 2nd edition, McGraw-Hill, New York (1966).
[15b] J. L. Latham and A. E. Burgess, *Elementary Reaction Kinetics,* 3rd edition, Butterworths, London (1977).
[15c] L. Zang, *Lecture Notes on Kinetics,* Department of Materials Science and Engineering, University of Utah, Spring (2015).
[16] K. J. Laidler and M. C. King, The Development of Transition-state Theory, *J. Phys. Chem.,* 87, 2657-64 (1983).

[17] K. I. Rybakov, E. A. Olevsky, and E. V. Krikun, Microwave Sintering: Fundamentals and Modeling, *J. Am. Ceram. Soc.*, 96, [4], 1003-20 (2013).
[18] D. Agrawal, Microwave Sintering of Ceramics, Composites, Metals, and Transparent Materials, *Journal of Materials Education*, Vol. 19 (4, 5 & 6), 49-57 (1997).
[19] K. H. Brosnan, G. L. Messing, and D. K. Agrawal, Microwave Sintering of Alumina at 2.45 GHz, *J. Am. Ceram. Soc.*, 86, [8], 1307-12 (2003).
[20] M. A. Janney and H. D. Kimrey, Diffusion-controlled Processes in Microwave-fired Oxide Ceramics, *MRS Proceedings: Microwave Processing of Materials II*, Vol. 189, 215-27, San Francisco, CA (1991).
[21] R. R. Thridandapani, The Effect of Microwave Energy on Sintering, *Ph.D. Thesis*, Virginia Polytechnic Institute and State University, Blacksburg, VA (2011).
[22] M. A. Janney, H. D. Kimrey, M. A. Schmidt, and J. O. Kiggans, Grain Growth in Microwave-annealed Alumina, *J. Am. Ceram. Soc.*, 74, 1675-81 (1991).
[23] K. Hirao, M. I. Jones, M. E. Brito, and M. Toriyama, Microwave Sintering of Silicon Nitride Ceramics, *Advances in Microwave and Radio Frequency Processing*, 533-40, Springer (2006).
[24] D. S. Wilkinson, *Mass Transport in Solids and Liquids*, Cambridge University Press, Cambridge, UK (2000).
[25] S. Somiya and Y. Moriyoshi, *Sintering – Key Papers*, Elsevier Science Publishers, Essex, England, UK (1990).
[26] A. Birnboim, J. P. Calame, and Y. Carmel, Microfocusing and Polarization Effects in Spherical Neck Ceramic Microstructures during Microwave Processing, *Journal of Applied Physics*, 85, 478-82 (1999).
[27] J. H. Booske, R. F. Cooper, S. A. Freeman, K. I. Rybakov, and V. E. Semenov, Microwave Ponderomotive Forces in Solid-state Ionic Plasmas, *Phys. Plasmas*, 5, 1664-70 (1998).

THERMAL STABILITY OF ELECTROMAGNETIC COMPRESSED FL-5305 PM PARTS

Daudi R. Waryoba
Penn State University, 1 College Place, DuBois, PA 15801

ABSTRACT

Powder metallurgy (PM) part manufacturers are increasingly faced with the demand for tighter dimensional tolerances and better mechanical properties. Due to the generic propensity of pores in the PM parts, the heat treatment process often creates considerable distortion, dimensional change, and residual stresses in these parts. This work examines the use of electromagnetic compression/processing for achieving better dimensional control and thermal stability. The investigation was carried out on sinter hardened FL-5305 PM steel, and results show reduced dimensional changes on electromagnetic compressed parts compared to as-sintered parts.

INTRODUCTION

Heat treatment of powder metal (PM) components creates considerable changes in their mechanical properties, dimensions, residual stresses, and metallurgical phase composition. The goal of heat treatment process is to enhance the characteristics of the metal so that the components meet pre-specified quality assurance criteria. For the production of good quality parts, parts manufacturers have to ensure that heat treated parts conform/meet desired microstructure, hardness, and strength level without undergoing cracking, distortion, or excessive dimensional changes. Dimensional changes may also arise from phase transformations during heat treatment that result from volumetric changes inherently associated with the crystal structure of parent and product phases during the phase transformations in the material [1].

In general, high dense materials undergo minimal dimensional change during heat treatment. Demand for high density and superior mechanical properties is one of the most important focal areas in powder metallurgy (PM) industry. Densification is a precursor to high performance levels in PM products. Unfortunately, a tradeoff has to be made because processing routes that promoted densification are processes that increase production cost [2,3]. High densities are possible via high temperature sintering, which requires high-temperature furnaces. High compacting pressure is also another means of achieving higher densities, but this requires high-pressure presses. Moreover, high compaction pressures result into high ejection forces. Other processing routes such as repressing, isostatic pressing, infiltration, and use of enhanced sintering techniques such as liquid phase sintering and field-assisted sintering, are all viable routes for densification but are expensive and/or require specialized tools.

In this study, a unique and nonconventional pressing process which uses electromagnetic formulation was used to achieve higher densification. Electromagnetic processing, which may be referred to as a 'non-contact deformation' process offers the advantage of providing pressing without any contact between the tool and the work piece as opposed to the traditional pressing method, which is a 'contact deformation' process. The investigation focused on the effects of electromagnetic pressing on the thermal stability of the microstructure and dimensional changes of the post-sintered FL-5305 through thermal cycles in the dilatometer. FL-5305 is a standard sinter-hardened PM steel pre-alloyed with chromium (Cr) and molybdenum (Mo), and has excellent performance enhancing ability both in the as-sintered and heat-treated states. Besides improving hardenability, alloying with Cr and Mo has a tendency to raise the eutectoid transformation point to higher temperatures thereby suppressing the formation of pearlite [4,5].

EXPERIMENTAL PROCEDURE

All materials for this investigation were supplied by Atlas Pressed Metals, and consisted of sinter hardened FL-5305 (3 wt% Cr, 0.5 wt% Mo, 0.5 wt% C). The parts were disks (see Figure 1a) with outer diameter of ϕ25.5 mm (1.0in), inner diameter of ϕ9.5 mm (0.375in), and thickness of 7.5 mm (0.3in). Compaction was done at 400 MPa (30 tsi), 550 MPa (40 tsi), and 690 MPa (50 tsi). Sintering was performed in a 90% nitrogen+10% hydrogen atmosphere at 1150°C (2100°F) for 30min using Abbott 7 (VariCool 60Hz) zone convective cooling furnace; and delubing was done at 400°C (750°F).

Electromagnetic pressing was carried out at the National High Magnetic Field Laboratory (NHMFL), Tallahassee, using a capacitor charged to 7kV and discharged through a coil wound around the sample (see Figure 1b). This creates a pulse of about 100kA and a magnetic field of ~40 Tesla around the coil for a very brief time (~4μs). The rapidly changing magnetic field induces an electric current in the sample and exerts a radial compressive Lorentz force on this sample.

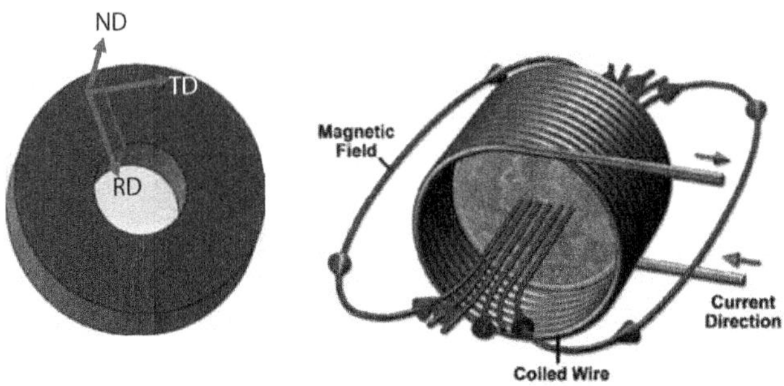

Figure 1. (a) Geometry of the sample and (b) direction of current and magnetic field for a winding used in the electromagnetic compaction. (Rendering by NHMFL Public Relations Dept.)

Densities were determined from five measurements of mass and dimensions of each sample before and after electromagnetic compaction. LECO LM300 AT microhardness test machine (with a 300g load) was used to measure the microhardness of the parts. Each microhardness value was the average of about ten diamond-pyramid indentations made across the normal (ND) and transverse (TD) sections of the sample. Microstructural investigations were performed on sections along the normal (ND) and transverse (TD) directions of the sample (see Figure 1a). All specimens were mechanically polished using Struers TegraPol-35/TegraForce-5 polishing machine, and etched by immersing in 5%Nital for about 20 - 35s.

Dilatometric measurement were performed in Anter 1161 Dilatometer. Each sample was heated at a controlled rate of 5°C/min to 1000°C, held at temperature for 1 hr., and cooled to room temperature at a controlled rate of 5°C/min. The heating and cooling cycle was repeated four times.

RESULTS AND DISCUSSION

Figure 2 shows typical dependency of sintered density on compacting pressure. The density of as-sintered FL-5305 increased from about 6.4 g/cm^3 at 400 MPa (30 tsi) to 6.9 g/cm^3 at 690 MPa (50 tsi). There was a notable increase in density after electromagnetic pressing. The effect was more pronounced for parts compacted at higher pressures. While there was no appreciable change in density for parts compacted at 30 tsi, there was about 2% increase in density for parts compacted at 50 tsi.

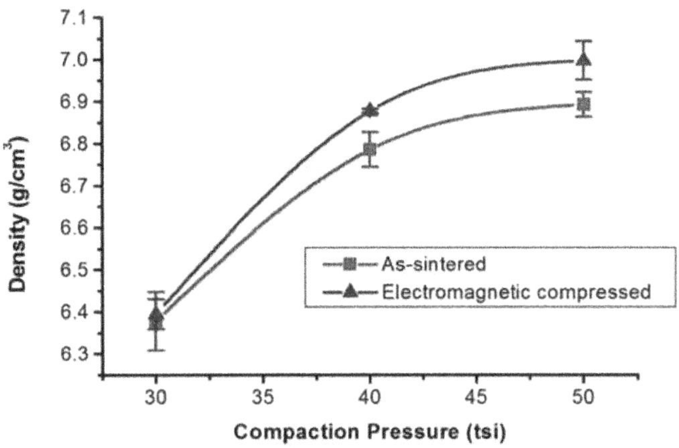

Figure 2. Densities of as-sintered and electromagnetic compressed FL-5305 PM parts.

The microstructure of as-sintered and electromagnetic compressed FL-5305 is shown in Figure 5. Evidently, both micrographs depict the familiar microstructure of sinter-hardened FL-5305 which consists of bainitic and martensitic structures [6,7]. However, the as-sintered microstructure exhibited more porosity compared to electromagnetic compressed parts. It was also noted that the bainitic and martensitic structures are sharper and finer in electromagnetic compressed parts than in as-sintered parts.

The microhardness of as-sintered and electromagnetic compressed FL-5305 is shown in Figure 4. Undoubtedly, there is an appreciable increase in the microhardness of the electromagnetic compressed parts compared to the as-sintered parts. This is supported by the observed refinement of the bainitic and martensitic structures after electromagnetic compression (Figure 3).

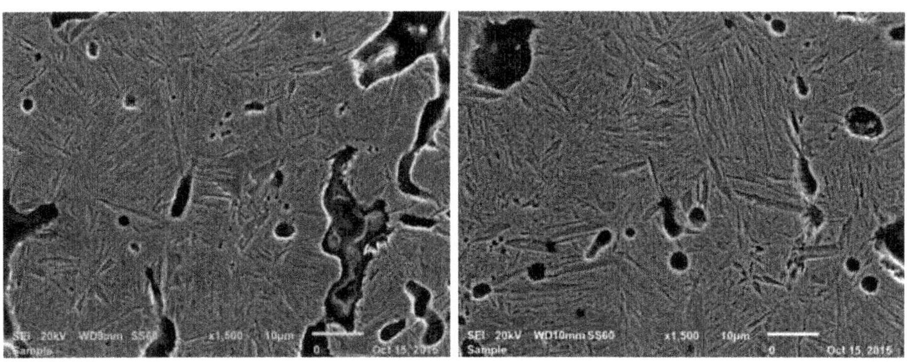
(a) (b)

Figure 3. SEM micrographs of (a) as-sintered, and (b) electromagnetic compressed FL-5305 compacted at 50 tsi and sintered at 1150°C.

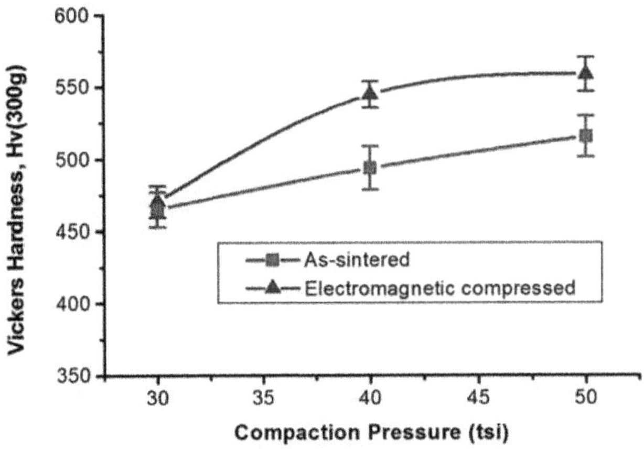

Figure 4. Microhardness of as-sintered and electromagnetic compressed FL-5305 PM Parts.

Figure 5 presents the microstructure of the as-sintered and electromagnetic compressed samples after thermal cycling in the dilatometer. Evidently, after 4 thermal cycles at 1000°c, the two microstructures are identical; and consisted of large grain sizes of ferrite with primary and secondary Widmanstatten ferrite, allotriomophic ferrite, and some pearlite [8].

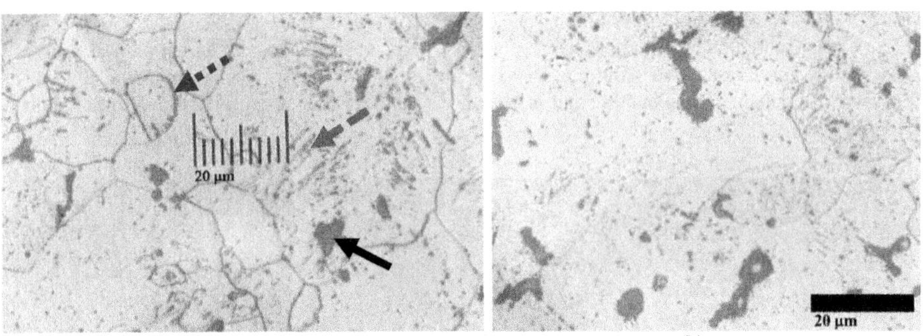

(a) (b)

Figure 5. Optical micrographs of (a) as-sintered, and (b) electromagnetic compressed FL-5305 after thermal cycling in the dilatometer at 1000°C. Red dashed arrow shows Widmanstatten ferrite, blue dotted arrow shows allotriomophic ferrite, and black arrow shows pearlite.

Figure 6 shows dimensional change ($\Delta L/L$) with temperature during the four thermal cycles. The first thermal cycle differed significantly from subsequent thermal cycles, and is accompanied with a pronounced shrinkage. There are two familiar phase transformations. As the specimen is heated, the bainite and the body centered tetragonal (BCT) martensite (α') transform to carbide and body centered cubic (BCC) ferrite (α) [8]. At about 770°C, the ferrite starts to transform to face centered cubic (FCC) austenite (γ), and this transformation is accompanied by shrinkage and ends at about 800°C. Further heating is accompanied by the expansion due to grain growth, and at the holding temperature there is no further dimensional change (see Figure 7) except for the first cycle which showed shrinkage. This shrinkage can be attributed to further reduction of porosity due to grain boundary diffusion, and disappears on subsequent thermal cycles. Upon cooling, the specimen shrinks as expected except at about 730°C where there is a notable expansion, which is due to the transformation of austenite (γ) back to ferrite (α) and carbide. This transformation ends at about 650°C, and is followed by normal shrinkage of the specimen to room temperature.

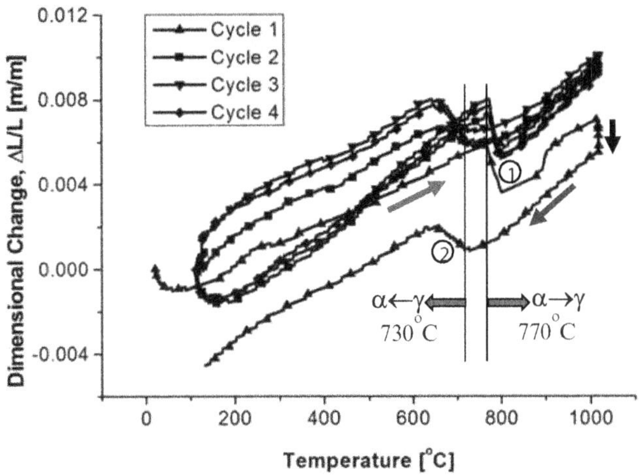

Figure 6. Dimensional change for as-sintered FL-5305 during thermal cyclic heat-treatment in the dilatometer.

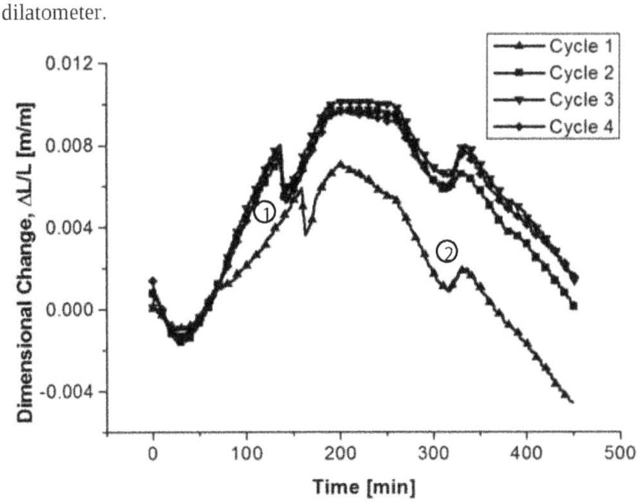

Figure 7. Rate of dimensional change for as-sintered FL-5305 during thermal cyclic heat-treatment.

Table 1 and Table 2 summarize the start and end temperatures for the transformation of ferrite to austenite, and vice versa, respectively, as well as the associated dimensional changes. The shrinkage from ferrite to austenite ranges from 0.2% to 0.25%, whereas the expansion from austenite to ferrite ranges from 0.09% to 0.18%.

Table 1: Dimensional Change (DC) for as-sintered FL-5305 during heating step of the thermal cycle.

	Cycle1		Cycle 2		Cycle 3		Cycle 4	
	DC (%ΔL/L)	Temperature (°C)	DC (%ΔL/L)	Temperature (°C)	DC (%ΔL/L)	Temperature (°C)	DC (%ΔL/L)	Temperature (°C)
Start	0.588	765	0.735	769	0.802	772	0.770	772
End	0.360	800	0.534	797	0.574	805	0.519	798
% Shrinkage	-0.228		-0.201		-0.229		-0.251	

Table 2: Dimensional Change (DC) for as-sintered FL-5305 during cooling step of the thermal cycle.

	Cycle1		Cycle 2		Cycle 3		Cycle 4	
	DC (%ΔL/L)	Temperature (°C)	DC (%ΔL/L)	Temperature (°C)	DC (%ΔL/L)	Temperature (°C)	DC (%ΔL/L)	Temperature (°C)
Start	0.859	728	0.575	735	0.648	726	0.598	724
End	0.202	665	0.665	670	0.802	642	0.774	648
% Expansion	0.116		0.089		0.154		0.177	

A similar dimensional change behavior and characteristic was observed during thermal cycling of electromagnetic compressed specimens (Figure 8). However, there was minimal difference in the dimensional change between the first cycle and subsequent cycles (also illustrated in Figure 9). During the ramping step, the transformation from ferrite (α) to austenite (γ) starts at about 770°C and ends at about 800°C. This transformation is accompanied by a shrinkage that ranges from 0.07% to 0.14%, which is significantly lower than for the as-sintered specimens. Upon cooling, the transformation from austenite (γ) to ferrite (α) and carbide starts at about 720°C and ends at about 700°C. This transformation is accompanied by an expansion that ranges from 0.03% to 0.09%, which is also significantly lower than for the as-sintered parts. Table 3 and Table 4 present summarized data for these transformations.

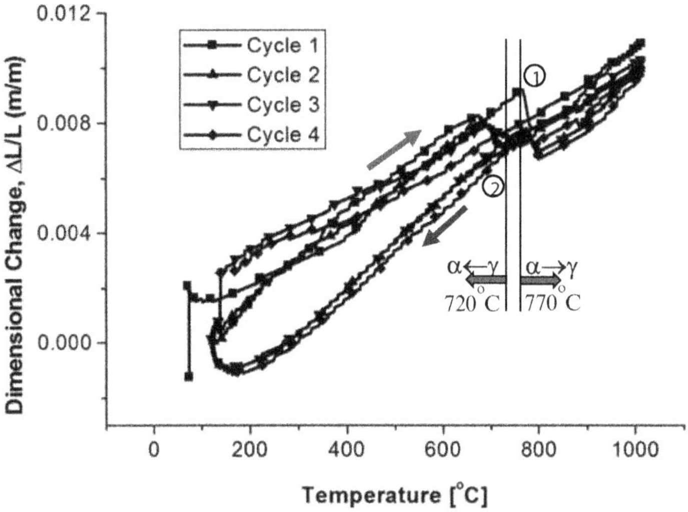

Figure 8. Dimensional change for electromagnetic compressed FL-5305 during thermal cyclic heat-treatment.

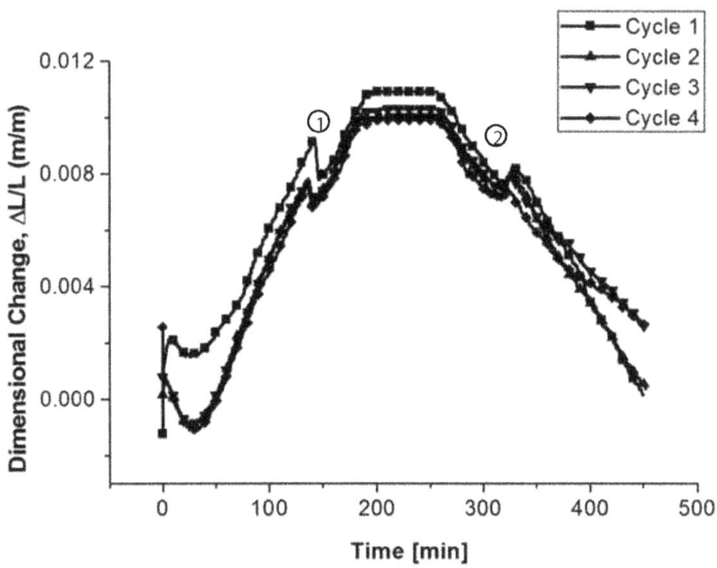

Figure 9. Rate of dimensional change for electromagnetic compressed FL-5305 during thermal cyclic heat-treatment.

Table 3: Dimensional Change (DC) for electromagnetic compressed FL-5305 during heating step of the thermal cycle.

	Cycle 1		Cycle 2		Cycle 3		Cycle 4	
	DC (%ΔL/L)	Temperature (°C)	DC (%ΔL/L)	Temperature (°C)	DC (%ΔL/L)	Temperature (°C)	DC (%ΔL/L)	Temperature (°C)
Start	0.922	765	0.768	773	0.782	774	0.757	776
End	0.779	786	0.695	797	0.698	800	0.675	804
% Shrinkage	-0.143		-0.073		-0.085		-0.083	

Table 4: Dimensional Change (DC) for electromagnetic compressed FL-5305 during cooling step of the thermal cycle.

	Cycle 1		Cycle 2		Cycle 3		Cycle 4	
	DC (%ΔL/L)	Temperature (°C)	DC (%ΔL/L)	Temperature (°C)	DC (%ΔL/L)	Temperature (°C)	DC (%ΔL/L)	Temperature (°C)
Start	0.756	715	0.709	722	0.750	723	0.721	721
End	0.830	675	0.801	673	0.795	697	0.747	704
% Expansion	0.073		0.093		0.046		0.026	

Figure 10 shows the comparison of the dimensional change of the as-sintered and electromagnetic compressed FL-5305 during the first and fourth thermal cycles. It is clear that in the first cycle the as-sintered specimen shrank considerably compared to the electromagnetic compressed specimen. However, the two specimen depict similar dimensional change on subsequent thermal cycling.

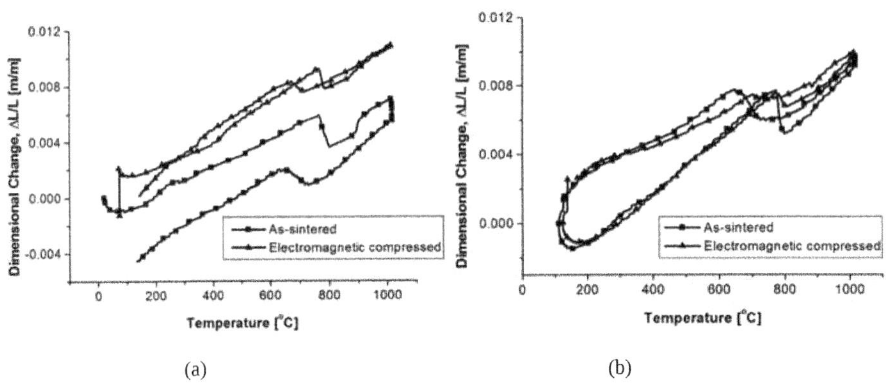

Figure 10. Comparison of the dimensional change of the as-sintered and electromagnetic compressed FL-5305 during (a) first thermal cycle, and (b) fourth thermal cycle.

CONCLUSIONS

This investigation has clearly shown that electromagnetic processing is a promising potential method that can be used to improve densification of PM parts. The results presented show evidence that parts processed by this method have better dimensional control and thermal stability compared to as-sintered parts. It should, however, be noted that the results presented here are preliminary assessments. This study is in its infancy, but the results so far are intriguing, and may have future practical applications.

ACKNOWLEDGEMENT

Funding by the Powder Metal Initiative (PMI) through the Department of Community and Economic Development (DCED), and that of the DuBois Educational Foundation (DEF) is highly acknowledged.

REFERENCES

[1] H. K. D. H. Bhadeshia, Bainite in Steels: Transformations, Microstructure and Properties, 2001, Institute of Materials.
[2] R.M. German, Powder Metallurgy & Particulate Materials Processing, 2005, Metal Powder Industries Federation (MPIF), Princeton, NJ.
[3] A. Upadhyaya and G.S. Upadhyaya, Powder Metallurgy: Science, Technology, and Materials, CRC Press, 2011.
[4] MPIF Standard 35, Materials Standards for PM Structural Parts, 2007, Metal Powder Industries Federation (MPIF), Princeton, NJ.
[5] B. Maroli, S.Berg, P. Thorne, and U.Engström, "Sinter-Hardening and Heat Treatment of Materials based on Astaloy CrM", Proceedings of PM²TEC 2003, Las Vegas, NV, June 8-12, 2003.
[6] R. M. German, Powder Metallurgy Science, 2nd Ed., 1994, Metal Powder Industries Federation (MPIF), Princeton, NJ.
[7] Handbook of Sintered Components, Volume 6: Metallography, North American Hoganas (NAH), 2004.
[8] H. K. D. H. Bhadeshia, Some phase transformations in steels, Materials Science and Technology 15, 22 – 29, 1999.

Advances in Composites

A NEW PRODUCTION PROCESS FOR THERMAL BARRIER COATING MATERIAL

Yunsheng Wang*, Wenzhong Tao, Decheng Pan, Zuxiong Chen
Jiangsu Lida Hi-Tech Special Material Co., Ltd, Changshu, Jiangsu, China 215500
*corresponding author, innova2010@126.com

ABSTRACT

We have developed a series of environmental friendly production process to manufacture specialized zirconia, yttria stabilized zirconia powder and suspension. The products are sulfur and chlorine free, very low impurities, and super fine particle (d50<20nm). Stable suspension products contain no aggregation. Completely solid with adjustable specific surface area enhances coating strength, good columnar crystal structure possesses superior anti thermal shock and low wear properties. The parameters are: ZrO_2 99.95%; Yttria-stabilized zirconia 92.5±0.3+7.5±0.3% Y_2O_3; Al2O3, SiO_2, Fe_2O_3, TiO_2, Na_2O, CaO are all <0.001-0.005%; Velocity of flow s/50g=15sc; Hollowness 90%; Zirconia thermal barrier coating has excellent high temperature phase stability, can be cool down to room temperature. Excluding monoclinic, they are fully structurally tetragonal and cubic phases. The 1250 °C thermal shock validation test has been performed more than 2800 times. Rare earth oxide Zirconia composite thermal spray powders and suspensions have been tested for 1400~1500 °C.

INTRODUCTION

Jiangsu Lida Hi-tech Special Materials Co., Ltd. specializes in zirconia, yttria powder production. Products cover areas of structural ceramics and corrosion wear-resistant of high temperature-resistant ceramic coating materials, such as ultrafine nano-zirconia powder, yttria-stabilized zirconia thermal spraying powder, yttria stabilized zirconia suspension, ultra-high temperature rare earth oxide thermal barrier coatings, and cerium based composite zirconia catalyst.

METHODS AND RESULTS

We have used patented environmentally friendly technologies to produce all our products[1]. Especially, we have used the environmentally friendly technology to produce high-purity zirconia powder products, while addressing problems of unable to remove the sulfur ions in the products by traditional production process using Sulfuric Acid Technologies. The produced zirconia powders have small diameter and contain no sulfur and chlorine ions. Moreover, the technologies allow us to reduce the atmosphere ammonia pollution of the conventional precipitation method.

Figure 1. Zirconia particle size by electron microscopy. The top figure shows a plurality of particle images, the bottom is an image of a single particle.

As shown in Figure 1 by transmission electron microscopy (TEM) test, the average particle size is less than 20nm. The nanoparticles appear polygonal, with regular and clear lattice structure.

The powder products possess controllable specific surface area. The fresh specific surface area of composite zirconium oxide powder used for cerium-based oxygen storage material can be as high as up to 400 ~ 500 m^2/g. The new technologies employ low temperature synthesis production process to produce none aggregated powders with no waste emissions, and process water can be recycled.

Jinan Winner

测试信息

样品名称：7.5wt%钇稳定氧化锆【平均】送样单位：试测 送样时间：2014-07-13

分 散 剂： 分散介质：水 超声时间：1s

测量范围：0.1μm~40μm 样品浓度：1.5 分析模式：标准

粒径μm	体积%	累积%	粒径μm	体积%	累积%	粒径μm	体积%	累积%
0.12	1.40	1.40	1.83	4.94	89.78	27.63	0.00	100.00
0.15	1.72	3.12	2.22	3.91	93.69	33.54	0.00	100.00
0.18	2.09	5.21	2.70	2.84	96.53	40.00	0.00	100.00
0.22	2.53	7.74	3.28	1.81	98.35	49.43	0.00	100.00
0.26	3.02	10.76	3.98	1.04	99.39	60.00	0.00	100.00
0.32	3.58	14.34	4.83	0.48	99.86	72.84	0.00	100.00
0.39	4.26	18.59	5.86	0.14	100.00	88.42	0.00	100.00
0.47	6.14	24.74	7.11	0.00	100.00	107.33	0.00	100.00
0.57	9.41	34.14	8.64	0.00	100.00	130.29	0.00	100.00
0.69	13.26	47.40	10.48	0.00	100.00	158.17	0.00	100.00
0.84	13.56	60.96	12.73	0.00	100.00	192.00	0.00	100.00
1.02	10.94	71.89	15.45	0.00	100.00	233.07	0.00	100.00
1.24	7.33	79.23	18.75	0.00	100.00	282.93	0.00	100.00
1.51	5.61	84.84	22.76	0.00	100.00	300.00	0.00	100.00

分析结果

D10=0.25μm D50=0.72μm D90=1.85μm

D99=3.72μm

分析结果

S/V=113367.50cm²/cm³ Dav=0.92μm

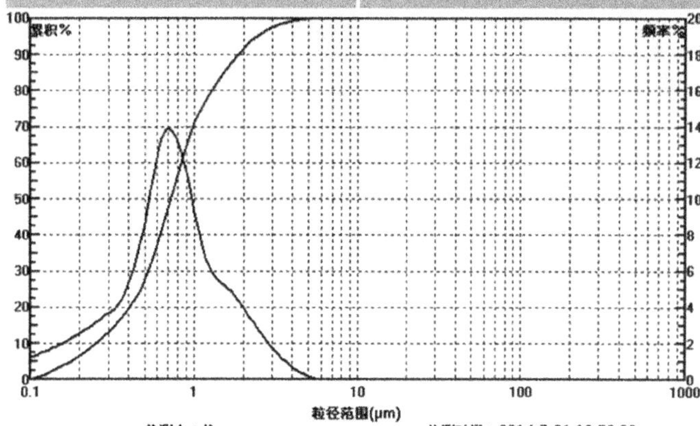

检测人：徐 检测时间：2014-7-21 10:53:30

仪器型号：Winner2000激光粒度分析仪

Figure 2. Particle size distribution chart of 7.5% Yttria stabilized zirconia by Winner2000 Laser Particle Analyzer. The test report shows a uniform particle size distribution and a narrow particle size range. Sample concentration is 1.5. Particle size measurement ranged from 0.1μm to 40μm, laser time 1 second. In the table, Columns 1, 4, 7 from the left are particle size (μm), columns 2, 5, 8 are volume (%) and columns 3, 6, 9 are accumulation (%). In bottom figure, x-axis is particle size (μm), left y-axis is accumulation (%), and right y-axis is frequency (%).

The powder products have high qualities; Figure 2 showed a particle size distribution chart of 7.5% yttria stabilized zirconia by Winner2000 Laser Particle Analyzer. The test report demonstrated a narrow particle size range less than 6 μm and evenly particle size distribution, the average particle diameter (D_{av}) is 0.92 μm, and D10= 0.25μm, D50= 0.72μm, D90= 1.85μm, D99= 3.72μm.

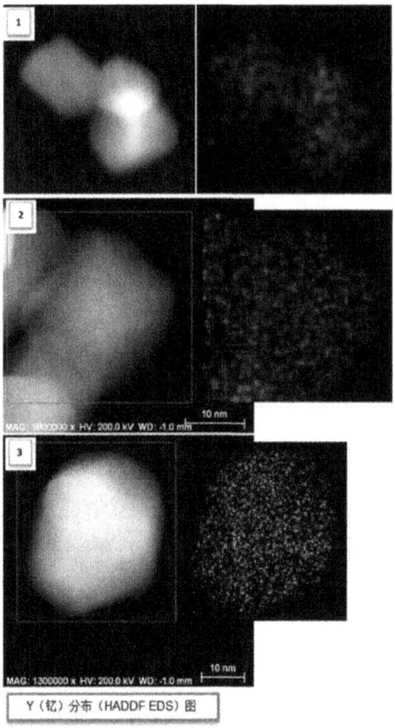

Figure 3. Yttria distribution in yttria stabilized zirconia. The image of high resolution transmission electron microscopy showed that yttrium and zirconium material became completely solid solution. Top, middle and bottom panels are shown in different resolutions.

As shown in Figure 3, the yttria is distributed evenly in the product yttria stabilized zirconia; yttrium and zirconium material has become completely solid solution.

The dehydration process of company's patented technology uses azeotropic distillation methods, the water molecules wrapped in colloids of prepared zirconia were removed azeotropically. The hydroxyl groups between adjacent particles surfaces were Zr-O-Zr bond substituted, therefore resulted no hard aggregates formed in the subsequent drying and calcination process.

Our patented production process is different from other methods of green synthesis. Zirconia crystal nuclei have initially formed during the mixing and reaction process, through heat treatment they can grow to form stabilized tetragonal (cubic) phase zirconia nano powders.

The yttria-stabilized zirconia thermal barrier coating materials have excellent high-temperature phase stability, when they were cooled to room temperature, showing completely tetragonal and cubic phases, excluding monoclinic.

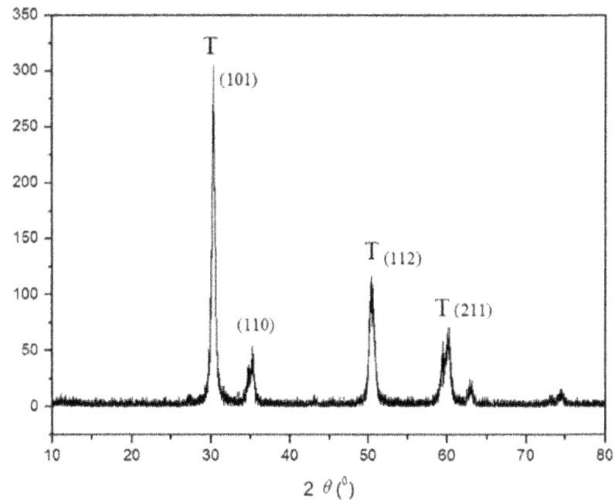

Figure 4. The X-ray diffraction test report of zirconia particles. Results showed that with the standard card (PDF # 50-1089) comparative analysis, crystal planes (101) (110) (112) (211) are the tetragonal phase. It indicated that the zirconia particles were 100% tetragonal crystal phase.

The zirconia powder products were examined by x-ray diffraction (XRD), as shown in Figure 4, by comparison with the standard card (PDF # 50-1089) analysis, crystal planes of peaks (101), (110), (112), and (211) are all tetragonal phase, no other peaks appeared prior to peak (101), it indicated that the zirconia particles is 100% tetragonal crystal phase.

Table 1. Products key performance index. All three products had low impurities and high performance indicators.

Product	Nano-zirconia powder	Nano-zirconia suspension	Yttria stabilized zirconia
ZrO_2%	99.95	99.95	92.5±0.3
Y_2O_3%	--	--	7.5±0.3
Al_2O_3 <%	0.002	0.002	0.005
SiO_2<%	0.002	0.002	0.005
Fe_2O_3<%	0.003	0.003	0.005
TiO_2<%	0.001	0.001	0.005
Na_2O<%	0.001	0.001	0.005
CaO<%	0.002	0002	0.002
*Hall Flow (s/50g)	--	--	<40
Density (g/cm^3)	0.7	--	1.2
Specific surface area m^2/g	15±2	--	15±2
Average particle size (μm)	<1.5	<70nm	20~70

* Thermal spray powder ball's velocity from a Hall Flowmeter measurement.

Unlike the products produced by the liquid phase methods in which sulfur and chlorine ions exist in common, the product produced by using new technologies contains no these ions, therefore it is easy to clean. As shown in Table 1, the powder products characterized as low impurities, loose powder by Hall Flowmeter testing (Hall flow of yttria stabilized zirconia <40 s/50g and the two powders bulk density were 0.7 and 1.2 g/cm^3), high reactivity and good thermal stability[2].

The suspension product has no settlement in a long time; its good columnar crystal structure improves coating strength, and possesses more excellent thermal shock resistance and thermal stability[2]. To the samples that were thermal spray coated with our yttria stabilized zirconia suspension materials, we performed 1250 °C thermal shock tests more than 2800 times to observe severe segment cracking.

With this new technology we have developed and prepared ultra-high temperature resistance up to 1400 ~ 1500 °C rare earth oxide composite zirconia thermal spray powders and suspensions. Such materials are mainly used in aviation to provide insulation protection for turbine engine blades, combustion chambers, heat shields and other hot section under conditions of high temperature hot air. It can also be used for the base and the high-temperature gas isolation, so that the metal or alloy substrates avoid high temperature oxidation, corrosion and erosion of the gas.

Using our powder materials we have manufactured structure ceramic products with excellent performance. The products have high resistance to abrasion, corrosion and impact resistance, with a life time 5 to 10 times longer than that of the metal products.

CONCLUSION

This new production process has a number of advantages in the production, no emissions and low power energy consumption. The products contain low impurities; have a small powder particle size, specific surface area large and controllable and stable crystal phase. The samples coated by our thermal barrier coating powder have characteristics of high strength, thermal stability and thermal shock resistance.

REFERENCES

[1]Yunsheng Wang and Ye Tao: A large specific surface area of stabilized nano zirconia powder preparation methods. Chinese Patent No. ZL 2013 1 0060335.8, Patentee Jiangsu Lida Hi Tech Special Materials Co. Ltd, inventor Wang Yunsheng, cultivate. Authorized Date June 11, 2014.

[2]Shunyan Tao, Xinghua Zhong, Huayu Zhao, Yuexing Zhao, Dachuan Li, Chuanxian Ding: Columnar structured thermal barrier coatings deposited by suspension plasma spraying. Thermal Spray 2014: Proceedings of the International Thermal Spray Conference (May 21–23, 2014, Barcelona, Spain) Pages: 745 - 750 (6), 2014.

SIMULTANEOUS SYNTHESIS AND SINTERING OF DENSE B_4C/CNF COMPOSITES USING A PULSED ELECTRIC-CURRENT PRESSURE SINTERING AND EVALUATION OF THEIR THERMAL PROPERTIES

Naoki Goto[1], Mitsuhiro Shima[1], Xiaolei Chen[1],
Masaki Kato[1], Ken Hirota[1*] and Toshiyuki Nishimura[2]

[1] Department of Molecular Chemistry & Biochemistry, Faculty of Science & Engineering, Doshisha University, Kyo-tanabe, Kyoto 610-0321, Japan.
[2] National Institute for Materials Science, Tsukuba, Ibaraki 305-0044, Japan.

ABSTRACT

Dense B_4C/CNF composites have been synthesized and sintered simultaneously using pulsed electric-current pressure sintering at 2173 K for 6.0×10^2 s (10 min) under 50 MPa in a vacuum. They showed moderate mechanical properties at room temperature, however, they revealed extraordinary high bending strength $\sigma_b \geq 600$ MPa at 1673~1873 K in an inert atmosphere, especially, σ_b of 830 MPa at 1773 K (12.5 vol% CNF added composite) and 800 MPa at 1873 K (10.0 vol% CNF). Furthermore, they also showed much improved high-temperature fracture toughness by a factor of around 20, in comparison with monolithic B_4C at 1773 K. These extraordinary high strength and toughness might be explained by the suppression of grain growth by CNF at elevated temperatures, an inverse σ_b temperature-dependence for carbides, and anchor effect induced by CNF.

In addition, their thermoelectric properties have been evaluated. Seebeck coefficient S, electric conductivity σ_e and thermal conductivity λ also have been measured from room temperature to 1073 K in air. With increasing the CNF content and decreasing temperature, the value of S decreased gradually. However, σ_e value increased with increasing both the CNF content and temperature, as if there is a "trade-off" relation between S and σ_e. On the other hand, λ value at room temperature changed from around 22 to 34 $W\cdot m^{-1}K^{-1}$ with increasing the CNF content, and reduced gradually to 10~12 $W\cdot m^{-1}K^{-1}$ at 1273 K. As the result, the highest dimensionless figure of merit ZT of monolithic B_4C reached ~2.3×10^{-2} at 1073 K.

INTRODUCTION

Until now, various kinds of studies and works concerning about utilization of carbon nano-materials, *i.e.*, fullerene (C_{60}), carbon nano-tube (CNT), and carbon nano-fiber (CNF) have been performed [1-3]. Especially, due to its easy mass-production and consecutive cost reduction, CNF has attracted much attention as reinforcements of materials because of their exceptional mechanical properties (tensile strength σ_t of ~2.20 GPa and Young's modulus E of 100 ~ 300 GPa for 0.15 μm diameter nanofibers) and a high electrical conductivity σ_e of 1×10^6 $S\cdot m^{-1}$ [4, 5]. However, many studies on the fabrication of ceramics/CNF composites have not been so much succeeded; their mechanical properties were often decreased by the CNF addition. This has been explained by the lack of good dispersion of CNF [6, 7] and the poor cohesion between CNFs and matrix [8]. Therefore, the fabrication of homogeneous nano-composites dispersed with CNF has been remained still as a technical challenge.

Under these circumstances, we have investigated the fabrication of ceramics/CNF nanocomposites by adopting pulsed electric-current pressure sintering (PECPS) [9, 10] or spark plasma sintering (SPS) from the mixed elemental powder compacts dispersed with CNF; for examples, α-Al_2O_3/CNF [11], β-SiC/CNF [12], β-SiAlON/CNF [13] and (B_4C with 2.5 vol% Al_2O_3)/CNF [14].

Previous our study [14] on CNF dispersed B_4C composites with sintering additive of 2.5 vol% Al_2O_3 showed high mechanical properties at room temperature; three-point bending strength σ_b ~710 MPa, Vickers hardness H_v ~36 GPa, and fracture toughness K_{IC} ~7.9 $MPa\cdot m^{1/2}$. However, their bending strength at elevated temperatures was not so high (σ_b ~590 MPa at 1873 K). Recently we noticed that the reduction in σ_b at high temperatures might be originated from thin aluminum oxide (Al-O) film formed between CNF and B_4C [14]; the rupture and/or the creep at high temperatures would originate at this Al-O film. In order to improve the high temperature strength, we tried to fabricate dense B_4C/CNF composites without sintering additives by utilizing both self-propagating high temperature synthesis (SHS) [15-17] between amorphous boron and carbon powders and pulsed electric-current pressure sintering (PECPS). The former SHS has been used to synthesize monolithic and composite materials with high formation enthalpy. However, in general, it is very difficult to fabricate dense bulk materials due to the poor control of their heating process. On the other hand, PECPS can produce dense materials from poor sinterable powders.

In these studies, we combined these techniques on the fabrication of pure B_4C/CNF composites and evaluated their mechanical properties [18]. The present paper deals with the mechanical and thermoelectric properties at elevated temperatures, in which the latter properties are much attractive from the viewpoints of energy saving, of densely consolidated B_4C/CNF composites.

EXPERIMENTAL PROCEDURE

High purity amorphous B powder (H.C. Starck-V Tech Ltd., average particle size P_s of ~30 nm, purity of 98.5%, theoretical density D_x of 1.73 $Mg \cdot m^{-3}$) [19], amorphous C powder (Mitsubishi Chemical Co. Ltd., Tokyo Japan, "Carbon Black #30", P_s ~30 nm, 98.5%, D_x of 1.80 $Mg \cdot m^{-3}$) [20], and carbon nano-fiber (CNF) "VGCF-H" (Showa Denko K. K., Tokyo, Japan, ~150 nm in diameter and 4~5 μm in length, ~99%, D_x 2.0 $Mg \cdot m^{-3}$) [5] were used as starting materials. These powders were weighed into B/C=4:1 (molar ratio) + 5.0, 7.5, 10, 12.5, and 15 vol% CNF; their compositions are presented as an inclusive manner. The weighed powders (4B/C) were mixed in regent grade ethanol using an alumina mortar and pestle for 1.8×10^3 s (30 min). Before mixing CNF with these mixed powders, the former CNF were pre-dispersed into 2-propanol using an ultrasonic vibrating homogenizer with a power output of 300 W at frequency of 20 kHz (*US-300T*, Nihon-Seiki Ltd., Tokyo, Japan) for 1.8×10^3 s (30 min). After the mixed 4B/C powders were added into the CNF dispersed solution, the mixed slurries with the compositions of B_4C/CNF=100/0~85/15 vol%, [here, the compositions are denoted as the final composite materials; as will be described in the later, the mixed powder of B/C=4:1 in molar ratio changed into chemical compound B_4C by simultaneous synthesis and sintering with a PECPS], were homogenized using the same ultrasonic vibrator for 1.8×10^3 s (30 min). Then, powder mixtures were prepared by drying the slurries with stirring at room temperature in air.

The mixtures were uniaxially pressed into disks with a 20 mm diameter and a ~5 mm thickness at 98 MPa and then cold isostatically pressed at 245 MPa. Their green densities were 1.13 $Mg \cdot m^{-3}$ for mixed powder (4B+C) corresponding to the final compound B_4C, 1.05 $Mg \cdot m^{-3}$ for the final composition of B_4C/CNF=87.5/12.5vol% powders, and 0.981 $Mg \cdot m^{-3}$ for the B_4C/CNF =85/15 vol% powders; these values corresponded to 64.5, 60.3 and 56.0% of their theoretical densities, 1.745, 1.747 and 1.751 $Mg \cdot m^{-3}$, respectively, which were calculated using the theoretical densities of 1.73 for amorphous B [19], 1.80 for amorphous C [20], and 2.0 $Mg \cdot m^{-3}$ for CNF [5]. From this, it is clear that their relative densities decreased gradually from 65 to 56% with increasing CNF content, which explained that both hollow cores inside of CNF and vacant space around needle like CNF in powder compacts.

The compressed powder compacts were pre-heat-treated for 3.6×10^3 s (2 h) at 1223 K (950°C) under reduced pressure to eliminate a small amount of water and oxygen adsorbed on the surfaces of amorphous boron and carbon nano-powders without particle-sintering during heating [14]. After elimination of adsorbed gas, they were put into a cylindrical mold (20mmϕ-40mmϕ-40mm^h, high-density graphite carbon) that was used for PECPS (*SPS-5104A*, SPS SYNTEX INC., Tokyo, Japan). Using this apparatus, sintering was performed at 2173 K (1900°C) for 6.0×10^2 s (10 min) under an uniaxial pressure of 50 MPa in a vacuum (~10 Pa) with a heating rate of 1.66 $K \cdot s^{-1}$ (100°$C \cdot min^{-1}$) and a cooling rate of ~8.33×10^{-1} $K \cdot s^{-1}$ (~50°$C \cdot min^{-1}$), applying DC pulse electric-current (rectangular wave current with a on/off interval=12(39.6ms):2(6.6ms) ratio). Heating temperature above ~873K (600°C) was monitored at the center position of the outside wall of carbon mold using a monochrome pyrometer. Sintering temperature was decided on the fabrication of monolithic B_4C. From the results of the preliminary experiments, a 2173 K (1900°C)-sintering gave a high density (relative density of 99.9%) of single-phase B_4C ceramics and high mechanical properties of Vickers hardness H_v of

33.5 GPa and fracture toughness K_{IC} of 4.98 MPa·m$^{1/2}$ which value was determined by IF method [21]; Crystalline phase of the sintered materials were identified by X-ray diffraction (XRD) analysis (Cu$K\alpha$ radiation with a graphite monochromator, *Rint 22000*, Rigaku, Osaka, Japan). Hereafter, CNF-dispersed B_4C composites were also sintered at the same sintering conditions of 2173K/6.0×10^{2}s/50MPa/vacuum (1900°C/ 10min/50MPa/vacuum).

Microstructural observation by both field emission-type scanning electron microscopy (FE-SEM, *JSM-7001FD*, JEOL Ltd., Tokyo, Japan) and high-resolution transmission electron microscopy (HR-TEM, *JEM-2100*, JEOL Ltd.) was performed on the sintered materials. Bulk densities of sintered materials were measured by Archimedes' method. After phase identification, test bars for mechanical-property measurements were cut from the sintered materials with a diamond cutting-blade and then their two (upper and lower) surfaces were polished to mirror surface with a diamond paste (nominal particle size 1~3 μm); the left two sides were grinded with a diamond abrasive tool containing #400 grid-pass diamond particles. And then three-point bending strength σ_b was evaluated with a cross-head speed of 8.33×10^{-3} mm·s^{-1} (0.5 mm·min^{-1}) and an 8 mm-span length using WC jigs. Vickers hardness H_v and fracture toughness K_{IC} were evaluated with an applying load of 19.6 N and a duration time of 15 s for the former, and the indentation fracture method (IF) with Niihara's equation [21] for the latter. Three-point bending strength at high temperatures up to 1973 K (1700°C) in N_2 was measured with a cross-head speed of 8.33×10^{-3} mm·s^{-1} (0.5 mm·min^{-1}) and a 16 mm-span length using graphite jigs (*Instron 4505*, Instron Japan Co. Ltd., Tokyo, Japan).

Thermoelectric properties (Seebeck coefficient S and electric conductivity σ_e) of B_4C/CNF composites with the sample size of 2 by 2 by 12 mm were measured using "Thermoelectric Property-Measurement System" (*RZ2001i*, Ozawa Science, Nagoya, Japan) from room temperature to 1073 K (800°C) by 25 K step in air. To evaluate their dimensionless figure of merit ZT, thermal conductivity λ was also measured at room temperature in air using "Thermal Conductivity Measurement System" (*LFA447 Nano Flash,* Netzsch Japan, Yokohama, Japan) based on the laser-flash method under the conditions of a pulse width of 0.2 m·s and intensity of 270 V. Furthermore, thermal conductivity λ of composites from 673 to 1273 K in Ar was also investigated from thermal diffusivity α, specific thermal capacity C_p, and bulk density D_{obs} with the equation of $\lambda=\alpha*C_p*D_{obs}$, using a laser flash thermal analysis (*LFA 457 Micro Flash,* Netzsch Japan).

RESULTS AND DISCUSSION

MICROSTRUCTURE

Figure 1 shows the SEM photographs of fracture surfaces of ceramics with the compositions of B_4C/CNF=(a) 100/0 (monolithic), (b) 95/5, (c) 90/10, and (d) 85/15 vol%, fabricated using PECPS under the conditions of 2173K/6.0×10^{2}s/50MPa/vacuum.

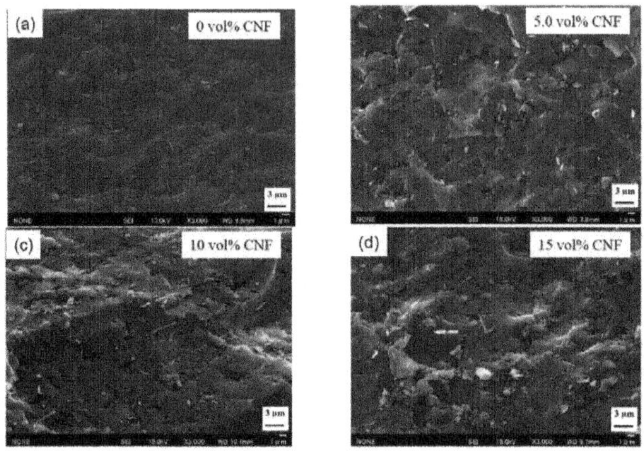

Figure 1. SEM photographs for the room temperature-fractured surfaces of B_4C/CNF composites sintered at 2173K/10min/50MPa: B_4C/CNF =(a) 100/0, (b) 95/5, (c) 90/10 and (d) 85/15 vol%.

Dense homogeneous microstructure with undistinguishable grain boundary and isolated dispersed CNF within the grains are observed in both materials and the latter for the composites, respectively. As it is very difficult to measure the grain size G_s of B_4C, the G_s is roughly estimated to be around 1 ~ 3 μm. After the confirmation by XRD that their phases were B_4C for the monolithic and B_4C+carbon (CNF) for the composite materials, they were characterized in terms of density (microstructural) and mechanical properties: their results were displayed in Table I. Their bulk density decreased gradually from 2.52 to 2.45 $Mg \cdot m^{-3}$ with increasing CNF content up to 15 vol%, however, their relative densities, determined from the theoretical densities calculated using the values of $D_x(B_4C)$=2.515 $Mg \cdot m^{-3}$ [PDF#35-0798] and D_x(CNF)=2.0 $Mg \cdot m^{-3}$ [5] are almost constant of ≥99.5%.

Table I Some properties of B_4C/CNF composites sintered at 2173 K

B_4C/CNF (vol %)	D_{obs} ($Mg \cdot m^{-3}$)	D_x* ($Mg \cdot m^{-3}$)	D_{obs}/D_x (%)	σ_b (MPa) R.T.	H_v (GPa)	K_{IC} ($MPa \cdot m^{1/2}$)
100/00	2.52	2.515	99.9	455	33.5	4.98
95/05	2.49	2.494	99.8	570	34.0	4.94
92.5/7.5	2.48	2.484	99.7	605	31.7	5.27
90/10	2.47	2.474	99.5	610	29.6	5.39
87.5/12.5	2.46	2.463	99.9	630	30.0	5.67
85/15	2.45	2.453	99.8	635	28.8	5.29

MECHANICAL PROPERTIES AT ROOM TEMPERATURE

Three-point bending strength σ_b measured at room temperature (R.T.) increased from 455 to 635 MPa with increasing CNF content, even though there are some variations among the values. On the other hand, Vickers hardness H_v and fracture toughness K_{IC} showed reduction (33.5 to 28.8 GPa) and increase (4.98 to 5.67 $MPa \cdot m^{1/2}$), respectively, with increasing CNF content. The reduction of H_v might be explained in terms of the addition of soft CNF and the accession of K_{IC} due to the crack deflection induced by CNF.

MECHANICAL PROPERTIES AT HIGH TEMPERATURES

Then, bending strength σ_b at high temperature was measured in an inactive atmosphere; at the same time load-displacement (f- δ) curves of the ceramics were also evaluated. Figure 2 displays the variations of σ_b of monolithic and composite ceramics from room temperature to 1973 K. The monolithic B_4C ceramics revealed a standard high-temperature σ_b curve, *i.e.*, with increasing temperature σ_b value decreases gradually. CNF homogeneously dispersed B_4C composites show also a standard high-temperature σ_b curve like monolithic B_4C around 773 to 1473 K, however, they show an extraordinary high values of $\sigma_b \geq 600$ MPa at 1673 to 1873 K. The former gradual decrease might be explained by little improvement effect on σ_b, which effect may be induced from the thermal expansion difference between CNF (its coefficient α: 0.3~1.0 $\times 10^{-6}$ K^{-1}) and B_4C (α: 2.3×10^{-6} K^{-1}) at low temperatures. From these curves, the optimum amount of CNF content might be about 10.0~12.5 vol%, the B_4C composites with the 87.5/12.5 vol% composition shows 830 MPa at 1773 K, in addition, the 90/10 vol% composites gave the σ_b value of 830 MPa at 1873 K; those values have not been reported yet. To investigate this improvement in high-temperature strength of CNF-dispersed composites, the microstructures of the test pieces after high-temperature bending test were observed using FE-SEM. Figure 3 shows the representative SEM photographs for the fractured surfaces of B_4C/CNF composites: (a) 100/0, (b) 95/5, (c) 90/10, and (d) 85/15 vol%, after bending test conducted at 1873 K.

Monolithic B_4C ceramics showed well-developed microstructure, *i.e.*, much grain-grown texture even after 1273 K-heating, though, this SEM photograph is not shown here, in comparison with those of composites. Now, it should be noted that the microstructures of B_4C/CNF composites reveal little change in their textures, grain size, CNF distributions, and porosity, after the heating at 1873 K. In focusing on B_4C/CNF composites with the optimum composition of 87.5/12.5 vol%, their microstructural developments with increasing heating temperature were investigated precisely, as shown in Fig. 4: (a) room temperature, (b) 1273 K, (c) 1573 K, (d) 1773 K and (e) 1973 K.

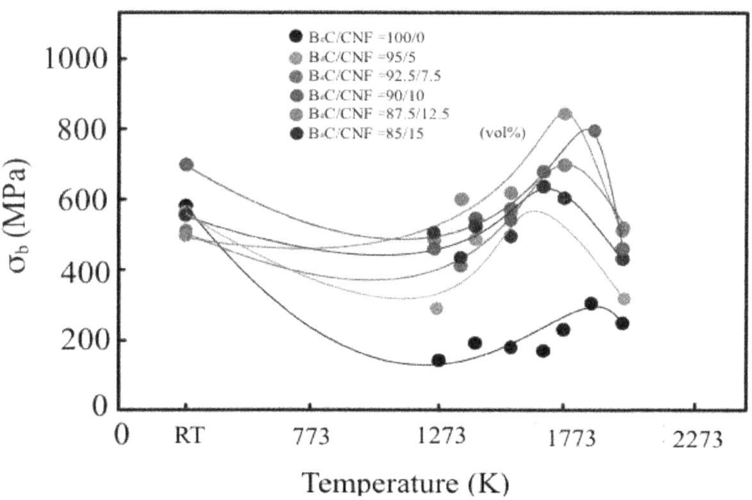

Figure 2. Bending strength σ_b of the present B_4C/CNF composites at high temperatures.

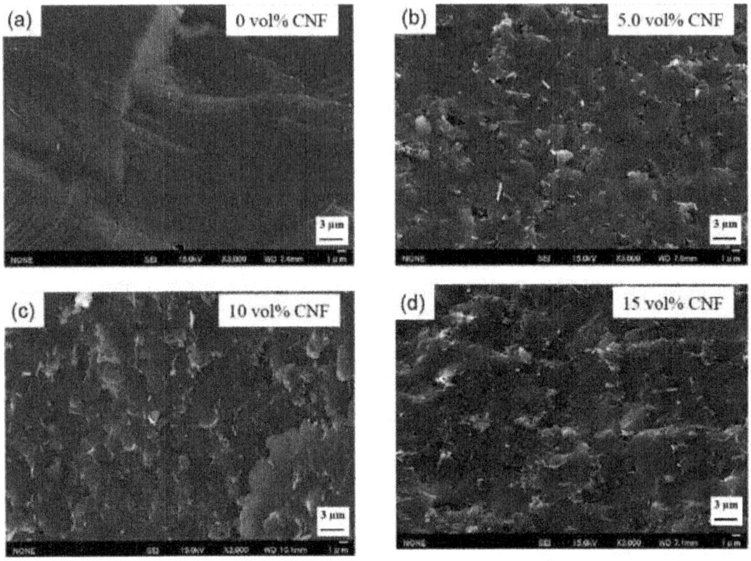

Figure 3. SEM photographs for the fractured surfaces of B_4C/CNF ceramics at 1873 K: B_4C/CNF= (a) 100/0, (b) 95/5, (c) 90/10 and (d) 85/15 vol%.

These SEM photographs also proved little change in microstructures. In addition to FE-SEM, precise observation using FE-TEM was conducted as shown in Fig. 4 (f). This photograph shows the interface between B_4C grain and CNF. Nevertheless CNF's lack of affinity, the direct bonding between CNF and B_4C is recognized. This strong covalent bonding among carbides might induce extraordinary high-strength by "Anchor-effects by CNF".

Then, other mechanical properties at high temperatures were evaluated. It is very difficult to *in situ* evaluation of the fracture toughness at elevated temperatures, especially higher than 1273 K. Then, we tried to measure load–displacement (stress-strain) curves during high-temperature bending test and estimate their areas inside the curves. This area was thought to correspond to the elastic strain energy approximately. Figure 5 displays the load–displacement curves of monolithic and composite ceramics measured at various temperatures.

It is very clear that the curves of monolithic B_4C are smaller than those of composites at all temperatures. Furthermore, 12.5 vol% CNF added composites display large curves even at 1973 K. These elastic strain energies of the monolithic and composite ceramics were summarized. Figure 6 shows the normalized elastic strain energy, *i.e.*, the elastic strain energy density of the materials, as a function of bending strength test temperature.

From these, it is cleared that the composites revealed high toughness at elevated temperatures, and their toughness reached a peak at 1773 K and it should be noted that 12.5 vol% CNF added composite showed higher toughness than the monolithic by a factor of around 20, however, the composites show moderate mechanical properties at room temperature.

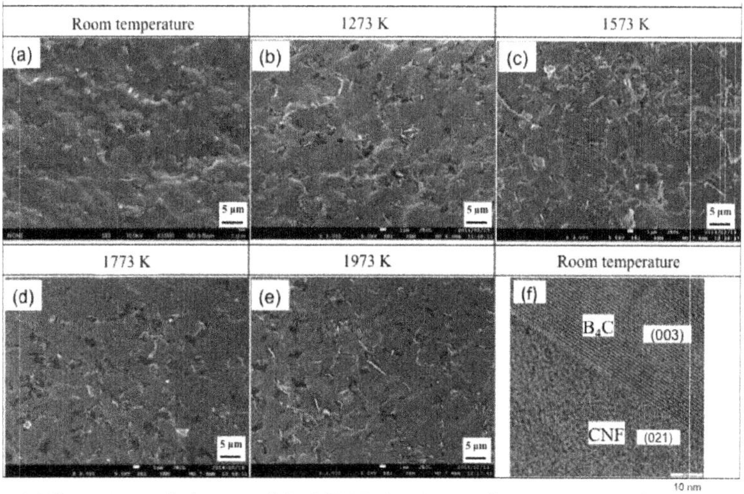

Figure 4. Microstructural changes of B_4C/CNF=87.5/12.5 vol% composites after the high-temperature bending test.

THERMOELECTRIC PROPERTIES

On the other hand, boron carbides have been reported to have high Seebeck coefficient S (~220 $\mu V \cdot K^{-1}$at 360 K) [22] and moderate electric conductivity σ_e (1.3~3.3 $\times 10^2$ S·m) [23]. Therefore, thermoelectric properties of B_4C/CNF were evaluated because high σ_e was expected by the addition of highly electrical conductive CNF [5]. Figure 7 represents the S value of composites with various compositions as a function of measuring temperature; monolithic B_4C shows high S about ~2.2×10^2 $\mu V \cdot K^{-1}$, this value is the same as reported for fine-grained B_4C ceramics [22]. On the other hand, once CNF added to B_4C, S values at room temperature were reduced to ~2.5 $\times 10^1$ $\mu V \cdot K^{-1}$, however they increased gradually to ~1.7 $\times 10^2$ $\mu V \cdot K^{-1}$ with increasing temperature; this might be explained in terms of the higher σ_e, as will be described in the latter.

Electrical conductivity σ_e is presented in Fig. 8; the values are increased from nearly 0.2 $\times 10^3$ to ~ 7.5 $\times 10^3$ $S \cdot m^{-1}$ with increasing temperature. The present σ_e value (1.2×10^3 $S \cdot m^{-1}$) at room temperature is a little higher than the reported values (1.3~3.3 $\times 10^2$ $S \cdot m^{-1}$) [23].

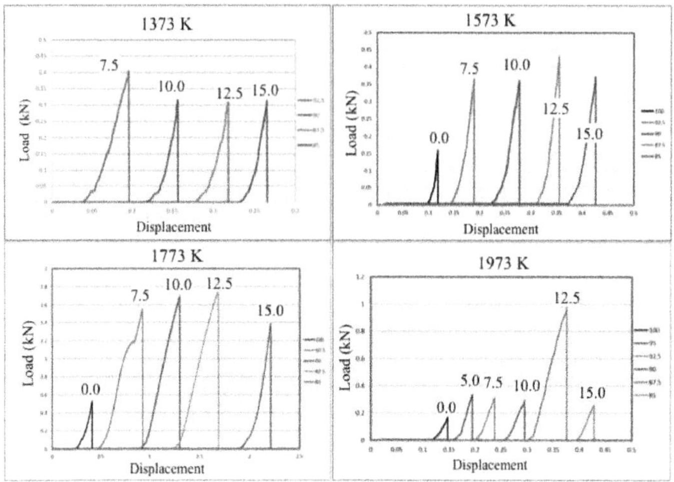

Figure 5. Load-displacement curves for the B_4C/CNF composite ceramics at 1373, 1573, 1773 and 1973 K.

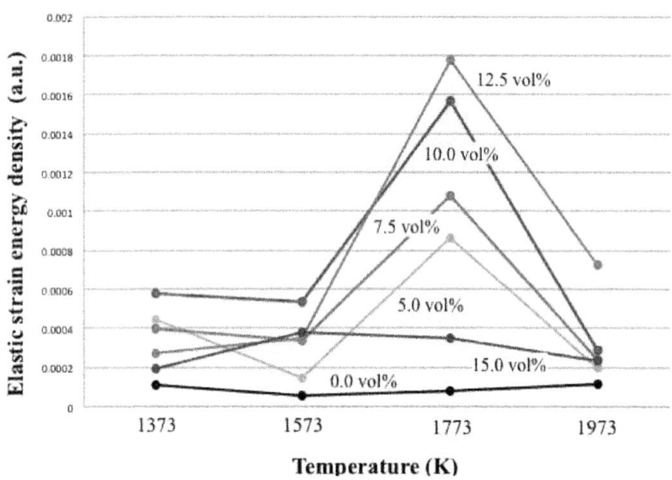

Figure 6. Elastic strain energy density of B_4C/CNF composites evaluated using the areas formed by the load-displacement curves measured at high temperatures.

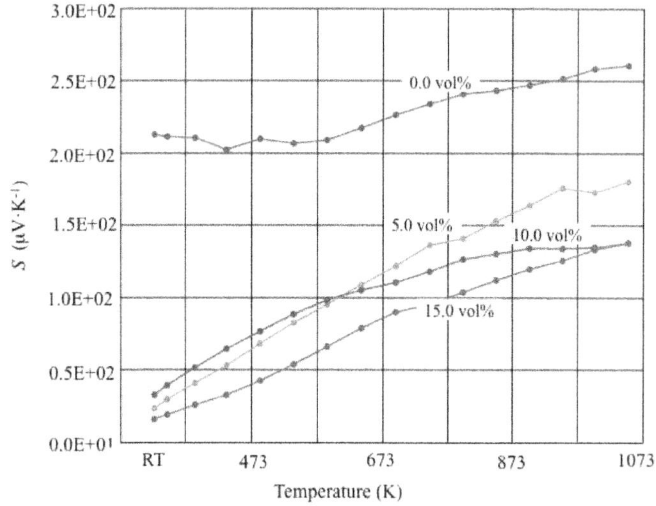

Figure 7. Seebeck coefficient S of B_4C/CNF composites as a function of temperature.

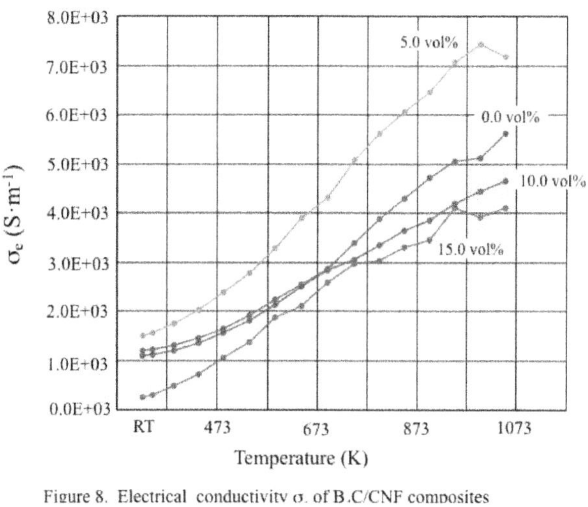

Figure 8. Electrical conductivity σ_e of B_4C/CNF composites with various compositions: as a function of temperature.

And as expected before, these graphs proved that the addition of CNF improved σ_e and there is the optimum content for CNF, *i.e.*, 5.0 vol%. 5.0 vol% CNF added B_4C revealed ~7.5×10^3 $S\cdot m^{-1}$ at ~950 K.

Then, thermal conductivity λ was evaluated from room temperature to 1273 K in order to calculate dimensionless figure of merit $ZT=(S^2*\sigma_e)T\cdot\lambda^{-1}$. Figure 9 reveals thermal conductivity λ of composites as a function of temperature, where, λ at R.T. was determined with a "*LFA447 Nano Flash*" in air and λ from 673 to 1273 K "*LFA 457 Micro Flash*" in Ar.

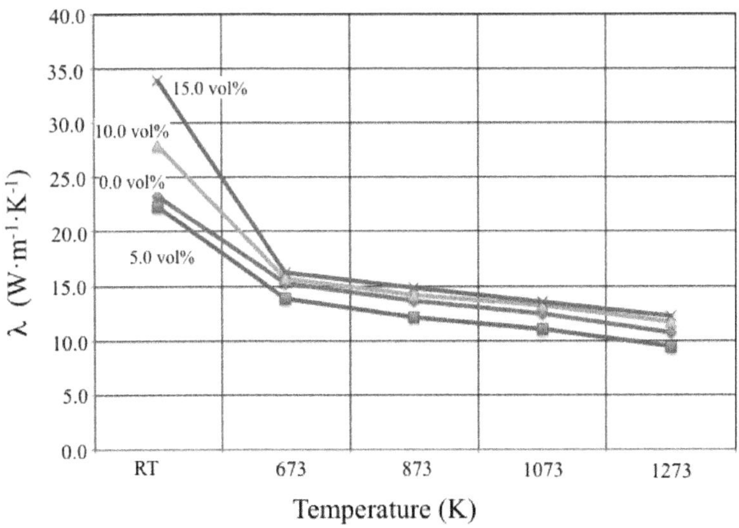

Figure 9. Thermal conductivity λ of B_4C/CNF composites as a function of temperature.

λ at R.T. changed from 22 to 34 $W\cdot m^{-1}K^{-1}$ are the same as the value λ=28.8 $W\cdot m^{-1}K^{-1}$ [24] reported on B_4C ceramics. When we consider the λ(CNF)=1200 $W\cdot m^{-1}\cdot K^{-1}$ [20], it should be noted that the experimental data (22 ~ 34 $W\cdot m^{-1}\cdot K^{-1}$) are almost constant nevertheless, increasing CNF content. Although σ_e was increased as mentioned above, λ values were not increased with increasing CNF content; this reason might be explained by that in the former case of σ_e, their carrier are electrons, on the other hand, in the latter, λ, their main carrier is phonons and not electrons. Therefore, σ_e could be improved by addition of highly electrical conductive CNF based on the percolation theory [25], however, in the latter case, main carrier of phonon is brought from the highly elastic B_4C matrix and CNF could not supply a lot of phonons. Using these data, dimensionless figure of merit ZT of B_4C/CNF at high temperatures were calculated (Fig.10).

The values of ZT reached 0.7 ~ 2.3×10^{-2} at the compositions of 0 ~ 15 vol% CNF and 1073 K; these values are not enough high for practical applications. However, under the situation that it is very difficult to increase both S and σ_e, furthermore, to decrease λ, or at least to keep constant in monolithic materials, it should be noted that by adopting the best combination of the high S-material and the high σ_e-material without increasing λ, totally high ZT composites could be fabricated.

Now, we should have described the dimensions of these B_4C/CNF composites, which can be fabricated using the large-scale pulsed electric-current pressure sintering (PECPS), would be outer diameter of 40 ~ 60 mm and thickness of 10 ~ 20 mm.

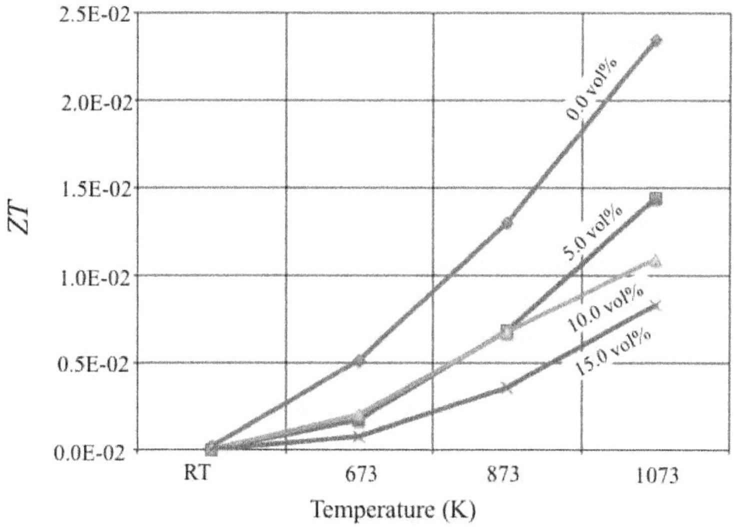

Figure 10. Dimensionless figure of merit ZT of B_4C/CNF composites as a function of temperature.

CONCLUSIONS

Now we have fabricated fully dense pure B_4C/CNF composites at relatively low temperature of 2173 K (1900°C) for only 6.0×10^2 s (10 min) by utilizing both SHS (self-propagating high-temperature synthesis) between amorphous fine B and C powders, and PECPS (pulsed electric-current pressure sintering) heating. The following results have been cleared.

1. Their mechanical properties at room temperature are;
   i) Bending strength σ_b: 455 ~ 635 MPa, ii) Vickers hardness H_v: 28.8~ 34.0 GPa, and iii) fracture toughness K_{IC}: 4.94 ~5.67 MPa·m$^{1/2}$.
2. High-temperature strength σ_b are;
   i) σ_b 830 MPa at 1773 K (12.5vol% CNF), ii) σ_b 800 MPa at 1873 K (10.0vol% CNF)
3. High-temperature elastic strain energy density;
   i) CNF dispersed B_4C composites showed much improved values,
   ii) Especially, 12.5 vol% CNF dispersed B_4C revealed higher values by a factor of ~ 20 at 1773 K, in comparison with monolithic B_4C.
4. These improved mechanical behaviors at high temperatures might come from
   i) High density and small grain sizes suppressed by CNF,
   ii) Anchor-effects induced by CNF,
   iii) Inverse temperature dependence of strength for carbon-based materials.
5. Thermoelectric properties were evaluated;
   i) Seebeck coefficient S was decreased a little with increasing CNF content.
   ii) As expected, electric conductivity σ_e was increased.

iii) However, thermal conductivity λ was almost constant.
iv) Dimensionless figure of merit ZT reached 0.7 ~ 2.3×10^{-2} at the compositions of 0 ~ 15 vol% CNF and 1073 K.

From these continuous studies, it has been cleared that dense B_4C/CNF composites thus fabricated are expected to be used for the key parts of airplane and marine engines which requires high bending strength and toughness at high temperatures around 1773 ~ 1873 K in the future.

ACKNOWLEDGMENT

This work was supported by a grant to Research Centre for Advanced Science and Technology at Doshisha University from the Ministry of Education, Culture, Sports, Science and Technology, Japan (The project S1311036). The authors thank Mr. O. Tsukamoto in Netzsch Japan for measuring thermal conductivity of B_4C/CNF composites at high temperatures. And also thank to Ms. M. Toda and Ms. J. Morita in Doshisha University Research Centre for Interfacial Phenomena for FE-SEM and FE-TEM observations of the samples.

REFERENCES

[1]E. Hammel, X. Tang, M. Trampert, T. Schmitt, K. Mauthner, A. Eder, and P. Pötschke,Carbon nanofibers for composite applications, *Carbon,* **42**, 1153-1158 (2004).

[2]J. W. An, D. H. You, and D. S. Lim, Tribological properties of hot-pressed alumina-CNT composites, *Wear,* **255**, 677-681 (2003).

[3]R. W. Siegel, S. K. Chang, B. J. Ash, J. Stone, P. M. Ajayan, R. W. Doremus, and L.S. Schqdler, Mechanical behavior of polymer and ceramics matrix nanocomposites, *Scripta Mater.,* **44**, 2061-2064 (2001).

[4]M. Endo, Y.A. Kim, T. Hayashi, K. Nishimura, T. Matsushita, K. Miyashita, and M.S. Dresselhaus, Vapor-grown carbon fiber (VGCFs)—Basic properties and their battery application, *Carbon,* **39**, 1287-1297 (2001).

[5]CNF Catalogue from supplier (Showa Denko K. K.). (http://www.sdk.co.jp/contents/product/index.htm)

[6]L. Jiang, L. Gao, and J. Sun, "Production of aqueous colloidal dispersions of carbon nanotubes," *J. Colloid and Interface Sci.,* **260**, 89-94 (2003).

[7]J. Sun, and L. Gao, "Development of a dispersion process for carbon nanotubes in ceramic matrix by heterocoagulation," *Carbon,* **41**, 1063-1068 (2003).

[8]Ch. Laurent, A. Peigney, O. Dumortier, and A. Rousset, Carbon-nanotubes-Fe-alumina nanocomposies, Part II: Microstructure and mechanical properties of the hot-pressed composites, *J. European. Ceram. Soc.,* **18**, 2005-2013 (1998).

[9]K. Hirota, K. Mitani, M. Yoshinaka, and O. Yamaguchi, Simultaneous synthesis and consolidation of chromium carbides (Cr_3C_2, Cr_7C_3 and $Cr_{23}C_6$) by pulsed electric-current pressure sintering. *Mater. Sci. Eng. A*, **399**, 154-160 (2005).

[10]M. Tokita, Trend in Advanced SPS Spark Plasma System and Technology, *J. Soc. Powder Tech. Jpn*, **30**, 790-804 (1993).

[11]K. Hirota, Y. Takaura, M. Kato, and Y. Miyamoto, Fabrication of Carbon Nanofiber (CNF)-Dispersed Al_2O_3 Composites by Pulsed Electric-Current Pressure Sintering and their Mechanical and Electrical Properties, *Mater. Sci*, **42**, 4792-4800 (2007).

[12]K. Hirota, H. Hara, and M. Kato, Mechanical Properties of Simultaneously Synthesized and Consolidated Carbon Nanofiber (CNF)-Dispersed SiC Composites by Pulsed Electric-Current Pressure Sintering, *Mater. Sci. & Eng., A* **458**, 216-225 (2007).

[13]K. Hirota, T. Shibata, M. Kato, and T. Nishimura, Fabrication of carbon nanofiber (CNF)/SiAlON composites by pulsed electric-current pressure sintering and their mechanical and electrical properties, *Materials Science and Technology (MS&T)*, 309-320 (2007).

[14]K. Hirota, Y. Nakayama, S. Nakane, M. Kato, and T. Nishimura, The Study on Carbon Nanofiber (CNF)-Dispersed B_4C Composites, *Int. J. Appl. Ceram. Technol.,* **6**, 607-616 (2009).

[15]A.G. Merzhanov, and I.P. Borovinskaya, Self-Propagating High-Temperature Synthesis of Refractory Inorganic Compounds. *Dokl Acad Nauk SSSR*, **204**, 429-432 (1972).

[16]Z.A. Munir, Synthesis of High Temperature Materials by Self-Propagating Combustion Methods, *Am. Ceram. Soc. Bull*, **67**, 342-349 (1988).
[17]J. Subrahmanyam, and M. Vijayakumar, Review Self-Propagating High-Temperature Synthesis, *J. Mater. Sci.*, **27**, 6249-6273 (1992).
[18]K. Hirota, Y. Nakayama, S. Nakane, M. Kato, and T. Nishimura, The study on carbon nanofiber (CNF)-dispersed B_4C Composites, *Int. J. Appl. Ceram. Technol.* **6** [5], 607-616 (2008).
[19]*Dictionary for Physics and Chemistry 3rd* (Tokyo: Iwanami Shoten Publisher, 1981), 1267.
[20]Information from the supplier (Mitsubishi Chemical Co. Ltd., Tokyo, Japan). (http://www.mkagaku.co.jp/about/mcc/division/carbon/carbon_b/index.htm)
[21]K. Niihara, R. Morena, and D. P. H. Hasselman, Evaluation of K_{Ic} of Brittle Solids by the Indentation Method with Low Crack-to-Indent Ratios, *J. Mater. Sci. Lett.*, **1**, 13-16 (1982).
[22]T. Goto, Thermoelectric conversion function materials and process. Borides system high temperature thermoelectric materials, *Kinzoku (metals)*, **68**, 1086-1092 (1998) in Japanese.
[23]*Handbook of Materials & Engineering 3rd* (New York: CRC Press, 2001), 956.
[24]*Handbook of Materials & Engineering 3rd* (New York: CRC Press, 2001), 404.
[25]D.S. McLachlan, M. Blaszkiewicz, and R.E. Newnham, Electrical resisitivity of composites. *J. Am. Ceram. Soc.*, **73**, 2187-2203 (1990).

Innovative Processing

ADVANCED MICROSTRUCTURAL STUDY OF NICKEL-TITANIUM ROTARY ENDODONTIC INSTRUMENT TIPS

Rahnuma Chowdhury,[1] Matthew R. Wheeler,[1] William A.T. Clark,[1] William A. Brantley[2] and John M. Nusstein[3]

[1]Department of Materials Science and Engineering, Ohio State University, Columbus, OH, USA

[2]Division of Restorative Science and Prosthodontics and [3]Division of Endodontics, College of Dentistry, Ohio State University, Columbus, OH, USA

ABSTRACT

Background: Nickel-titanium rotary instruments are used worldwide in endodontics. This investigation employed advanced microscopy techniques to obtain new information about microstructures at the tip region of importance for clinical performance. Representative instruments (HyFlex® CM™) from one manufacturer (Coltène Whaledent) were selected. SEM examination and EDS analyses were performed (Quanta 200, FEI, Hillsboro OR). Instruments were serial-sectioned (Helios NanoLab 600 FIB, FEI) for 3D image reconstruction. Results: The tip regions contained nanoscale NiTi phases, along with other phases and porosity. Special crystallographic orientation was observed for the NiTi phases. Conclusions: Novel insight was obtained about the complex microstructures of these instruments in the tip region, which may contribute, along with overload and/or excessive fatigue cycles, to occasional rotary instrument failure under clinical conditions. TEM examination is necessary for detailed phase identification and determination of specific orientation relationships. Additional examination of as-received instruments and clinically used instruments is needed.

INTRODUCTION

The use of nickel-titanium rotary instruments with slow-speed dental handpieces to enlarge and shape root canals has revolutionized endodontics,[1] and a review article has recently discussed the impact of nickel-titanium instrumentation on clinical outcomes.[2] In their pioneering study, Walia et al.[3] pointed out that the much lower elastic modulus of near-equiatomic NiTi, compared to the austenitic stainless steel alloy that had been used, would allow endodontic instruments to negotiate curved root canals with much greater facility.

The traditional manufacturing technique involves machining the NiTi rotary instruments from starting wire blanks in the pseudoelastic condition[4] with substantial transformable austenitic NiTi.[5] Because occasional instrument fractures occur under clinical conditions,[6] there has been considerable interest in modifying the NiTi alloy to yield improved properties, notably fatigue performance.[7] The recent introduction of a new process for the NiTi wire blanks that yields a heavily cold-worked martensitic structure[8,9] has led to a new-generation of rotary endodontic instruments with improved performance[10-17].

While there have been several investigations of these new rotary NiTi instruments and the starting wire blanks using scanning electron microscopy, transmission electron microscopy, differential scanning calorimetry and X-ray diffraction,[8,10,11,13,15,17] much work remains to understand the complex microstructures of these instruments. The present study has employed advanced microscopic techniques to examine the tip region of clinical interest for a recently introduced next-generation rotary instrument. This initial investigation has provided new insight into how the manufacturing process influences the microstructure and introduction of microstructural defects that should have pivotal roles for clinical performance of the instruments.

EXPERIMENTAL PROCEDURES

A representative rotary instrument (HyFlex® CM™) from one manufacturer (Coltène Whaledent), similar to those analyzed by Micro-X-ray diffraction in our previous study,[17] was selected. Scanning electron microscopic (SEM) examination and X-ray energy-dispersive spectrometric (EDS) analyses were performed (Quanta 200, FEI, Hillsboro OR). The instrument was serial-sectioned (Helios NanoLab 600 FIB, FEI) for three-dimensional image reconstruction of the analysis results. This initial microscopy study focused on the tip region of the instrument, which is of considerable clinical interest.

RESULTS

The tip regions of the HyFlex® CM™ rotary instruments have complex microstructures, containing NiTi phases and other intermetallic particles, along with porosity. EDS analyses revealed that Ti-rich carbide precipitates were present and precipitate compositions were complex. Some porosity was also observed. A representative cross-section near the tip is shown in Figure 1, and a series of such sections is presented in Figure 2.

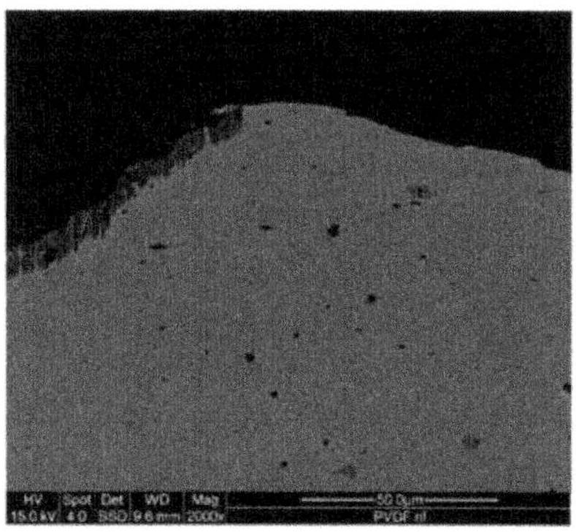

Figure 1. SEM image showing intermetallic phases and precipitates in cross-section near tip.

Figure 2. Series of FIB cross-sections obtained at instrument tip. (Note 15 μm scale bar.)

A higher-magnification SEM image of the surface of a tip region cross-section (FIB reconstruction) is shown in Figure 3, and two different intermetallic phase particles can be seen. The small linear features near the center of the image and large intermetallic particle are pores.

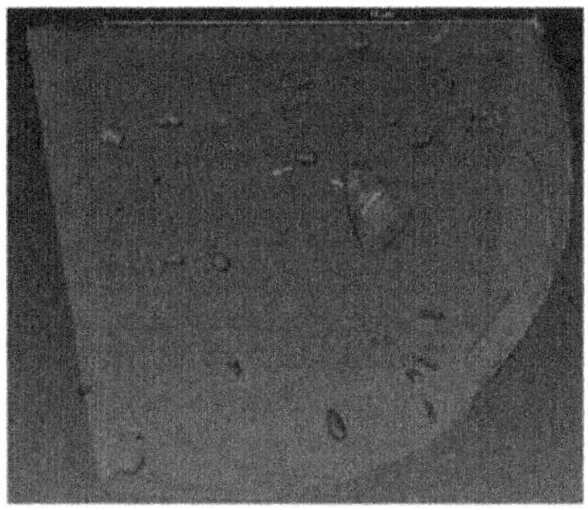

Figure 3. SEM image of reconstructed volume element at tip region (scale bar = 15 μm), revealing intermetallic phases and some porosity.

Figure 4 presents an extended reconstructed image from the FIB cross-sections, showing the two types of precipitates seen in Figure 3, along with oriented grains parallel to the instrument axis. The nanoscale dimensions of these grains is evident from the 15 μm scale bar at the top of this figure.

Figure 4. Nanoscale NiTi grains in tip region oriented parallel to instrument axis. Intermetallic phases, similar to those shown in Figure 3, can also be seen. (Scale bar = 15 μm)

DISCUSSION

Transmission electron microscopic (TEM) examination and differential calorimetric analyses show that the new-generation NiTi rotary endodontic instruments have a heavily work-hardened martensitic ultrastructure, with higher austenite-finish temperatures than conventional NiTi rotary instruments.[8,11,13] These new instruments have improved flexibility and fatigue life, which are considerable benefits for in vivo use.[7,12,16]

HyFlex® instruments have previously been investigated by conventional X-ray diffraction[15] and Micro-X-ray diffraction[17]. The latter can provide analyses at regions of approximately 100 µm dimensions, whereas conventional X-ray diffraction only yields information about phases for the overall instrument. Micro-X-ray diffraction has shown that the tip region of HyFlex instruments contain both austenitic and martensitic NiTi phases, but the peak intensities are much lower than for analysis sites at 3, 6, 9 and 9 mm from the tip region because of substantially greater permanent deformation. [17] Very similar Micro-X-ray diffraction peaks were observed for the Vortex Blue® (Dentsply Tulsa Dental Specialties) next-generation rotary instruments. [17]

The present investigation, employing three-dimensional reconstruction of SEM images from FIB-derived cross-sections, has provided novel information about the microstructures of the HyFlex instruments. The reconstructed images show the existence of an oriented nanostructure in the tip region, along with the presence of other intermetallic phases and some porosity. The oriented microstructure cannot be established by conventional metallographic examination of polished and etched rotary instrument specimens, which show a classic lenticular martensitic structure.[10,11] The oriented microstructure in the tip region and the intermetallic phases provide strengthening, but effects of the relatively small amount of porosity and these phases on the crack-microstructure interactions involved in fatigue failure remain to be determined.

This study has been limited to the tip region of an instrument from one manufacturer, and additional regions along the working lengths of instruments from several manufacturers should be investigated using the present methodology. Future TEM investigations of the complex ultrastructures of as-received and retrieved instruments are needed to determine the orientation relationships within the martensitic structure and their evolution with clinical use. Such studies should include the new-generation (R-phase) rotary instruments that are fabricated by twisting, rather than machining.[7] Electron diffraction is needed to identify the locations of austenitic NiTi, martensitic NiTi, and the intermediate R-phase. Energy-dispersive analyses with the TEM are also needed to determine the compositions of the intermetallic phase particles. The metallurgy of the NiTi instruments is complex,[18] and proprietary heat treatments now being performed by manufacturers on the instruments yield further unknown changes in the ultrastructure.

CONCLUSIONS

The present investigation has shown that as-received rotary endodontic instruments from one manufacturer have complex microstructures and some porosity in the tip region. These microstructures are the result of the NiTi alloy composition and processing by manufacturer. Advanced microscopy techniques utilizing three-dimensional reconstruction of SEM images from FIB sections have new insight into how these microstructures affect instrument properties.

ACKNOWLEDGMENTS

We thank Colténe Whaledent for providing the sample instruments. Expert assistance was provided by John Sosa and Brian Welk of CEMAS (Center for Electron Microscopy and Analysis, Ohio State University) in developing the three-dimensional image reconstruction program and performing the EDS analyses, respectively.

REFERENCES

[1]W.A. Brantley, Introduction of Nickel-Titanium Alloy to Endodontics, In: J.I. Ingle, L.F. Bakland, and J.C. Baumgartner (Editors), Ingle's Endodontics (6th Edition), Chapter 26B, B.C. Decker, Hamilton, Ontario, Canada (2008).

[2]E. Schäfer and S. Bürklein, Impact of Nickel–titanium Instrumentation of the Root Canal on Clinical Outcomes: A Focused Review, *Odontology*, **100**, 130-36 (2012).

[3]H. Walia, W.A. Brantley, and H. Gerstein, An Initial Investigation of the Bending and Torsional Properties of Nitinol Root Canal Files, *J. Endod.*, **14**, 346-51 (1988).

[4]S.A.Thompson, An Overview of Nickel-Titanium Alloys Used in Dentistry, *Int. Endod. J.*, **33**, 297-310 (2000).

[5]W.A. Brantley, T.A. Svec, M. Iijima, J.M. Powers, and T.H. Grentzer, Differential Scanning Calorimetric Studies of Nickel Titanium Rotary Endodontic Instruments, *J. Endod.*, **28**, 567-72 (2002).

[6]S.B. Alapati, W.A. Brantley, T.A. Svec, J.M. Powers, J.M. Nusstein, and G.S. Daehn, SEM Observations of Nickel-Titanium Rotary Endodontic Instruments that Fractured During Clinical Use, *J. Endod.*, **31**, 40-3 (2005).

[7]J.L. Gutmann and Y. Gao, Alteration in the Inherent Metallic and Surface Properties of Nickel–Titanium Root Canal Instruments to Enhance Performance, Durability and Safety: A Focused Review, *Int. Endod. J.*, **45**, 113-28 (2012).

[8]S.B. Alapati, W.A. Brantley, M. Iijima, W.A.T. Clark, L. Kovarik, C. Buie, J. Liu, and W.B. Johnson, Metallurgical Characterization of a New Nickel-Titanium Wire for Rotary Endodontic Instruments, *J. Endod.*, **35**, 1589-93 (2009).

[9]W.B. Johnson, Fatigue-resistant Nitinol Instrument, US Patent 2012/0231414 A1 (September 13, 2012).

[10]Y. Shen, H. Zhou, Y. Zheng, L. Campbell, B. Peng, and M. Haapasalo, Metallurgical Characterization of Controlled Memory Wire Nickel-Titanium Rotary Instruments, *J. Endod.*, **37**, 1566–71 (2011).

[11]W.A. Brantley, J. Liu, F. Zheng, S.R. Schricker, J.M. Nusstein, W.A.T. Clark, L. Kovarik, M. Iijima, and S.B. Alapati. Next-Generation Rotary Endodontic Instruments Fabricated from Special Nickel-Titanium Alloy, In: R. Narayan, S. Bose, and A. Bandyopadhyay (Editors), Biomaterials: Science, Processing, Properties, and Applications II, *Ceram. Trans.*, **237**, 11-8 2012.

[12]O.A. Peters, A.K. Gluskin, R.A. Weiss, and J.T. Han, An in vitro Assessment of the Physical Properties of Novel Hyflex Nickel-Titanium Rotary Instruments, *Int. Endod. J.*, **45**, 1027-34, (2012).

[13]W.A. Brantley, J. Liu, S.R. Schricker, F. Zheng, J.M. Nusstein, M. Iijima, W.A.T. Clark, and S.B. Alapati, Characterization of Next-Generation Nickel-Titanium Rotary Endodontic Instruments, In: S. Bose, A. Bandyopadhyay, and R. Narayan (Editors), Biomaterials Science: Processing, Properties, and Applications III, *Ceram. Trans.*, **242**, 11-8 (2013).

[14]E.Ninan and D.W. Berzins, Torsion and Bending Properties of Shape Memory and Superelastic Nickel-Titanium Rotary Instruments, *J. Endod.*, **39**, 101-4 (2013).

[15]Y. Shen, J.M. Coil, H. Zhou, Y. Zheng, and M. Haapasalo, HyFlex Nickel-Titanium Rotary Instruments after Clinical Use: Metallurgical Properties, *Int. Endod. J.*, **46**, 720-9 (2013).

[16]H.H. Nguyen, H. Fong, A. Paranjpe, N.M. Flake, J.D. Johnson, and O.A. Peters, Evaluation of the Resistance to Cyclic Fatigue among ProTaper Next, ProTaper Universal, and Vortex Blue Rotary Instruments, *J. Endod.*, **40**, 1190-3 (2014).

[17]W.A. Brantley, M. Iijima, W.A.T. Clark, S.R. Schricker, J.M. Nusstein, and I. Mizoguchi, Micro-X-ray Diffraction Study of New Nickel-Titanium Rotary Endodontic Instruments, In: R. Narayan,

S. Bose, and A. Bandyopadhyay (Editors), Biomaterials Science: Processing, Properties, and Applications V, *Ceram. Trans.*, **254**, 47-54 2015.
[18]K. Otsuka and X. Ren, Physical Metallurgy of Ti–Ni-Based Shape Memory Alloys, *Progr. Mater . Sci.*, **50**, 511-678 (2005).

SYNTHESIS OF $TiC-TiB_2$ COMPOSITE POWDERS FROM CARBON COATED TiO_2 PRECURSORS

Zhezhen Fu, Rasit Koc
Department of Mechanical Engineering and Energy Processing, Southern Illinois Univerisyt, 1230 Lincoln Drive, Carbondale, Illinois, US

ABSTRACT

Submicron $TiC-TiB_2$ composite powders were synthesized by carbo/borothermal reduction from carbon coated precursor. TiO_2 directly mixed with carbon black and B_4C were also investigated as comparison. Precursors and traditionally mixed powders were reacted from 1100 to 1500°C for 2 hours in flowing argon at 1L/min for both methods. Phase evolution, microstructures and surface areas were characterized by X-ray diffraction (XRD), transmission electron microscope (TEM) and Brunauer-Emmett-Teller (BET) surface area analyzer, respectively. Products synthesized from precursors had high purity and single phases of $TiC-TiB_2$ mixture, fine particle size, narrow size distribution, regular shape, loose agglomeration, and high surface area. Composite powders with different content of TiB_2, which were controlled by changing of ratio between TiO_2, C and B_4C, were synthesized. The sintering properties were tested on synthesized powders with 5 wt. % nickel through pressureless sintering at 1550°C in argon atmosphere, which lead to a relative density of 99.6% theoretical density (TD).
Keywords: carbon coated precursor; TiC; TiB_2; carbothermal reduction; borothermal reduction

INTRODUCTION

Both titanium carbide(TiC) and titanium diboride(TiB_2) combine the advantages of high melting point, good mechanical properties, high electrical conductivity, and high resistance to oxidation and corrosion. These materials have many potential applications in key technologies from mechanical and chemical industries to electronics[1,2]. Compared with monolithic TiC or TiB_2, $TiC-TiB_2$ composites appear better properties, including high thermal shock resistance, stability under high temperature, and high fracture toughness[1]. The most widely studied methods to synthesize $TiC-TiB_2$ composites are self-propagating high-temperature synthesis (SHS)[3,4] and high energy ball milling[5,6,7]. Using titanium, B_4C/B, and carbon as starting materials, reactions are as followings:

$$3Ti + B_4C \rightarrow TiC + 2TiB_2 \quad (1)$$
$$4Ti + B_4C + C \rightarrow 2TiC + 2TiB_2 \quad (2)$$
$$2Ti + 2B + C \rightarrow TiB_2 + TiC \quad (3)$$

However, very fine titanium and boron powders are needed. Large scale production is also not possible for both methods. Substoichiometric TiC_x (x is much smaller than 1)[4], which has demonstrated a deleterious effect on both mechanical and electrical properties[8,9], may also form in final products. From equation 1, content of TiB_2 cannot be higher than 66.7 mol. %, which means non-controllable composition. Former research has demonstrated that fine microstructure ceramic materials have an advantage in mechanical properties[7]. Submicron or nano size $TiC-TiB_2$ composite powders can be synthesized through high energy ball milling or SHS methods. But either long time period (up to 72 hours)[6,10] or annealing is needed[4,6,10]. This also leads to high cost of production.

Commercial methods, which means large scale of products and economic on costs, to produce TiC[11] and TiB_2[2] are from carbothermal and borothermal reduction, respectively. Mixtures of TiO_2, carbon, and B_4C are reacted at temperature up to 1800°C in vacuum or with inert atmosphere protecting:

$$TiO_2 + 3C \rightarrow 3TiC + CO\ (g) \quad (4)$$
$$2TiO_2 + 3C + B_4C \rightarrow 2TiB_2 + 4CO\ (g) \quad (5)$$

Carbothermal and borothermal reduction methods have the advantages of large scale products and are cost effective compared with SHS and high energy ball milling method. Carbon coating TiO_2 precursor by carbothermal reduction to prepare TiC, TiN, Ti(C, N) was developed by Koc[12]. Instead of carbon black, the carbon is from the pyrolysis of propylene. This would eliminate the contaminations like iron, copper and other inorganic elements in carbon black. At the same time, pyrolysis resulted carbon would deposit on the surface of TiO_2 to ensure intimate contact of reactants. Because of those factors, reaction temperature can be lowered by up to 200°C. Also, the produced powders have the advantages of pure and single phase, submicron in particle size, narrow distribution, loose agglomerate and spherical particle shape. Carbon coating precursor method can also be applied in the synthesis of TiB_2[13], SiC[14], W[15], WC[16], and CaB_6[17]. Except TiC-Al_2O_3[18], no other composite has been prepared through the coating precursor method. Most of above reports concern synthesis only and sintering properties of produced powders are not clear yet.

The present paper aims to prepare high quality, varying and controlling TiB_2 content of TiC-TiB_2 composite powders in large scale using carbo/borothermal reduction from carbon coating precursor. Reaction equation is designed as following:

$$(x+2y)TiO_2 + yB_4C + (3x+3y)\,C \rightarrow xTiC + 2yTiB_2 + (2x+4y)\,CO\,(g) \quad (6)$$

Since this is the combination of equation 4 and 5, the content of TiB_2 can change from 0 to 100% theoretically. The content of TiB_2 is controlled by changing the ratio between TiO_2, C and B_4C in the mixtures. Intermediate phases of composite powders with 60 wt. % TiB_2 are investigated by XRD. Final products, which contain 40, 60 and 80 wt. % TiB_2, are characterized by TEM, XRD, and surface area analyzer. Sintering properties are examined through presureless sintering with nickel.

EXPERIMENT

The synthesis process consists of two steps. The first step is the coating of TiO_2 (Aeroxid P25, ACROS Organic, NJ) with carbon by decomposing a hydrocarbon gas (C_3H_6) at temperature of 600°C. This step was performed in a rotating chamber furnace consisting of a 10 cm ID×35 cm stainless steel cylinder. About 100 grams of TiO_2 powder was placed in the vessel. The vessel was evacuated, purged with argon, and evacuated again. The vessel was then heated to 600°C and filled with propylene gas until the pressure reached about 3 atm. After 5 minutes, the remaining gas was released and the vessel was filled with fresh propylene. The coating step was repeated until the desired amount of carbon was deposited. The weight percentage of the carbon in the precursor is determined using thermogravimetric analysis in air as 24.4 wt. % aiming for 60 wt. % TiB_2. In this case, equation 6 became:

$$2.29\,TiO_2 + 0.65\,B_4C + 4.95\,C \rightarrow TiC + 1.29\,TiB_2 + 4.58\,CO\,(g) \quad (7)$$

Precursors were then mixed with stoichiometric amounts B_4C (Grade HS, H.C Starck, Munich, German) in a plastic bottle with two methacrylate balls using a Spex Model 8000 Mixer/Mill (Spex Corp., Metuchen, NJ) for 20 minutes. In order to evaluate phase evolution, final mixtures were then heated at temperatures ranging from 1100 to 1500°C for 2 hours in a tube furnace (Model CTF 17/75/300, Carbolite, Sheffield, UK) with flowing argon gas (1L/min). Weights of samples before and after the reaction were recorded to calculate weight loss. Mixtures from conventional mixing of TiO_2, carbon black (Monarch 880, Cabot, Boston, MA) and B_4C aiming for 60 wt. % TiB_2 were also heated as comparison. About 15 mg of each mixture were applied for differential scanning calorimetry (DSC, 404C, Netzsch, German) characterization from room temperature to 1450°C with ramp rate of 20°C/min in flowing argon.

Another two batches carbon coated TiO_2 precursors were also prepared. Based on equation 6, coated carbon amount was determined as 26.9 and 21.6 wt. % respectively to prepare 40 and 80 wt. % TiB_2 composite powders. Weight percentage of coated carbon was controlled by cycles of coating process. Stoichiometric amounts of B_4C were then mixed with precursors at 1500°C for 2 hours in flowing argon.

Produced powders were then characterized by XRD (Rigaku D/Max-B, Tokyo, Japan) using Cu Kα (λ= 1.54056 Å) radiation and a scan rate of 1.8° 2θ/min, BET gas adsorption surface area analyzer (Micromeritics, Gemini 2360, Nocross, GA), and TEM (JEOL JEM-2100F). Mass fractions of produced samples were calculated based on the relative intensity of the strongest diffraction peak using the K value method as follows[19]:

$$x_j = \frac{1}{1 + I_R K_c^j / I_j K_C^R}$$

Where x_j is j phase's mass fraction, I_j and I_R are the relative intensities of j phase and R phase's strongest diffraction peak respectively, and K_C^j and K_C^R are the reference spectral intensities of j phase and R phase, respectively. Crystallite size was calculated from Scherrer equation[20]:

$$D = \frac{0.93\times\lambda}{\beta\times cos\theta}$$

Where D is the crystallite size, λ is the X-ray wavelength (λ=1.54056Å), β is line boarding at the half the maximum intensity and θ is the Bragg angle. TiC crystallite size was estimated based on the average values from (111)(200) (220) peaks and TiB$_2$ from (001)(100)(101) respectively.

Synthesized 40 wt. % TiC-60 wt. % TiB$_2$ powders from carbon coated precursors with 5 wt. % nickel powders(<1μm, 99.5% pure, CERAC, Milwaukee, WI) were mixed using the Spex Model 8000 Mixer/Mill for 1 hour. Pellets were made by single action, uniaxial pressing in a hardened steel die (ID=12.95 mm) under pressure of 200MPa. Sintering process was conducted at the temperature of 1550 °C for 2 hours with ramp rate of 4°C/min in the tube furnace under flowing argon. Densities of sintered pellets were determined by the Archimedes method. Scanning electron microscope (SEM) was used to investigate the polished surfaces. Vickers hardness values were measured 10 times for the sample with Shimazdu HSV-20 hardness tester using a load of 20 kg. The fracture toughness was determined using Vickers hardness indentation method.

RESULTS AND DISCUSSION

TEM of mixtures from carbon coated precursors and conventional mixing

TEM micrographs of mixtures of carbon coated TiO$_2$ precursors with B$_4$C and conventional mixing TiO$_2$, carbon black and B$_4$C were shown in Figure 1(a) and (b) respectively. As shown in Figure 1a , very uniform pyrolytic carbon layers were coated on the surface of nano TiO$_2$ particles. This will ensure the intimate contact between carbon and TiO$_2$. In this case, for carbon coating precursors, there would be sufficient and continues carbon source for reactions toward both TiB$_2$ and TiC. At the same time, uniform carbon layers prevent the direct contact between TiO$_2$ and B$_4$C. This may influence the formation path and intermediate phases. While in Figure 1b, reactants just mix together and TiO$_2$ particles do not have intimate contact with carbon. As a result, there is no continuous carbon source during reaction and would be considered as low reactivity. Meanwhile, direct contact between B$_4$C and TiO$_2$ in conventional mixing samples could result in different reaction paths.

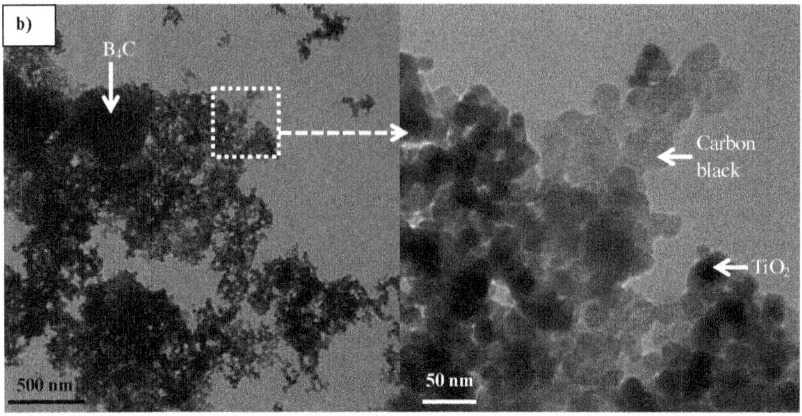

Figure 1. TEM micstructure of mixture from different sample preparation methods. (a) carbon coated precursors, (b) conventional mixing method.

DSC

DSC curves of two different mixtures are shown in Figure 2. Both sample preparation methods have two endothermic peaks. Peaks for carbon coated precursors are at 1236 and 1331°C respectively while conventional mixing samples at 1311 and 1377°C, are both at higher temperatures. This means higher reaction activity comparing with conventional mixing samples. Precursors with intimate contact between TiO_2 and C have higher carbonthermal reductions activity, which is consistent with former research about formation of TiC through carbon coated TiO_2 precursors[21]. At the same time, different contacting can also influence the activity of borothermal reductions. There is no extra peaks for conventional mixing sample although it was supposed to have more reactions because of the direct contact of TiO_2 and B_4C. This is probably because reactions happened at similar temperature range and can't form obvious split reaction peaks.

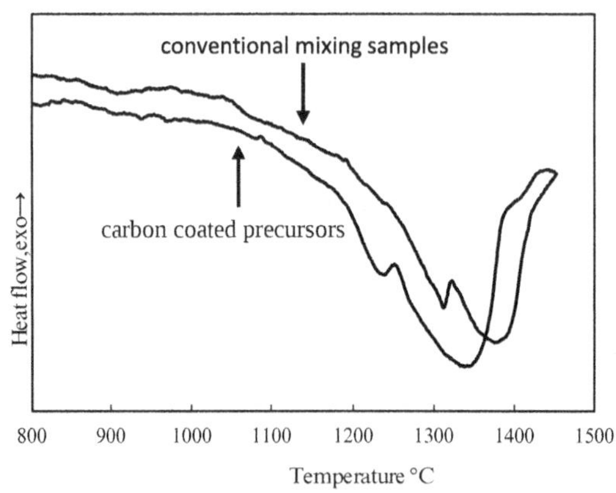

Figure 2. DSC curves for mixtures of carbon coated TiO_2 with B_4C and conventional mixing samples. Samples were heated in alumina crucible at 20 °C/min in flowing argon.

Weight loss, phase evolution and surface area

Based on equation 7, the theoretical weight loss of the reaction is 46.17%. In carbo/borothermal reduction reactions, by comparing the actual and theoretical weight loss, fractions of the reaction can be determined. From the results in Figure 3, when reaction temperature was 1100°C, either samples showed obvious weight loss. With temperature went up to 1200°C, carbon coated precursors method sample started weight loss while conventional mixed sample still had relative low weight loss. Samples prepared from carbon coated precursors had a weight loss of 46% at 1400°C, which means a near completed reaction by comparing with theoretical weight loss percentage. For directly mixed samples, the weight loss was just 39.4% at 1400°C. The conventional directly mixed method sample reached theoretical weight loss at 1500°C. This illustrates that conventional mixing samples have low reactivity because of high finalization reaction temperature. It should be noticed that weight loss of carbon coated precursors is close to linear between 1100°C to 1400°C. There is a slope between 1300°C and 1400°C for mixing samples,. This is because that sufficient carbon source in carbon coated precursors would ensure continuously reactions.

XRD patterns in Figure 4 and 5 show formation paths and phase evolutions at temperature range from 1100 to 1500°C. For carbon coated precursors method, phases were identified to be Ti_3O_5, B_2O_3 and TiC_xO_y at 1100°C; B_2O_3, TiC_xO_y, and TiB_2 at both 1200 and 1300°C. Intensities of B_2O_3 peaks decreased and TiB_2 increased between 1200 and 1300°C. Sharp TiC and TiB_2 peaks with no other impurity formed at 1400°C and 1500°C, which means pure phases TiC-TiB_2 composites were synthesized. For conventional mixed samples, more complicated reactions happened. Phases from carbon mixing method were determined to be $TiBO_3$, Ti_2O_3, Ti_3O_5, and B_2O_3 at 1100°C; $TiBO_3$, Ti_2O_3, B_2O_3, and TiC_xO_y at 1200 °C; $TiBO_3$, B_2O_3, and TiC_xO_y at 1300 °C; B_2O_3, TiB_2, and TiC_xO_y at 1400 °C. Sharp TiC and TiB_2 peaks can be observed at 1500°C finally. Based on XRD patterns, the calculated TiB_2 mass fraction of carbon coated precursors is 59 wt. %, comparing with 55 wt. % for mixing sample preparation method. This is the result of low reactivity for conventional mixing samples and lead to the evaporation of B_2O_3 during reaction.

Surface areas of samples from different preparation methods at temperature range between 1100 and 1500°C are shown in Figure 6. Both values increased at the lower temperature range. With reaction continues, surface areas decreased because the eliminating of carbon. The surface area of sample from carbon coated precursors at 1500°C is 5.8 m^2/g, comparing with 10.7 m^2/g from mixed samples. This is because mixed samples contain high level of unreacted carbon black.

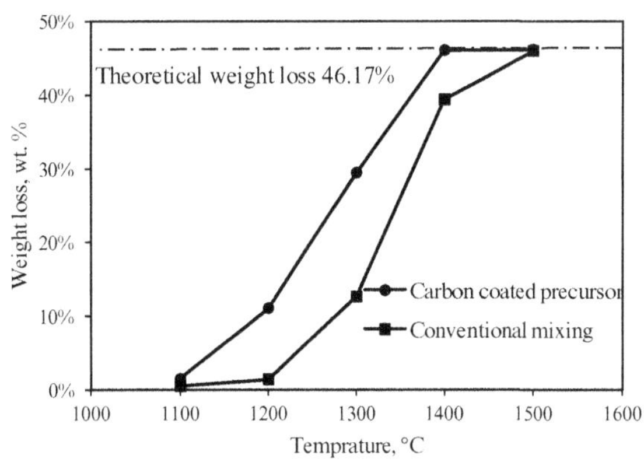

Figure 3. Weight loss of sample from different preparing method.

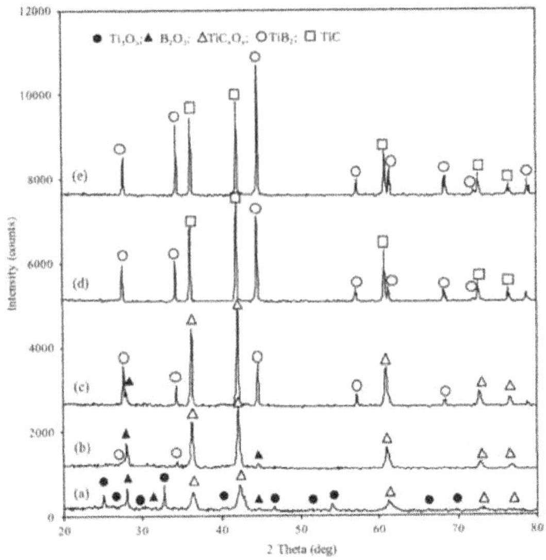

Figure 4. XRD patterns for samples from carbon coated precursors mixture at different temperature. (a) 1100°C, (b) 1200°C, (c) 1300°C, (d) 1400°C, (e)1500°C.

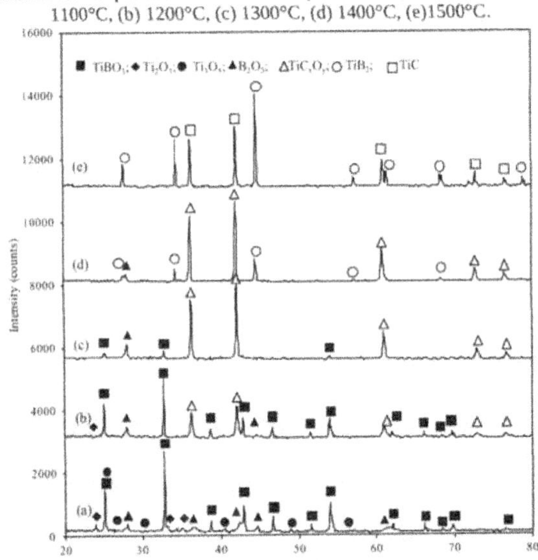

Figure 5. XRD patterns for samples from conventional carbon black mixing mixture at different temperature. (a) 1100°C, (b) 1200°C, (c) 1300°C, (d) 1400°C, (d)1500°C.

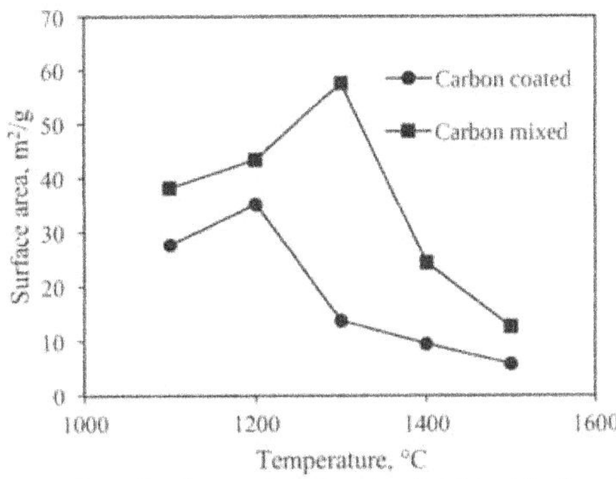

Figure 6. BET surface area of samples from carbon coated and carbon mixed precursors as a function of reaction temperature.

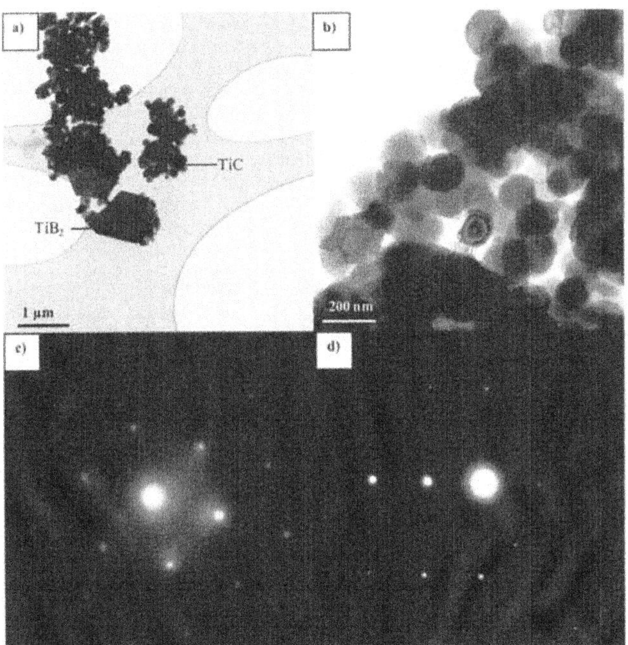

Figure 7. TEM micrographs of synthesized TiC-TiB$_2$ composite from carbon coated precursors at 1500°C for 2 hours and their diffraction patterns. (a)mixture of TiC and TiB$_2$, (b) enlarge view of TiC particles, (c) diffraction patterns of selected single particle TiB$_2$, (d) diffraction patterns of selected single particle TiC.

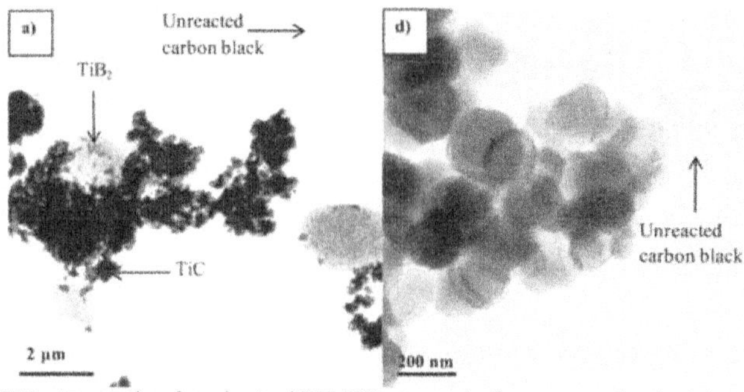

Figure 8. TEM micrographs of synthesized TiC-TiB_2 composite from conventional mixing at 1500°C for 2 hours. (a)mixture of TiC and TiB_2, (b)enlarge view of TiC particles.

TEM micrographs of final products

TEM micrographs of produced composites from carbon coated precursors are shown in Figure 7. TiB_2 particles have regular shape with size about 0.7 to 0.9 µm. Spherical shape TiC with loose agglomeration have particle size around 0.1 to 0.2 µm which means narrow distribution. Trace free carbon can be found through TEM. Diffraction patterns of selected single particles confirm the formation of TiC and TiB_2. However, samples from conventional mixing methods are much different as shown in Figure 8. Although fine TiB_2 and TiC particles with regular shape can also be observed, there are still some unknown particles. These might be intermediate phases such as $TiBO_3$, B_2O_3, or Ti(B,C) solid solution. At the same time, large amounts of unreacted carbon black are still left in the samples as shown in Figure 8 (b). TEM micrographs confirm former results and discussion from XRD patterns and BET surface areas. This means uncompleted reaction for samples from conventional mixing even at 1500°C. All of the above differences about micrographs are the result of reactivity of different sample preparation method. Carbon coated precursors mixing with B_4C are more reactive because of intimate contact and continues carbon source during reaction. Sample made from carbon coating precursors at 1500°C features with fine particle size and shape, high purity, and loose agglomeration, which has much better quality compared with conventional mixing samples.

Controlling of TiB_2 content

Another two samples, which contain 40 wt. % TiB_2 and 80 wt. % TiB_2 from different weight percent of coated carbon on TiO_2 precursors, were synthesized at 1500°C for 2 hours under flowing argon. Their XRD patterns combing 60 wt. % TiB_2 samples are shown in Figure 10. Similarly with former results and discussion, there is no extra containment phase in all final products. Sharp and strong TiC and TiB_2 peaks can be observed. Mass fractions and crystallite size calculated from XRD reflections, BET surface area of produced TiC-TiB_2 composite powders are shown in Table I. Mass fractions shown TiC-TiB_2 composites with variable content can be controlled by change ration of TiO_2, carbon and B_4C. The mass fraction of TiB_2 is close to the theoretical value because only trace amount B_2O_3 evaporate under high temperature. Calculated crystallite size is around 30 to 50 nm for each component, which is comparable with samples from 4 to 8 hours high energy milled[7]. Surface areas are related with composite compositions since TiC has smaller particle size which means higher surface area.

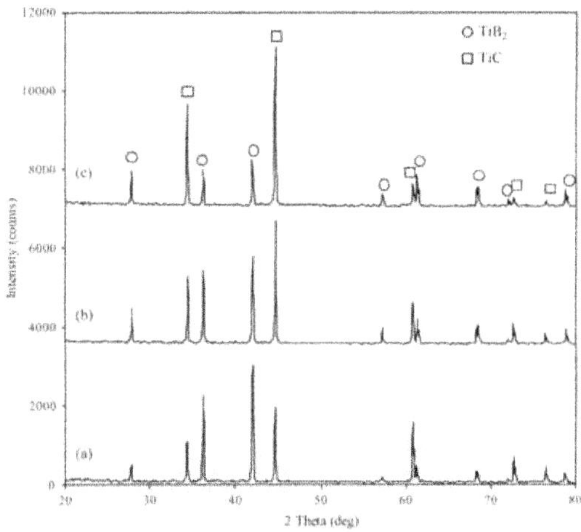

Figure 9. XRD patterns of composite powders with different content. (a) 40 wt. %TiB_2 and 60 wt. % TiC, (b) 60 wt. % TiB_2 and 40 wt. % TiC, (c) 80 wt. % TiB_2 and 20 wt. % TiC.

Table I Summary of mass fraction and crystallite size of synthesized TiC-TiB_2 composites

Target composites	Mass fraction	TiC crystallite size	TiB_2 crystallite size	Surface area
40 wt%TiB_2-60 wt % TiC	39 wt % TiB_2	42±3 nm	47±3 nm	7.2 m^2/g
60 wt%TiB_2-40 wt % TiC	59 wt % TiB_2	40±2 nm	47±1 nm	5.8 m^2/g
80 wt%TiB_2-20 wt % TiC	78 wt % TiB_2	31±5nm	39±2 nm	4.7 m^2/g

Sintering and mechanical properties of produced powders with nickel

Sintering studies were performed on produced powders from carbon coated precursors with content of 40 wt. % TiC and 60 wt. % TiB_2 with nickel. A dense sample with 99.6% TD can get through pressureless sintering(PS). Backscattered detector (BSD) micrographs of the polished sample surface is shown in Figure 10. TiC grains(gray ones) and TiB_2 grains(dark gray ones) are embedded in nickel binder with low porosity. Grain size of TiC and TiB_2 are measured as 1.5 μm and 2.3 μm respectively via line intercept method. The Vickers hardness is 24.7±0.6 GPa based on 10 times examinations. Fracture toughness was calculated by measuring the length of crack length using following equation derived by Anstis[22]:

$$K_{IC} = \delta \left(\frac{E}{H_V}\right)^{1/2} P C_0^{-\frac{3}{2}}$$

Where: K_{IC} is fracture toughness, δ is constant value of 0.016±0.004, E is the Young's modulus taken from literature and the law of mixture is applied for composites, H_V is the Vickers hardness, P is the indentation load, and C_0 is the crack length measured from the center of the indent. Calculated fracture toughness is 7.5±1.5 $MPa \cdot m^{1/2}$.

From previous reports, for TiB_2-TiC cermet, a sintering temperature of higher than 1600°C and with the assistance of pressure is needed to reach a higher TD. Ogwu reported commercial TiC and TiB_2 mixing and sintered with 5 wt. % Ni + X at 1550°C only reached a 85% TD[23], comparing 99.6 % TD with current research. Both Zou and Huang reported a sintering temperature of 1650°C with the assistance of pressure is need to prepare dense TiC-TiB_2 or TiC-TiB_2-WC with 8 wt. % Ni, with 22.54 to 23.5 GPa hardness and 7.5 to 7.8 $MPa \cdot m^{1/2}$ fracture toughness[24,25]. While for current research, at sintering temperature of only 1550°C without the assistant of pressure, high density $TiC-TiB_2$ composites can be sintered with only 5 wt. % Ni. Low content of Ni ensure the application of cermet at more extreme environment. Pressureless sintering at relative low temperature can greatly improve the production efficiency and lower the cost. At the same time, because of the high qualities of synthesized powders and high TD, sintered composite also shows outstanding combination of high hardness and toughness. Details of sintering and mechanical properties are conducted and will be reported in separate literature.

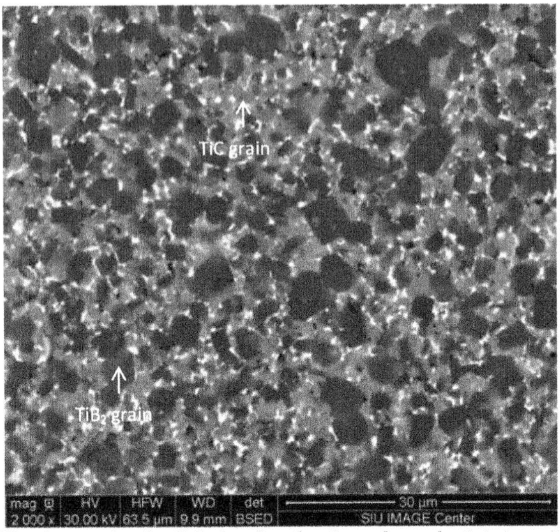

Figure 10. BSD micrographs of polished surface for sintered sample sintering with 5 wt. % Ni.

CONCLUSION

Submicron $TiC-TiB_2$ composite powders, with TiC particle size about 100 to 200 nm and TiB_2 about 700 to 900 nm, were successfully synthesized by carbo/borothermal reduction from carbon coated TiO_2 precursors at relative low temperature(1500°C for 2 hours). Final products combine the advantages such as pure phases, fine particle size, narrow size distribution, loose agglomerate and high surface area. Content of TiC and TiB_2 can be controlled with the change of ration between TiO_2, carbon and B_4C. Both TiC and TiB_2 have nano scale crystallite size. Sintering examination of produced powders with 5 wt.% nickel showed near full density can be achieved with outstanding Vickers hardness (24.7±0.6 GPa @20kg) and fracture toughness (7.5±1.5 $MPa \cdot m^{1/2}$).

ACKNOWLEDGEMENT

Partial financial supported by a grant (DE-FE0008864) from U.S. Department of Energy was gratefully acknowledged. The first author was also supported by China Scholar Council, Project of Study Aboard.

REFERENCE

1. Vallauri, Dario, IC Atias Adrian, and A. Chrysanthou. "$TiC–TiB_2$ composites: a review of phase relationships, processing and properties." Journal of the European Ceramic Society 28.8 (2008): 1697-1713.
2. Basu, B., G. B. Raju, and A. K. Suri. "Processing and properties of monolithic TiB_2 based materials." *International materials review* 51.6 (2006): 352-374.
3. Yeh, C. L., and Y. L. Chen. "Combustion synthesis of $TiC–TiB_2$ composites."*Journal of Alloys and Compounds* 463.1 (2008): 373-377.

4. Zou, Binglin, et al. "Combustion synthesis of TiC_x–TiB_2 composites with hypoeutectic, eutectic and hypereutectic microstructures." *Journal of the European Ceramic Society* 28.11 (2008): 2275-2279.
5. Li, Jianlin, et al. "TiB_2/TiC nanocomposite powder fabricated via high energy ball milling." *Journal of the European Ceramic Society* 21.16 (2001): 2829-2833.
6. Wang, Huihua, et al. "Characterization of the structure of TiB_2/TiC nanocomposite powders fabricated by high-energy ball milling." *Ceramics International* 37.7 (2011): 2689-2693.
7. Lee, J. W., Z. A. Munir, and M. Ohyanagi. "Dense nanocrystalline TiB_2–TiC composites formed by field activation from high-energy ball milled reactants." *Materials Science and Engineering: A* 325.1 (2002): 221-227.
8. Guemmaz, M., et al. "Elastic properties of sub-stoichiometric titanium carbides: comparison of FP-LMTO calculations and experimental results." *Solid state communications* 110.6 (1999): 299-303.
9. Fu, Z., Mondal, K. and Koc, R. (2015) Sintering, Mechanical, Electrical and Oxidation Properties of Ceramic Intermetallic TiC-Ti_3Al Composites from Nano-TiC Particles, in Advances in Solid Oxide Fuel Cells and Electronic Ceramics: A Collection of Papers Presented at 39th International Conference on Advanced Ceramics and Composites (eds N. P. Bansal, M. Kusnezoff and K. Shimamura), John Wiley & Sons, Inc., Hoboken, NJ, USA. doi: 10.1002/9781119211501.ch4
10. Zhezhen Fu, Kanchan Mondal and Rasit Koc, Sintering, Mechanical, Electrical and Oxidation Properties of Ceramic IntermetallicTiC-Ti_3Al Composites from Nano-TiC Particles, ICACC 2015.
11. Wang, Huihua, et al. "Characterization of the structure of TiB_2/TiC composites prepared via mechanical alloying and subsequent pressureless sintering."*Powder Technology* 217 (2012): 340-346.
12. Koc, R., and J. S. Folmer. "Carbothermal synthesis of titanium carbide using ultrafine titania powders." Journal of Materials science 32.12 (1997): 3101-3111.
13. Koc, Rasit, and Gregory C. Glatzmaier. "Process for synthesizing titanium carbide, titanium nitride and titanium carbonitride." U.S. Patent No. 5,417,952. 23 May 1995.
14. Koc, R., and D. B. Hodge. "Production of TiB_2 from a precursor containing carbon coated TiO_2 and B4C." *Journal of materials science letters* 19.8 (2000): 667-669.
15. Koc, Rasit, Gregory Glatzmaier, and Jack Sibold. "β-SiC production by reacting silica gel with hydrocarbon gas." *Journal of materials science* 36.4 (2001): 995-999.
16. Swift, Geoffrey A., and Rasit Koc. "Tungsten powder from carbon coated WO3 precursors." *Journal of materials science* 36.4 (2001): 803-806.
17. Koc, Rasit, and Suneel K. Kodambaka. "Tungsten carbide (WC) synthesis from novel precursors." *Journal of the European Ceramic Society* 20.11 (2000): 1859-1869.
18. Akkoyunlu, A., et al. "Synthesis of Submicron-Size CaB6 Powders using Various Boron Sources." *Nanostructured Materials and Nanotechnology V: Ceramic Engineering and Science Proceedings, Volume 32* (2011): 125-135.
19. Kaga, Hisashi, and Rasit Koc. "Formation of Al_2O_3–TiC Composite Nano-Particles Synthesized from Carbon-Coated Precursors." *Journal of the American Ceramic Society* 90.2 (2007): 407-411.
20. Ni, De-Wei, et al. "Synthesis of monodispersed fine hafnium diboride powders using carbo/borothermal reduction of hafnium dioxide." *Journal of the American Ceramic Society* 91.8 (2008): 2709-2712.
21. H.P. Klug, L.E. Alexander X-ray Diffraction Procedures Wiley (1974)
22. Swift, G. A., and R. Koc. "Formation studies of TiC from carbon coated TiO_2."*Journal of materials science* 34.13 (1999): 3083-3093.
23. Anstis, G. R., et al. "A critical evaluation of indentation techniques for measuring fracture toughness: I, direct crack measurements." *Journal of the American Ceramic Society* 64.9 (1981): 533-538.
24. Ogwu, A. A., and T. J. Davies. "The densification and mechanical properties of a TiC and TiB_2 hardmetal sintered with a reactive alloy binder." *Physica status solidi (a)* 153.1 (1996): 101-116.

25. Zou, Bin, et al. "Mechanical properties and microstructure of TiB_2–TiC composite ceramic cutting tool material." *International Journal of Refractory Metals and Hard Materials* 35 (2012): 1-9.
26. Song, Jinpeng, et al. "Microstructure and mechanical properties of TiB_2–TiC–WC composite ceramic tool materials." *Materials & Design* 36 (2012): 69-74.s

NICKEL NITRATE AND MOLYBDENUM OXIDE AS A YTTRIA-STABILIZED ZIRCONIA SYNERGISTIC SINTERING AID

Clay Hunt [a], David Driscoll [a], Adam Weisenstein [b], Stephen Sofie [a]

[a] Montana State University, Bozeman, MT, USA
[b] ZAF Energy Systems Inc, Kalispell, MT, USA

ABSTRACT

The sintering performance, microstructure, and crystal structure of yttria-stabilized zirconia stabilized with 8 mol% yttria (8YSZ) doped with both Mo and Ni together are compared to the sintering performance, microstructure, and crystal structure of pure 8YSZ, 8YSZ with only Ni added as a sintering aid, and 8YSZ with only Mo added as a sintering aid. Results suggest that the addition of both Mo and Ni as co-sintering aids work together synergistically. Enhanced densification and grain growth is observed in the Mo-Ni co-doped 8YSZ samples when compared to all other microstructures. Order of magnitude sintering rate increases are observed in the Mo-Ni co-doped 8YSZ over that of pure 8YSZ. With a maximum sintering temperature of 1200°C and a one hour dwell, sintered densities of 85% theoretical density $(5.02 \cdot gm/cm^3)$ are achieved with the co-doped samples: a 57% increase in density over pure 8YSZ sintered with the same sintering profile. X-ray diffraction (XRD) results indicate some destabilization of the cubic phase of YSZ with the addition of MoO_2. Finally, the spontaneous generation of nickel-molybdenum nano-rods was observed on the 5, and 10 mol% Mo-Ni infiltrated 8YSZ samples after being left under vacuum in a scanning electron microscope chamber, suggesting evaporation of a possible nickel-molybdenum compound from the sample fracture surfaces.

INTRODUCTION

The structural stabilization of the cubic and tetragonal polymorphs of zirconium oxide with the addition of yttrium has yielded an incredibly versatile oxide with numerous functional and structural applications. This material is currently used to produce kitchen knives, mechanical bearings, artificial gemstones, and medical and dental implants. More technically, yttria-stabilized zirconia (YSZ) is used in the production of thermal barrier coatings of turbine blades, oxygen sensors in fuel-injected engines, and electrolytes and anodes of solid oxide fuel cells (SOFCs)[1-3]. The tetragonal phase of zirconia is mechanically tough and is the phase of the polymorphic material used to make bearings, knives, and dental implants[4]. Because of its low electrical conductivity, single crystals of the cubic phase of YSZ are quite transparent, and are used to make artificial diamonds. The low electrical conductivity and the relatively high oxygen ion conductivity make polycrystalline cubic zirconia an excellent candidate for the SOFC electrolyte[5, 6].

Zirconia chemically stabilized with 8 mol% yttria (8YSZ) exhibits a maximum in oxygen ion conductivity with respect to yttria content[7]. 8YSZ also exhibits chemical stability in the caustic fuel cell environment at typical fuel cell operating temperatures of 850°C. However, 8YSZ can be difficult to sinter to 94% theoretical density; the density required for removing open porosity. Temperatures of 1400°C or more for more than two hours are required to sinter 8YSZ to a density that makes it useful as a SOFC electrolyte[5]. Because of the usefulness and difficulty of processing of 8YSZ, many studies have been undertaken to understand why 8YSZ sinters at such a high temperature and how to reduce the temperature at which densification occurs.

To better understand the phenomenon of sintering, and find ways to increase sintering performance, the effects of initial powder size and shape have been examined[8-11]. Although improvements in sintering performance have been realized through changing sintering pressure, a function of powder particle size and shape, the sintering performance achieved via the addition of impurities as sintering aids can eliminate nano-particle requirements[12, 13]. Al_2O_3 as a sintering aid has been shown to increase sintering performance by decreasing the activation energy of densification of 8YSZ[14, 15]. A small amount of Al_2O_3 added to 8YSZ has been shown to increase ionic conductivity according to electrical impedance spectroscopy (EIS)[16]. The presence of Al_2O_3 at the electrolyte-anode interface was found to form $NiAl_2O_4$ by means of chemical interactions with Al_2O_3 and NiO[17]. Although the study results of Knibbe et al. suggested the presence of $NiAl_2O_4$ did not negatively impact electrochemical cell performance at the operating temperature of 850°C, a study also performed by Knibbe et al. found that Al_2O_3 at the YSZ-$La_{1-x}Sr_xMnO_3$ (LSM) interface caused delamination of the LSM air electrode from the YSZ causing failure of the system[18].

The addition of some metal oxides may or may not act as a sintering aid, but have been shown to change the atomic structure of cubic zirconia. Adding TiO_2 to cubic zirconia has been found to introduce the tetragonal phase of YSZ[19, 20]. This observed phase change has been shown to decrease the ionic conductivity of YSZ[21]. The addition of Bi_2O_3 has also been found to destabilize the cubic phase of YSZ[22]. Nickel has been shown to enhance YSZ sintering performance[23]. However, nickel metal has been shown to precipitate out of YSZ solution in a reducing environment, irreversibly decreasing oxygen ion conductivity in the process[24-26]. Other effective sintering aids such as cobalt and manganese show promise as effective sintering aids that preserve the oxygen ion conductivity of 8YSZ[24, 27, 28].

The prior literature studies share the common theme of examining the effects of a single dopant as a sintering aid. Reported 8YSZ sintering experiments concerned with more than one impurity typically examine the inadvertent addition of alumina or silica, and the negative performance impacts those additions have on electrochemical performance[29-31]. An in-house study concerned with the sintering performance of $Ni(NO_3)_2$ infiltrated 8YSZ in an oxygen rich atmosphere revealed that the $Ni(NO_3)_2$ infiltrated samples densified greatly in the oxygen-rich atmosphere. Elemental investigation of the samples that demonstrated greater densification revealed the presence of large amounts of molybdenum, an impurity serendipitously present in the tube furnace from prior experimental work. It is thought that the molybdenum contamination of the furnace volatilized as molybdenum oxide in the oxygen rich atmosphere, providing a low level source of well-distributed Mo. The nickel in the $Ni(NO_3)_2$ infiltrated pellets appears to have facilitated absorption of the molybdenum from the atmosphere, and the combination of nickel and molybdenum in the sub-micron particulate 8YSZ samples yielded substantial densification near 1100°C, a temperature well below the onset of densification temperature of pure 8YSZ. This unexpected discovery inspired the study of the synergetic effects of nickel and molybdenum as sintering aids for 8YSZ discussed in this study.

EXPERIMENTAL PROCEDURE

Four sample groups were prepared for sintering studies: pure 8YSZ (1), $Ni(NO_3)_2$-infiltrated 8YSZ (2), MoO_2 doped 8YSZ (3), and MoO_2 doped $Ni(NO_3)_2$-infiltrated 8YSZ (4).

Control Group: Sample Set 1

8YSZ (TZ-8YS, Tosoh Corp., Tokyo, Japan) powder was uniaxially pressed as received in a cylindrical, stainless steel die under a pressure of 250 MPa. Samples were held under that pressure for 10 to 30 seconds to reduce damage from the powder springing back due to the storage of elastic energy in the compressed powder particles. Pressed pellets were then sintered in a vertical-load-frame dilatometer in a flowing oxygen atmosphere (L75, Linseis, Robbinsville, NJ) up to 1200°C with heating rates of 5°, and 10°C per minute, held at 1200°C for one hour, then allowed to cool to room temperature. One of the sintering samples was fractured, attached to an aluminum slide with carbon tape, and iridium sputter-coated. The microstructure of the pellet was then examined via scanning electron microscopy (SEM) (JSM-6100A SEM, Jeol Ltd., Tokyo, Japan). The remaining pellets were pulverized in a synthetic sapphire mortar and pestle. X-ray diffraction (XRD) (SCINTAG X1 Diffraction System) was used to examine the sample crystal structure. Scans were conducted from $25° - 78°\ 2\theta$ with $0.01°$ step size and 3 second dwell.

$Ni(NO_3)_2$ Infiltrated Group: Sample Set 2

Pellets pressed in the manner described in the *Control Group* section, above, were infiltrated with a polymeric nickel-nitrate ($Ni(NO_3)_2{\cdot}6H_2O$, Inframat Advanced Materials, Farmington, CT) solution prior to sintering. The addition of nickel in this manner helped to insure uniform distribution of the dopant particularly in the region of particle-particle interfaces. The polymeric nickel nitrate solution was prepared by dissolving $Ni(NO_3)_2{\cdot}6H_2O$ in water. Ethylene glycol was added to the $Ni(NO_3)_2$ solution. The water was evaporated by holding the mixture at 95°C while stirring. The solution was thinned with 2-n-butoxyethanol. The pellets were infiltrated with the polymeric $Ni(NO_3)_2$ solution by placing the pellet on a densified Al_2O_3 plate, and placing a few drops of the $Ni(NO_3)_2$ solution onto the pellet. The pellet was allowed to absorb this solution for 24 hours before it was sintered in the dilatometer in an atmosphere of flowing oxygen. This process resulted in a NiO content of approximately 0.85 mol% after sintering. The NiO content of the samples was determined by weighing each pellet before infiltration with $Ni(NO_3)_2$ solution, and after sintering. The increased mass of the sintered pellets was attributed to the presence of NiO. A Mathcad algorithm which computed the mole fraction of NiO in the infiltrated pellet was used to determine the mole fraction of NiO in each sample. The sintering profiles described in the *Control Group* section, above were used to sinter the $Ni(NO_3)_2$ infiltrated pellets. Samples were then examined via SEM and XRD according to the *Control Group* section, above.

Mechanically-Mixed MoO_2 8YSZ: Sample Set 3

Three MoO_2-doped 8YSZ powders were prepared by mechanical mixing of oxides by dry ball milling appropriate amounts of molybdenum oxide powder (Molybdenum(IV) oxide, Alfa Aesar, Ward Hill, MA) with 8YSZ in high density polyethylene bottles with zirconia milling media. The resulting powders contained 1.0, 5.0, and 10.0 mol% MoO_2. Each powder was then sieved through a screen, and the resulting powders were pressed into pellets according to the *Control Group* section, above. A pellet of each concentration of MoO_2 was then sintered in a flowing oxygen atmosphere using the sintering profiles described in the *Control Group* section, above. Samples were then examined via SEM and XRD according to the *Control Group* section, above.

Mechanically-Mixed MoO_2 8YSZ Infiltrated with $Ni(NO_3)_2$: Sample Set 4

8YSZ pellets with each concentration of MoO_2 were pressed according to the process described in the *Control Group* section, above. These pellets were additionally infiltrated with the polymeric $Ni(NO_3)_2$ solution according to the method described in the *$Ni(NO_3)_2$ Infiltrated Group* section, above. The pellet infiltration process resulted in pellets with 1, 5, and 10 mol% MoO_2 doped to about 0.85, 0.80, and 0.75 mol% NiO, respectively. The pellets were sintered in a flowing oxygen atmosphere, and then examined via SEM, and XRD as described in the *Control Group* section, above.

RESULTS AND DISCUSSION

The time-density curves determined from the dilatometry measurements of the sintering performance of 8YSZ (Sample Set 1), 8YSZ infiltrated with $Ni(NO_3)_2$ (Sample Set 2), MoO_2 doped 8YSZ (Sample Set 3) and , and MoO_2 doped 8YSZ infiltrated with $Ni(NO_3)_2$ (Sample Set 4) are shown as Figure 1. Each plot of Figure 1 directly compares the sintering performance of each sample set. Note that in all cases the sintering rate and sintered densities of Sample Set 4 were much greater than the sintering performance of any other sample set. The maximum sintering rates, green pellet densities, and sintered pellet densities are summarized in Table I. Note that densities are reported relative to the theoretical density of 8YSZ of 5.9 gm/cc. Sintering rates are reported in units of change in relative density per minute (1/min).

Table I: Green density (top left), final sintered density (top right), and maximum sintering rate (bottom); density measurements relative to 5.9 gm/cc, and sintering rate in change in relative density per minute (1/min)

Heating Rate (°C per Min)	Control	$Ni(NO_3)_2$ Inf	1% MoO_2	1% MoO_2 $Ni(NO_3)_2$	5% MoO_2	5% MoO_2 $Ni(NO_3)_2$	10% MoO_2	10% MoO_2 $Ni(NO_3)_2$
	50.7, 54.3	50.3, 59.3	49.9, 54.9	51.3 ,67.3	50.4, 56.5	50.4, 82.0	52.1, 54.5	52.1, 77.7
5	0.101	0.249	0.154	0.718	0.204	1.287	0.068	2.227
	50.0, 54.0	50.4, 61.2	47.2, 51.9	50.7, 69.0	49.6, 55.9	49.2, 85.9	47.7, 50.0	51.6, 82.4
10	0.156	0.293	0.277	1.208	0.317	3.696	0.103	4.003

Plots A and B of Figure 1 compare the sintering performance of Sample Set 1, Sample Set 2, 1 mol% MoO_2 doped 8YSZ, and 1 mol% MoO_2 doped $Ni(NO_3)_2$ infiltrated 8YSZ sintered up to and held for one hour at 1200°C with heating rates of 5° and 10°C per minute, respectively. Plots C, and D of Figure 1 compare the sintering performance of Sample Set 1, Sample Set 2, 5 mol% MoO_2 doped 8YSZ, and 5 mol% MoO_2 doped $Ni(NO_3)_2$ infiltrated 8YSZ sintered up to and held for one hour at 1200°C with heating rates of 5° and 10°C per minute, respectively. Plots E, and F of Figure 1 compare the sintering performance of Sample Set 1, Sample Set 2, 10 mol% MoO_2 doped 8YSZ, and 10 mol% MoO_2 doped $Ni(NO_3)_2$ infiltrated 8YSZ sintered up to and held for one hour at 1200°C with heating rates of 5° and 10°C per minute, respectively. The sintering temperature profile has been shown on each plot.

The sintering temperature of 1200°C was intentionally chosen because of the poor sintering performance of 8YSZ at that temperature. The sintering performance increase that is achieved with the addition of 5, and 10 mol% MoO_2 with $Ni(NO_3)_2$ is greater than the sum of the sintering performance increases achieved with the individual addition of MoO_2 and $Ni(NO_3)_2$. Further, it is shown that for each sample in Sample Set 4, the pellet sintered with a heating rate of 10°C per minute achieved a greater final density than the pellet sintered with the 5°C per minute heating rate.

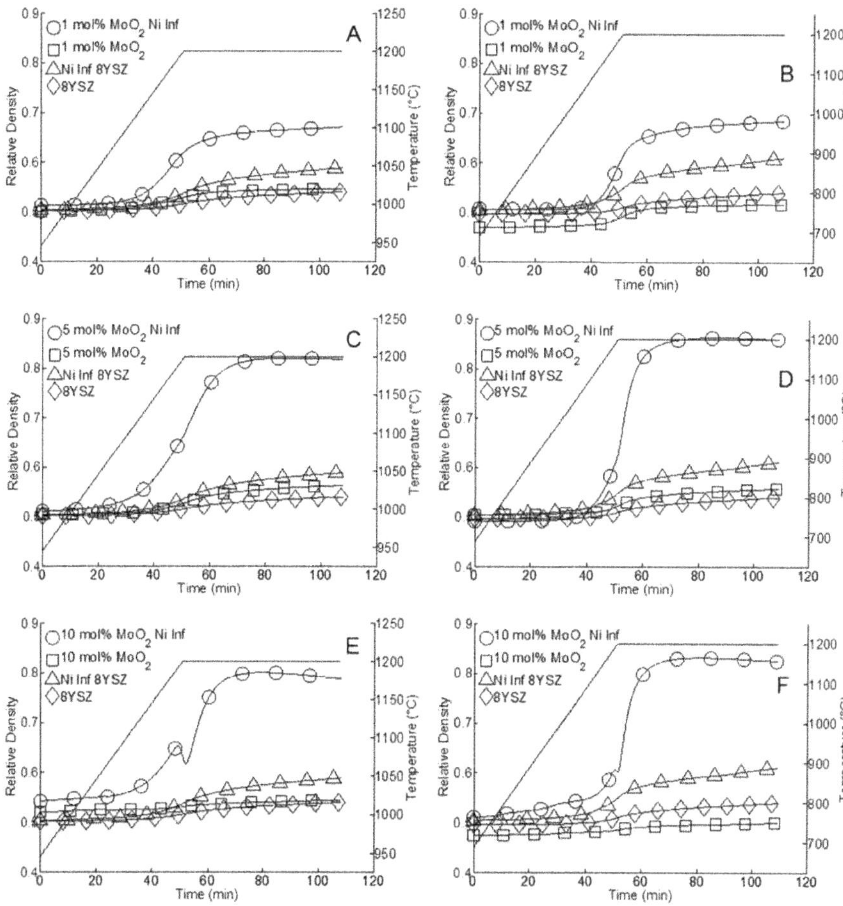

Figure 1: Sintering profiles of 8YSZ, $Ni(NO_3)_2$ infiltrated 8YSZ, MoO_2 doped 8YSZ, and MoO_2 doped $Ni(NO_3)_2$ infiltrated 8YSZ sintered up to and held at 1200°C for one hour with heating rates of 5°C per minute (A, C, and E), and 10°C per minute (B, D, and F) according to the temperature profile indicated in each figure by the solid black line (-)

The abrupt decrease in density of the 10 mol% MoO_2 $Ni(NO_3)_2$ infiltrated 8YSZ sample sintered at 5°C per minute, as shown in the densification curve, is not fully understood. However, closer examination of the Sample Set 4 samples sintered at 10°C per minute reveals similar, though less pronounced behavior. The XRD patterns of the cubic and tetragonal phases of zirconia with indexed peaks have been included as Figure 2(A) as a reference for comparison with Figure 2(B), the XRD patterns produced by the XRD examination of Sample Set 4[32].

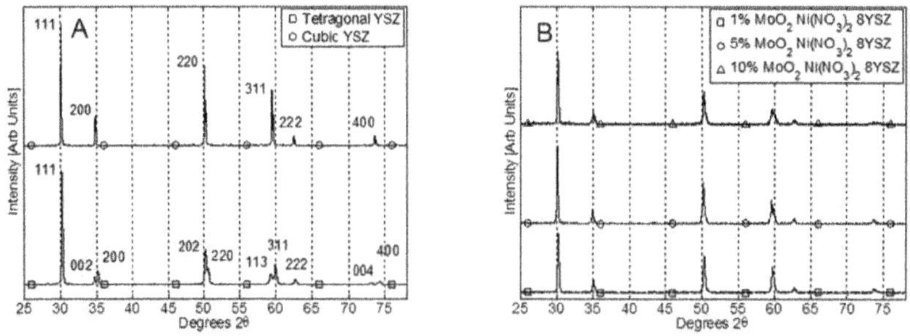

Figure 2: XRD patterns with indexed peaks of (A) Cubic (8YSZ) and tetragonal (3YSZ) YSZ sintered at 1200°C for one hour[32], and (B) XRD patterns of 8YSZ with the addition of 1, 5, and 10 mol% MoO_2 and $Ni(NO_3)_2$ also sintered at 1200°C for one hour

The peak locations and lattice parameter estimates from all XRD plots have been shown in Table II. Peak locations were determined using a peak-finding algorithm developed in Matlab. All members of Sample Sets 3 and 4 were examined with the peak finding algorithm as though they possessed cubic structure because of the overlap of the cubic and tetragonal peaks in the XRD patterns. The Matlab peak finder was used in preference over the standard *Jade* software approach due to the variations in peak location encountered when exploring different peak-finding options in *Jade*. The Matlab based peak finder overlays a plot of the parabola used to fit the XRD data defining each peak of the XRD pattern over the XRD data. These plots allow the user to visually evaluate the quality of the fit of each peak, and change the number of points used in the parabolic fit of the XRD data so that a more satisfactory peak fit might be achieved. Any increase in the accuracy of determining peak locations allows better evaluation of lattice parameter changes introduced by the addition of dopants. Lattice parameters of the cubic and tetragonal phases were calculated in the algorithm according to Cullity[33].

Table II: Peak locations and corresponding lattice parameters

	Cubic Peaks Index and Location (Degrees 2q)						Cubic Lattice Parameter
	1 1 1	2 0 0	2 2 0	3 1 1	2 2 2	4 0 0	Angstroms
8YSZ	30.0635	34.8459	50.1180	59.5712	62.5071	73.6111	5.144
1% MoO_2	30.1681	34.9343	50.2095	59.6418	62.6172	73.6426	5.135
1% MoO_2 with $Ni(NO_3)_2$	30.2702	35.0494	50.3282	59.7435	62.7244	73.7292	5.124
5% MoO_2	30.1022	34.8632	50.1474	59.5708	Fail	Fail	5.141
5% MoO_2 with $Ni(NO_3)_2$	30.1433	34.9201	50.1995	59.6185	62.6263	73.6711	5.136
10% MoO_2	30.3346	35.0780	50.3623	59.7837	62.8259	74.5049	5.111
10% MoO_2 with Ni(NO_3	30.2275	34.9996	50.2582	59.6899	62.7217	Fail	5.126

The cubic lattice parameter of 0.5144·nm is within 0.1% of the lattice parameter estimate of 0.5140·nm of White et al[34]. The tetragonal structure was found to have lattice parameters of a=b=0.5103·nm and c=0.5167·nm. The values of the tetragonal lattice parameter according to Schelling were calculated to be a=b=0.5142·nm and c=0.5186·nm after adjusting for thermal expansion from absolute zero[35]. These values are within 1% of each other. Note that the average lattice parameter of the tetragonal structure is 0.5135·nm, and that the lattice parameters of those samples that are thought to be composed of more of the tetragonal phase than the cubic phase are seen to have lattice parameters of 0.5135·nm or less.

The microstructure of Sample Sets 3 and 4 has been shown as Figure 3. Comparison of Figure 3(A), the microstructure of the 10 mol% MoO_2 doped 8YSZ, with Figure 3(B) shows that with the addition of $Ni(NO_3)_2$ to 1 mol% MoO_2 doped 8YSZ, the microstructure begins to change. Figure 3(C) shows that the 5 mol% MoO_2 $Ni(NO_3)_2$ microstructure is changed less subtly. With 10 mol% MoO_2, and $Ni(NO_3)_2$ the resulting microstructure, shown as Figure 3(D), is smooth, dense, and almost pore-free. The melting points of MoO_2 and MoO_3 are 1100°C and 795°C, respectively. Although the melting points of these molybdenum oxides are both below the sintering temperature of 1200°C used in this work, the sintering performance of the MoO_2-doped 8YSZ was not better than the sintering performance of pure 8YSZ, which suggests volatilization of the molybdenum oxide before the melting temperature was reached. Further, the porosity of the MoO_2-doped 8YSZ samples, shown in Figure 3, suggests no liquid-phase sintering. The densification enhancement was only achieved with the addition of molybdenum and nickel. The formation of a nickel-molybdenum compound has been considered to explain the increased densification. The only mention in the literature of such a compound is of $NiMoO_4$. The material safety data sheets (MSDS) for $NiMoO_4$ indicate that the melting point of this compound is not determined. However, the formation of this compound requires a nickel ion for every molybdenum ion. There is less than 1 mol% NiO in each $Ni(NO_3)_2$ infiltrated pellet. Energy dispersive X-ray spectroscopy (EDS) mapping (not shown) of the sample region shown in the backscattered electron image, shown as Figure 4(A), suggests an abundance of molybdenum and absence of nickel in the grain boundary. Stoichiometry suggests that even if $NiMoO_4$ has a melting temperature less than the 1200°C sintering temperature, there is not enough nickel in each sample to form this compound in MoO_2-rich grain boundaries. Thus, it is not possible to conclude that liquid-phase sintering is responsible for the observed increase in sintering performance.

The images of the 1, and 5 mol% MoO_2 doped 8YSZ microstructure were not shown because of the similarity of the 1, 5, and 10 mol% MoO_2 doped 8YSZ microstructure. The grain boundaries of the fracture surface of the 10 mol% MoO_2 $Ni(NO_3)_2$ infiltrated 8YSZ pellet visible in the back scattered electron image, shown as Figure 4(A), indicate that grain size is on the order of $25 \cdot \mu m$, indicating exaggerated grain growth. Contrasting the microstructure of Sample Set 3 with Sample Set 4 indicates that, by itself, the addition of MoO_2 to 8YSZ does little to affect the microstructure. This indication is confirmed by the similarity of the 1, 5, and 10 mol% MoO_2 images. The microstructure of the $Ni(NO_3)_2$ infiltrated 8YSZ, not shown, is similar to that of the 1, 5, and 10 mol% MoO_2 doped 8YSZ microstructure.

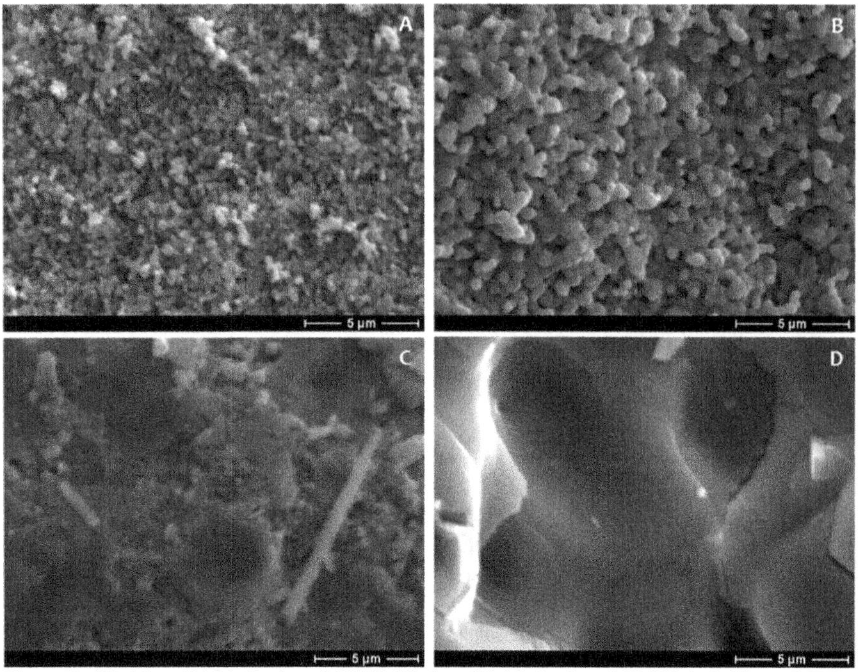

Figure 3: (A) 10 mol% MoO_2 doped 8YSZ (B) 1 mol% MoO_2 doped $Ni(NO_3)_2$ infiltrated 8YSZ (C) 5 mol% MoO_2 doped $Ni(NO_3)_2$ infiltrated 8YSZ (D) 10 mol% MoO_2 doped $Ni(NO_3)_2$ infiltrated 8YSZ, where all samples were sintered up to and held at 1200°C for one hour

Figure 3(C) and Figure 4(B) show the presence of tiny rods on the surface of the 5, and 10 mol% MoO_2 doped $Ni(NO_3)_2$ infiltrated samples, respectively. Elemental analysis of these rods indicates they are composed largely of nickel and molybdenum. The presence of these rods was not observed during the initial SEM analysis of the MoO_2 doped $Ni(NO_3)_2$ infiltrated samples. The samples were left under high vacuum overnight to continue characterization the following day. These tiny structures were observed as SEM study continued. This observation suggests that the formation of these rods was entirely spontaneous.

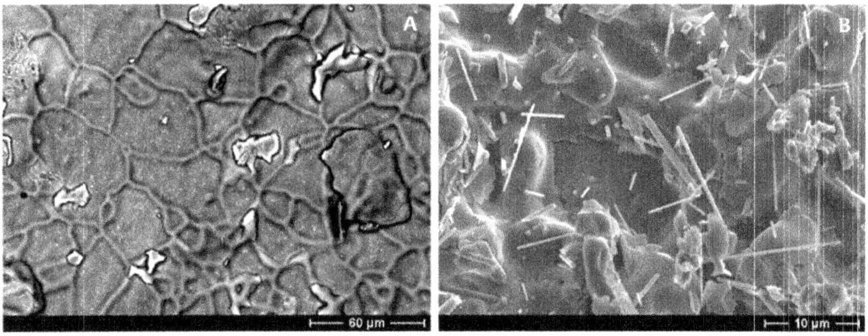

Figure 4: (A) Grain structure and (B) formed nano rods on surface of 10 mol% MoO_2 doped $Ni(NO_3)_2$ infiltrated 8YSZ sintered at 1200°C for one hour

CONCLUSIONS

The densification behavior, microstructure, and atomic structure of MoO_2 doped $Ni(NO_3)_2$ infiltrated 8YSZ was examined with dilatometry, XRD, and SEM. Although the sintering performance of 8YSZ with either MoO_2 or $Ni(NO_3)_2$ by itself was affected by the individual dopant addition, the effect on sintering of both of these materials together produced changes in the sintering performance that was seen to be greater than the sum of the performance increase of each individual dopant. Sintering rate increases of a factor of ten are observed in the MoO_2 doped $Ni(NO_3)_2$ infiltrated 8YSZ over that of pure 8YSZ. Sintered densities of 85% theoretical density ($5.02 \cdot gm/cm^3$) are achieved with the MoO_2 doped $Ni(NO_3)_2$ infiltrated samples at 1200°C. This reflects a 57% increase in density over pure 8YSZ sintered with the same sintering profile.

The microstructural differences between 5 and 10 mol% MoO_2 doped 8YSZ and the 5 and 10 mol% MoO_2 doped $Ni(NO_3)_2$ infiltrated 8YSZ suggest increased densification and grain growth of the MoO_2 doped $Ni(NO_3)_2$ infiltrated 8YSZ over 8YSZ, or 8YSZ doped with only $Ni(NO_3)_2$ or MoO_2. Although a chemically-induced phase change of 8YSZ may have been observed during the sintering of MoO_2 doped $Ni(NO_3)_2$ infiltrated 8YSZ, the reasons for the enhanced sintering performance are not known. The serendipitous observation of the nickel-molybdenum nano-rods on the surface of the 5 and 10 mol% MoO_2 doped $Ni(NO_3)_2$ infiltrated 8YSZ fracture surfaces indicate evaporation of molybdenum and nickel from the surface of the MoO_2 doped $Ni(NO_3)_2$ infiltrated 8YSZ.

References

1. Thermal Barrier Coatings for Gas-Turbine Engine Applications. Padture, Nitin P., Gell, Maurice and Jordan, Eric H. s.l. : Science, 2002, Science, Vol. 296, pp. 280 - 284.
2. Phase Evolution in Yttria Stabilized Zirconia Thermal Barrier. Witz, Gregoire, et al., et al. 9, s.l. : Journal of the American Ceramic Society, 2007, Vol. 90.
3. Demonstration of a Highly Efficient Solid Oxide Fuel Cell Power System Using Adiabatic Steam Reforming and Anode Gas Recirculation. Powell, Mike, et al., et al. s.l. : Journal of Power Sources, 2012, Journal of Power Sources, Vol. 205, pp. 377 - 384.
4. Transformation Toughening in Zirconia-Containing Ceramics. Hannink, Richard H.J., Kelly, Patrick M. and Muddle, Barry C. 2000, Journal of the American Ceramic Society, Vol. 83, pp. 461-487.
5. Fabrication, Microstructure and Properties of a YSZ Electrolye for SOFCs. Han, Minfang, et al., et al. s.l. : Journal of Power Sources, 2007, Journal of Power Sources, Vol. 165, pp. 757 - 763.
6. Grain Boundary Resistivity in Zirconia-Based Materials: Effect of Sintering Temperatures and Impurities. Badwal, S. P.S. s.l. : Sold State Ionics, 1995, Vol. 76. 67 - 80.
7. Zirconia-Based Solid Electrolytes: Microstructure, Stability and Ionic Conductivity. Badwal, S. P.S. s.l. : Solid State Ionics, 1992, Solid State Ionics, Vol. 52, pp. 23 - 32.
8. Sintering Kinetics at Isothermal Shrinkage: Effect of Specific Surface Area on the Initial Sintering Stage of Fine Zirconia Powder. Matsui, Koji, et al., et al. s.l. : Journal of the American Ceramic Society, 2007, Vol. 90. 44 - 49.
9. Analysis of Nanocrystalline and Microcrystalline ZnO Sintering Using Master Sintering Curves. Ewsuk, Kevin G., Ellerby, Donald T. and DiAntonio, Christopher B. s.l. : Journal of the American Ceramics Society, 2006, Journal of the American Ceramics Society, Vol. 89, pp. 2003 - 2006.
10. Sinterability of Commercial 8 mol% Yttria-Stabilized Zirconia Powders and the Effect of Sintered Density on the Ionic Conductivity. Gibson, I. R., Dransfield, G. P. and Irvine, J. T.S. s.l. : Journal of Materials Science, 1998, Journal of Materials Science, Vol. 33, pp. 4297 - 4305.
11. Master Sintering Curves of a Nanoscale 3Y-TZP Powder Compacts. Mazaheri, Mehdi, et al., et al. s.l. : Ceramics International, 2009, Ceramics International, Vol. 35, pp. 547 - 554.
12. German, Randall M. Sintering Theory and Practice. New York : John Wiley & Sons, 1996.
13. A Modified Model for the Sintering of Tungsten with Nickel Additions. Gessinger, G. H. and Fischmeister, H. F. s.l. : Journal of the Less-Common Metals, 1972.
14. Sintering Kinetics at Constant Rates of Heating: Effect of Al2O3 on the Initial Sintering Stage of Fine Zirconia Powder. Matsui, Koji, et al., et al. 12, s.l. : Journal of the American Ceramic Society, 2005, Vol. 88.
15. Activation Energy for the Sintering of Two-Phase Alumina/Zirconia Ceramics. Wang, Jenqdaw and Raj, Rishi. 8, s.l. : Journal of the American Ceramic Society, 1991, Vol. 74. 1959 - 1963.
16. Effect of Alumina Additions Upon electrical Properties of 8 mol% Yttria-Stabilised Zirconia. Feighery, A. J. and Irvine, J. T.S. 1999, Solid State Ionics, Vol. 121, pp. 209 - 216.
17. Effect of Alumina Additions on the Anode | Electrolyte Enterface in Solid Oxide Fuel Cells. Knibbe, R., et al., et al. s.l. : Journal of Power Sources, 2008, Journal of Power Sources, Vol. 179, pp. 511 - 519.
18. Effect of Alumina Additions in YSZ on the Microstructure and Degradation of the LSM-YSZ Interface. Knibbe, Ruth, Drennan, John and Love, Jonathan G. s.l. : Solid State Ionics, 2009, Solid State Ionics, Vol. 180, pp. 984 - 989.

19. Stability of Cubic ZrO2 (10 mol.% Y2O3) when Alloyed with NiO, Al2O3, or TiO2: Implications to Solid Electrolytes and Cermets. Chen, Shueiyuan, Deng, Wenhow and Shen, Pouyan. 1994, Materials Science and Engineering, Vol. B22, pp. 247-255.
20. Diffusion of Niobium in Yttria-Stabilized Zirconia and in Tatania-doped Yttria-Stabilized Zirconia Polycrystalling Materials. Kowalski, K., et al., et al. s.l. : Journal of the european Ceramic Society, 2006, Vol. 26. 3139 - 3143.
21. Structural and Electrical Characterization of Titania-Doped YSZ. Traqueia, L. S. M., Pagnier, T. and Marques, F. M. B. Great Britain : Journal of the European Ceramic Society, 1997, Journal of the European Ceramic Society, Vol. 17, pp. 1019 - 1026.
22. The Effect of Bi2O3 on the Microstructure and Electrical Conductivity of ZrO2-Y2O3 Ceramics. Kim, Y. H. and Kim, H. G. s.l. : Journal of Materials Science: Materials in Electronics, 1994, Journal of Materials Science: Materials in Electronics, Vol. 5, pp. 260 - 266.
23. Densification and Electrical Conductivity of Yttria-Stabilized Zirconia Containing NiO Additions. Batista, R. M. and Muccillo, E. N. S. s.l. : Electrochemical Society Transactions, 2008, ECS Transactions, Vol. 13, pp. 47 - 54.
24. Conductivity of Mn and Ni-Doped Stabilized Zirconia Electrolyte. Van herle, Jan and Vasquez, Ruben. s.l. : Journal of the European Ceramic Society, 2004, Journal of the European Ceramic Society, Vol. 24, pp. 1177 - 1180.
25. Conductivity Degradation of NiO-Containing 8YSZ and 10YSZ Electrolyte During Reduction. Coors, W. G., O'Brien, J. R. and White, J. T. Coorstek : Solid State Ionics, 2009, Solid State Ionics, Vol. 180, pp. 246 - 251.
26. Effect of NiO-to-Ni Transformation on Conductivity and Structure of Yttria-Stabilized ZrO2. Linderoth, Soren, et al., et al. s.l. : Journal of the American Ceramics Society, 2001, Journal of the American Ceramic Society, Vol. 84, pp. 2652 - 2656.
27. Chemical and Structural Changes in Manganese-Doped Yttria-Stabilized Zirconia Wtudied by Electron Energy Loss Spectroscopy Combined with Electron Diffraction. Appel, Charlotte C., et al., et al. s.l. : Journal of the American Ceramics Society, 1999, Journal of the American Ceramic Society, Vol. 82, pp. 429 - 455.
28. Cobalt Additive for Lowering the Sintering Temperature of Yttria-Stabilized Zirconia. Lewis, G. S., Atkinson, A. and Steele, B. C. H. s.l. : Journal of Materials Science Letters, 2001, Journal of Materials Science Letters, Vol. 20, pp. 1155 - 1157.
29. Silicon Volatility from Alumina and Aluminosilicates Under Solid Oxide Fuel Cell Operating Conditions. Gentile, Paul S., et al., et al. 2011, International Journal of Applied Ceramic Technology, Vols. doi: 10.1111/j.1744-7402.2011.00688.x, pp. 1 - 14.
30. Synergetic Effect of NiO and SiO2 on the Sintering and Porperties of 8 mol% Yttria-Stabilized Zirconia Electrolytes. Zhang, T. S., et al., et al. s.l. : Electrochimica Acta, 2009, Electrochimica Acta, Vol. 54, pp. 927 - 934.
31. Effect of Mn Addition on the Densification, Grain Growth and Ionic Conductivity of Pure and SiO2-Containing 8YSZ Electrolytes. Zhang, T. S., et al., et al. s.l. : Solid State Ionics, 2009, Vol. 180.
32. Ionic Conductivity Studies of Ultrafine-Grained Yttria Stabilized Zirconia Polymorphs. Ramamoorthy, R., Sundararaman, D. and Ramasamy, S. 1999, Solid State Ionics, Vol. 123, pp. 271-278.
33. Cullity, B. D. Elements of X-Ray Diffraction. pp 297 - 304 : Addison-Wesley Publishing Company, Inc, 1956.
34. The Enhanced Stabilization of the Cubic Phase in Yttria-Stabilized Zirconia with the Addition of Nickel Oxide. White, Josh, et al., et al. Golden, Colorado : Journal of American Ceramics Society, 2011, Journal of the American Ceramics Society, Vol. 94, pp. 2030 - 2036.

35. Mechanism of the Cubic-to-Tetragonal Phase Transition in Zirconia and Yttria-Stabilized Zirconia by Molecular-Dynamics Simulation. Schelling, Patrick K., Phillpot, Simon R. and Wolf, Dieter. 7, s.l. : Journal of the American Ceramic Society, 2001, Vol. 84. 1609 - 1619.

Surface Protection for Enhanced Performance

MODELING AND PREDICTION OF THE EFFECTIVE THERMAL CONDUCTIVITY OF THERMAL BARRIER COATINGS USING FFT AND FE APPROACHES

N. Ferguen, Y. Lahmar, Y. Fizi, R. Lakhdari
Centre de Développement des Technologies Avancées, Division des Milieux Ionisés et Lasers, Algiers, Algeria

W. Leclerc, M. Guessasma
Université de Picardie Jules Verne, Laboratoire des Technologies Innovantes, St-Quentin, France

ABSTRACT

This work aims to investigate the relationship between the porosity and the effective thermal conductivity (ETC) of alumina coating produced by air plasma spraying (APS). A novel approach for modeling the ETC based on Fast Fourier Transform (FFT) approach in conjunction with image analysis is developed. A statistical tool analysis was also used to assess the porosity content within the coating and to evaluate the pore distributions. In addition, the ETC was evaluated with OOF2 software based on finite element (FE) approach for comparison between both approaches. Numerical predictions of ETC based on FFT approach are in good agreement with FE approach and the results show that the inhomogeneity of pore orientations within the coating contribute to the anisotropic thermal properties behavior.

INTRODUCTION

Thermal barrier coatings (TBCs) are used to protect the metallic components from the aggressive environment at high temperatures so as to improve the durability of the components and enhance the engine efficiency. Due to their low thermal conductivity, TBCs allow to maintain a higher temperature drop across their thickness [1]. One of the protective means to solve the problem of hot corrosion is to coat the super alloy with a protective layer. The usual configuration of TBCs system consist of an intermediate layers with a bond coat between the super alloy substrate made of NiCrAlY or CoNiCrAly [2] and the ceramic topcoat that provides the oxidation resistance such as the alumina coatings. However, the thermomechanical properties of coatings and their insulation performance are mostly related to the void architecture [4]. The quantification of this later is hence fundamental and usually include several characteristics such as porosity content, size distribution and the pore network connectivity [5]. This pore architecture highly affects the ETC of coatings. Several investigations have shown the effect of the coating pore structure on its effective thermal conductivity [6-8]. In particular, image-based numerical analysis was widely applied using either real images or artificial coating models during the last 10 years [9]. This technique can be combined with use of finite element (FE) based codes such as OOF [10] to assess the ETC of coatings from a real microstructure images. Wang et al. [11] and Kulkarni et al. [12] applied image analysis to study the effect of the pores on ETC of plasma sprayed coatings, in their approach the artificial porous structure is built from the collective microstructural information by using small angle neutron scattering (SANS) [13]. Qaio et al. [9] have developed a 3D image-based finite difference (FD) model to study the heat transfer mechanisms through a porous structure, the ETC results measured by their FD model are in good agreement with the experimental value, but the comparison with the FE calculations shows that there is a small discrepancies. In the present work, we propose another way of investigation based on Fast Fourier Transform (FFT) approach in conjunction with image analysis for numerical predictions of ETC for TBCs. This approach is developed by Leclerc et al. [14] to study the effect of the microstructure of particle reinforced composites on the thermal properties, this approach is extended and adapted to investigate the area of plasma sprayed coatings.

The image used as sample for an input to the model is produced with SEM (scanning electron microscope) micrographs of polished alumina coating cross-section, Figure 1. From this figure we can observe the microstructure defects: globular pores, fine pores, and cracks, which are inherent features of plasma sprayed coatings. In this study we assume that the pore phase represents the total pores given by the globular/fine pores and cracks.

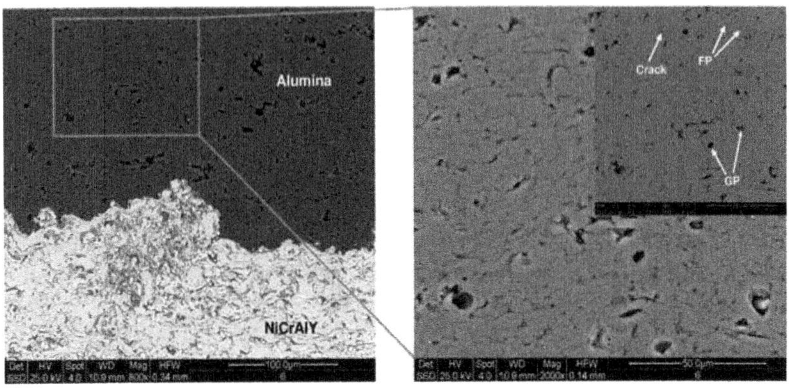

Figure 1: a) SEM cross-section micrograph of alumina coating on the NiCrAlY layer sprayed by APS. b) The magnified image at 2000x showing the different defects within the alumina, more details can be seen on the upper right of the image on globular pores (GP), fine pores (FP) and interlamellar crack.

IMAGE PROCESSING

The image produced with SEM micrographs of polished alumina coating cross-section, as shown in Figure 2, in grayscale level provides a large amount of information on pixel brightness such as pores and cracks inside the alumina coating. The original SEM image has a resolution of 1024 x 943 pixels which represents approximately 187 x 112 μm^2. A square domain is considered as sample for an input to the model have a resolution of 512 x 512 pixels. This sample was extracted from the original SEM image as seen in Figure 2b). In order to separate pores from the area filled with alumina material, the image was converted into a binary image, Figure 2 c), and the pixel values are written in a form of matrix. This matrix contains information about the pixel intensity value (0 or 255) and its pixel coordinates x and y in the grid. This step of image processing is achieved by using OpenCV libraries [15]. In this step, The image was segmented to two phases using threshold value of a grey level, so the pixels is divided into two groups matrix (white) and pores (black). The white color (255) corresponds to the area filled with alumina phase while the black color (0) stands for the pore phase. Several threshold values are selected allow to generate different level of porosity and the corresponding porosity values were then measured using a statistical tool (covariance) which will be discussed in the following section. These values of porosity are then compared with other ones determined by ImageJ (U.S. National Institutes of Health, Bethesda, USA), as shown in Figure 3. This figure shows the measurements of the porosity level inside the alumina coating determined by both covariance (COV) and imageJ, the ones measured by the covariance are in good agreement with those determined by imageJ.

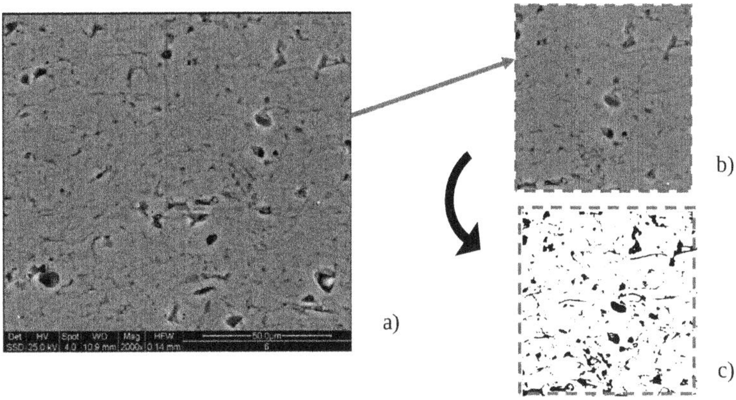

Figure 2: a) SEM image of alumina coating at 2000X magnification. b) Final image was extracted and used as sample for an input into the model (512 x 512 pixels, 0.132 µm/pixel) and c) Conversion into the binary image.

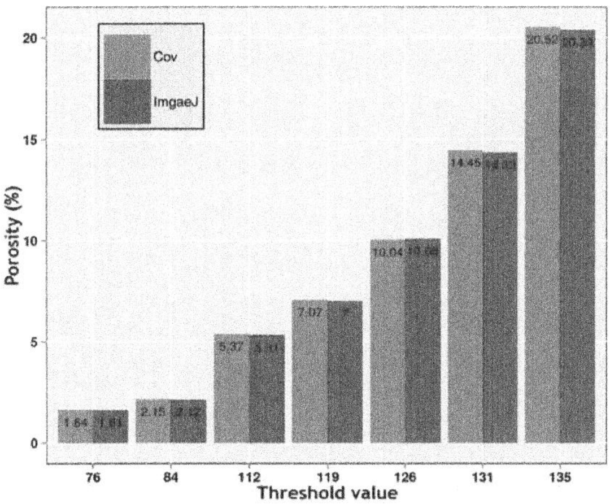

Figure 3: Comparison of porosity measurements for different values of threshold determined by both covariance and imageJ.

COVARIANCE

The covariance or two-probability function $S_2(r)$ [16-17] is a mathematical tool widely used to study heterogeneous materials. From this tool we can extract useful statistical information like the spatial distribution of heterogeneities, voids and the average size of the inclusions. The covariance describes the probability that 2 points of heterogeneous domain belong to the same phase [17], which directly depends on a distance r which is the norm of the vector connecting two random points.

Several numerical methodologies exist for evaluating the function $S_2(r)$. In the present work, we consider the approach introduced by Yeong and Torquato [18] and has been adapted and used to evaluate the porosity within the coatings from a digitized images. From the binary image, the process of calculation starts first by reading information about pixels values and their coordinates. Then the distance is defined in units of pixels so that both extremities of the random vector are located at the pixels centers. Based on 2D microstructures we consider only 2 directions of the Cartesian coordinate system (x and y), and r is limited to the two third of the total number of pixel (n) in one direction of the investigated domain. Finally, $S_2(r)$ is then simply estimated by translating the vector of one pixel at a time and increasing it by increment of one pixel after spanning the whole domain for each calculated direction. Two particular values of $S_2(r)$ can be exhibited in the case of coating contains only pores with different shapes. First, when the two points coincide ($r = 0$), the function is equal to the area fraction of pores (Φ). Second, the function is equal to the square of the volume fraction (Φ^2) this means that the distance between the two points tends to infinity and the heterogeneities (pixel phases) are well-dispersed. This last point provides a good criterion for investigating whether the spatial distribution of pores is homogeneous or heterogeneous within the coating. Figure 4 shows the evolution of $S_2(r)$ as function normalized distance (n/r) of the alumina microstructure. From this figure, $S_2(r)$ function gives an interesting results on the porosity content. First, the case of $r = 0$ the function $S_2(r)$ is equal to the area fraction of the porosity Φ and equal to 20.52 %. Second, when the function $S_2(r)$ is equal to the square of the area fraction of porosity when both the distance between the two points tends to infinity. We can observe a small oscillations about the asymptote with a delay for both directions which reflects the inhomogeneity of the pores orientation and distribution within the microstructure which is typical of APS coatings. The half-period is roughly equal to the radius a of the pores.

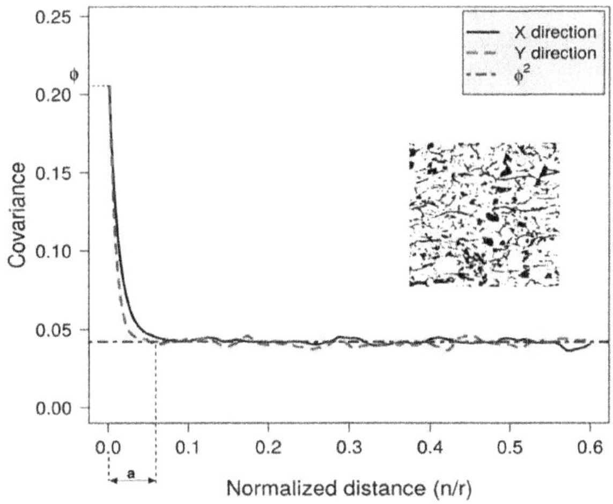

Figure 4: Covariance of the alumina microstructure (binarized image) with porosity of 20.52%. The COV is measured in tow orthogonal directions (*OX:* parallel, *OY*: vertical to the coating).

OBJECT-ORIENTED FINITE ELEMENT (OOF)

Langer et al. have developed an image-based finite element code called OOF [10,19,20]. This code can performs physical computations on microstructures starting from a micrograph or other image like simulated microstructures to determine the macroscopic properties. Based upon colors of an image, OOF allows to generate finite element meshes, this mesh adapts to the boundaries of different phases within the microstructure. To do this, OOF provides a number of methods for mesh adaptation based on adaptation algorithms and should meet criteria of the homogeneity or the shape quality [19] in order to get a better reproduction of the real shape of the image (the homogeneity of the mesh) or on a good shape of its elements (avoiding sharp angles or high aspect ratios). The mesh adaptation routines are oriented towards lowering the energy of the mesh based on eq. 1. Adaptation algorithms in OOF consists of 2 parts: the homogeneity energy (E_{hom}) which measures how well the mesh matches the pixel regions, and the second to account for shape energy (E_{shape}) which quantifies the quality of the shapes of elements.

$$E = \alpha\, E_{hom} + (1 - \alpha)\, E_{shape}$$

Where α is an adjustable parameter between 0 and 1, it is adjusted according to whether the priority is improving shape quality or homogeneity.

This code it is becoming an increasingly popular tool for the computation of properties like thermal properties with complex microstructure and used to predict properties of plasma sprayed YSZ coating [11,22,23]. The analysis technique of OOF can be summarized as follows: first, OOF reads an image and create a pixel groups based on colors in an image or on the grayscale contrast for SEM images. Then, a skeleton defines the geometry of the mesh is created, specifying nodes positions and elements edges only. A full finite element mesh, including information about element type are then created from the skeleton. Finally, fields are defined, applying boundary conditions, which equations are being solved to perform the calculation. Figure 5 shows the mesh of the binarized image generated by OOF2D, white and red areas denote alumina and pore phases respectively. A finer mesh is applied at the boundary of pixels, which takes account of small microstructural details and it is required for higher accuracy of the predicted results. The mesh in total had 266070 nodes and 93323 elements. Homogeneity index = 0.9934 it is very close to 1, this means that the mesh gives a good reproduction of the microstructure.

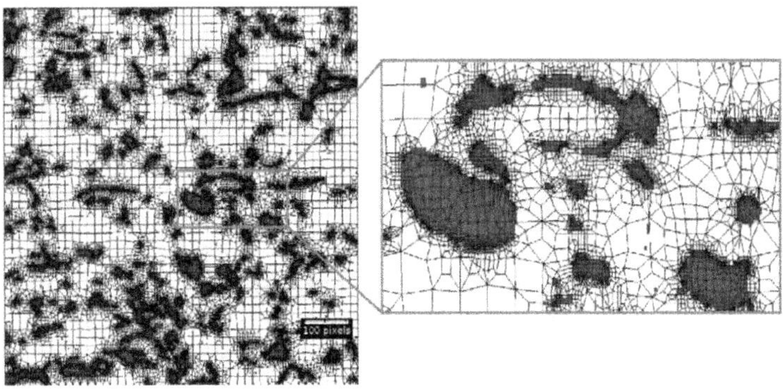

Figure 5: Finite element mesh generated by OOF2D. White and red areas denote alumina and pore phases respectively. A finer mesh and higher node density (266070 nodes and 93323 elements) along the interface between two phases can be seen in the zoomed view. Homogeneity index = 0.9934.

FFT

The FFT approach is an efficient numerical technique typically provided in the form of pixel or voxel-based geometry for microstructural characterization arising in homogenization of periodic media. This technique is based on the Lippmann–Schwinger equation which related to a classic linear problem via Green's operator expressed in the Fourier space in which the convolution operator becomes a simple classical product. Numerous iterative schemes have been proposed to compute efficiently the convolution product appearing in the Lippmann–Schwinger equation with periodic boundary and the most successful solver was proposed by Moulinec and Suquet [24,25]. Based on the recent work of Moulinec et al. [26] they have demonstrated that Eyre–Milton scheme [Eyre1999] gives the better efficiency. The FFT-based approach has the great advantage to not require any mesh but under periodic boundary conditions. Leclerc et *al.* [14] developed an FFT-based approach using Eyre-Milton scheme [27] implemented with the modified Green's operator provided in [28] for the fast assessment of the thermal conductivity in composites materials. In this work we used the approach developed by Leclerc et al. [14] to compute the ETC of TBCs. The ETC calculations are performed under steady-state heat flux. The coating bottom and top were assigned constant temperatures of $T1$ and T2 (> $T1$), respectively, while other two boundaries were kept insulated. Then the ETC K_{eff} is computed with Fourier's equation as follows:

$$K_{eff} = \frac{Q.h}{\Delta T.l}$$

Where Q, h, ΔT and l represent the heat flux per unit thickness through the transverse cross-section, the thickness, the temperature difference and the width of the coating respectively. The thermal conductivities of each phase are summarized in Table 1.

Table1: Thermal properties of the alumina coating phase constituents at the room temperature.

Pixel Group	Thermal conductivity (W/m.K)
Alumina	25
Pores (filled with air)	0,05

ETC

The evolution of the of alumina coating ETCs measured as function of the porosity fraction with both numerical and analytical models are shown in Figure 6. It is clear that thermal conductivity decreases when the porosity inside the coating increases. The ETCs measured by FFT approach are very close to those obtained with FE approach. When comparing results obtained by both numerical and analytical [30-33] models, there is a discrepancies between them and become more significant when the porosity is larger this is due that the analytical models only require knowledge of the volume fraction of the porosity for each phase and the respective thermal conductivities, but do not take into consideration microstructural details.

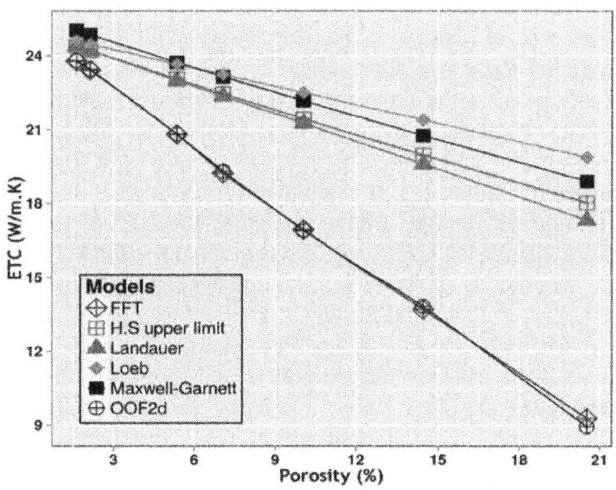

Figure 6: Influence of the porosity fraction on ETC of alumina coating

ANISOTROPIE

The anisotropic pore orientations of the microstructure as investigate by the covariance affect highly the properties of the coating. Figure 7 compares the computed ETCs as function of the porosity fraction for both directions of the coating x and y, x is parallel to the substrate (longitudinal) and *y* through the thickness of coating (transverse). It can be seen that both models exhibit anisotropic properties behavior.

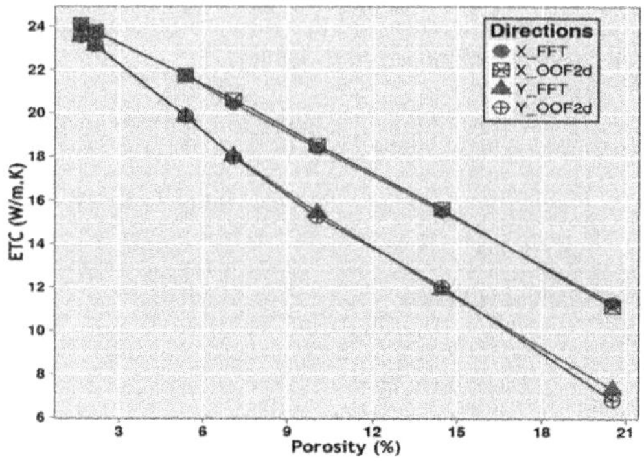

Figure 7: Influence of the porosity fraction of alumina coating on ETC for both directions, *x* is parallel to the substrate (longitudinal) and *y* through the thickness of coating (transverse).

COMPUTATIONAL EFFICIENCY

For the purpose of efficiency comparison between FE and FFT approaches we have performed simulations by these two approaches to compute the same ETC for a given binary image. FFT-based approach has the great advantage it does not require any mesh since the unit cell is discretized into a regular pixel then a discretization of 512 x 512 is considered. OOF simulations were performed with a refined mesh in order to get a good reproduction of the microstructure. Table 2 summarizes the simulation parameters for both approaches. The simulations were carried out on the Linux cluster at the CDTA center. The machine uses Intel®Xeon (R) CPU E5-2620, 2.1 GHz with 16 GB of memory (RAM). The comparison is based on the CPU (process time) consumption in seconds and the corresponding memory usage in Mega octets. The efficiency results show that the fast calculation using FFT is at least 30 times faster than the FE. It is well known that the FE simulations depend mainly on the complexity of the geometrical features and the mesh resolution which lead to the significant memory consuming (3100 Mo vs 12.3 Mo for FFT).

Table 2: Simulation parameters.

Approach	Discretization	Error Tolerance	Porosity (%)	ETC (W/m.K)	CPU (s)	RAM (Mo)
FFT	512 * 512 (262144 pixels)	10^{-14}	10.04	16.62	49	12.3
OOF2	NOE: 93323 NON: 266070	10^{-14}	10.04	16.89	1620	3100

CONCLUSION

This work presents a numerical model based on FFT approach using image processing for evaluating the effective thermal conductivity of thermal barrier coatings produced by air plasma spraying. The FFT approach was used to investigate the relationship between the porosity of the coating and the ETC. The FFT results are in good agreement with those obtained with FE approach. It was shown also that the porosity defects cause considerable reduction of the ETC and the homogeneity of pore orientations contribute to the anisotropic thermal properties behavior. In the future work, the proposed numerical model must be validate with experimental result, improve the model for defect detections (pores, cracks, splat interfaces, etc.), and investigate the effects of the anisotropic microstructural features on the thermal and mechanical properties of the coatings.

REFERENCES

[1] N.P. Paduture, M. Gell, and E.H. Jordan, Thermal barrier coatings for gas-turbine engine applications. Science, 296(5566), 280-284 (2002).
[2] E. Bahadori, S. Javadpour, M.H. Shariat et al., Preparation and properties of ceramics Al2O3 coatings as TBCs on MCrALY layer applied on Inconel alloy by cathodic plasma electrolytic deposition, Surface & Coatings Technology, 228, S611-S614(2013).
[3] S. Sobhanverdi, and A. Akbari, Prosity and microstructural features of plasma sprayed Yttria stabilized Zirconia thermal barrier coatings, Ceramics Int., 41(10) Part B, 14517-28(2015).
[4] P. Fauchais, J.V.R. Heberlein and M.I. Boulos, Thermal Spray Fundamentals, Ed. Springer (2014)
[5] R.W. Rice, Porosity dependence of physical properties of materials: a summary review, Key Eng Mater 115, 1-20(1995).
[6] G. Antou, G. Montavon, F. Hlawka et al., Modification of thermal barrier coating architecture by in situ laser remelting, J.Eur. Soc., 26, 3583-97(2006).
[7] A.F. Renteria, B. Saruhan, U. Schulz et al., Effect of morphology on thermal conductivity of EB-PVD PYSZ TBCs, surf. Coat. Technol., 201, 2611-20(2006).
[8] C. Petorak, J. Ilavsky, H. Wang et al., Microstructural evolution of 7 wt.% Y2O3-ZrO2 thermal barrier coatings due to the stress relaxation at elevated temperatures and the concomitant changes in thermal conductivity, surf. Coat. Technol., 205, 57-65(2010).
[9] J.H. Qiao, R. Bolot, H. Liao and al., A 3D finite-difference model for the effective thermal conductivity of thermal barrier coatings produced by plasma spraying, Int. J. of Thermal Sciences, 65, 120-26(2013).
[10] S.A. Langer, E.R. Fuller, and W.C. Carter., OOF: An Image-Based Finite-Element Analysis of Material Microstructures, *Computing in Science and Engineering* **3:3**, 15 (2001).
[11] Z. Wang, A. Kulkarni, S. Deshpande and al., Effects of pores and interfaces on effective properties of plasma sprayed zirconia coatings, Acta Materiala, 51, 5319-34(2003).
[12] A. Kulkarni, Z. Wang, T. Nakamura and al., Comprehensive microstructural characterization and predictive property modeling of plasma-sprayed zirconia coatings, 51, 2457-75(2003).
[13] S. Deshpande, A. Kulkarni, A. Sampath and al., Application of image analysis for characterization of porosity in thermal spray coatings and correlation with small angle neutron scattering

[14] W. Leclerc, N. Ferguen, C. Pélegris, and al., An efficient numerical model for investigation the effects of anisotropy on the effective thermal conductivity of alumina; Al composites, Advances in Engineering Software, 77, 1-12(2014).
[15] opencv.org
[16] K. Lochman, L. Oger, and D. Stoyan, statistical analysis of random sphere packings with variable radius distribution. Solid State Sci., 8, 1397-413(2006).
[17] S. Torquato, Random heterogeneous materials. Ed. Springer (2002).
[18] CLY. Yeong and S. Torquato, Reconstruction random media II: Three-dimensional from two dimensional cuts. Phys. Rev. E., 58,224-233(1998).
[19] A. C.E. Reid, A. S. Langer, R. C. Lua, and al., Image-based finite element mesh construction for material microstructures, *Computational Materials Science* **43**, 989-999 (2008).
[20] A. C.E. Reid, R. C. Lua, R. E. García, and al., Modelling Microstructures with OOF2, International Journal of Materials and Product Technology **35**, 361-373(2009).
[21]A. Goel, K.K. Chawla, U.K. Vaidya, and al., Mater. Sci. Technol., 24,864–869((2008).
[22] A.D. Jadhav, N.P. Padture, E.H. Jordan, and al. ActaMater. 54 3343–3349(2006).
[23] P. Michlik , C.C. Berndt,Image-based extended finite element modeling of thermal barrier coatings. Surf. Coat. Technol. 201, 2369–2380(2008).
[24] H. Moulinec, and P. Suquet, A numerical method for computing the overall response of nonlinear composites with complex microstructure. Comput. Methods Appl. Mech. Eng., 157:69–94 (1998).
[25] J. Michel, H. Moulinec, P. Suquet, Effective properties of composite materials with periodic microstructure: a computational approach. Comput. Methods Appl. Mech. Eng., 172,109–43(2001).
[26] H. Moulinec, and F. Silva, Comparison of three accelerated FFT-based schemes for computing the mechanical response of composite materials. Int. J. Numer. Methods Eng. 97(13):960–85 (2014).
[27] D. Eyre, G. Milton, A fast numerical scheme for computing the response of composites using grid refinement. Eur. Phys. J. Appl. Phys., 6, 41–47(1999).
[28] F. Willot, B. Abdallah, and Y.P. Pellegrini, Fourier-based schemes with modified Green operator for computing the electrical response of heterogeneous media with accurate local fields. Int. J. Numer. Methods Eng., 98(7), 518–33 (2014).
[30] R. Landuaer, The lectrical resistance of binary metallic mixtures, J. Appl. Phys., 23,779-84(1952)
[31] A.L. Loeb, Thermal conductivity: VIII, A theory of thermal conductivity of porous materials, J. of American Ceramic society, 37(2),96-99(1954).
[32] Z. Hashin and S. Shtrikman, J. of Mech. Phys. Solids, 11,127-40(1963).
[33] J.C. Maxwell-Garnett, Philosophical Transaction of the Royal society of London, Series A, 203,385-20(1904).

MATERIAL DESIGN OF CERAMIC COATING FOR JET ENGINE BY ELECTRON BEAM PVD

Hideaki Matsubara
Tohoku University, Graduate School of Environmental Studies
Sendai, Miyagi, Japan
(Japan Fine Ceramics Center, Nagoya, Aichi, Japan)

ABSTRACT

For the application of thermal barrier coating (TBC), microstructure (columnar, feather-like, porous, etc) of yttria stabilized zirconia layers prepared by electron beam physical vapor deposition was investigated, focusing on the effect of processing parameters and La_2O_3 content. Substrate rotation enhances formation of columnar structure with inter-columnar gaps and feather-like structure containing a large number of nano-pores and gaps. Computer Simulation and Design for Ceramics Coating Materials is also studied to predict microstructure change, mechanical properties degradation and peeling.

EXPERIMENTAL STUDIES ON TBC BY EB-PVD

At first, microstructure such as columnar, feather-like, porous, etc. of yttria stabilized zirconia layers prepared by electron beam physical vapor deposition was studied. We are focusing on the effect of processing parameters and La_2O_3 or other elements content. We have developed a technique for high speed deposition of nano-structure controlled layers of refractory oxide ceramics and next generation TBC materials[1-13].

Substrate rotation enhances formation of columnar structure with intercolumnar gaps and feather-like structure containing a large number of nano scale pores and gaps. Substrate arrangement contributes to increasing porosity by widening intercolumnar gaps. The flux shadowing effect caused by oblique vapor incidence is an important contributor to structure formation, because EB-PVD is line-of-sight process.

The addition of small amount of La_2O_3 gives the YSZ coating a precisely controlled nanoscale structure by suppressing sintering during deposition, thereby contributing a reduction in the thermal conductivity. La_2O_3 doped coatings also show improved high temperature stability of thermal conductivity due to high resistance to sintering. Because of low thermal conductivity and high resistance to sintering, ZrO_2-Y_2O_3- La_2O_3 coating has great potential as a new TBC.

COMPUTER SIMULATION STUDIES FOR CERAMIC COATING

The Monte Carlo (MC) method and the finite element method (FEM) have been used for the simulations of microstructure design[14-24]. The MC simulation was applied to the microstructural development of sintering and grain growth at the solid state and under the presence of a liquid phase. The compound effects of the second particle, liquid phase and anisotropy were successfully analyzed by the MC simulations in order to design complex and important microstructures in sintered materials. The MC simulation has currently remarkable development for the capability to design complex three-dimensional microstructure involving continuous or network structure of particles or pores in sintered materials. Furthermore, the combination method between MC and

FEM has been developed to predict precise shrinkage behaviors during sintering process.

In the Japanese national project, "Structural Material for Inovation, Cross-ministerial Strategic Innovation Promotion Program (SIP)", computer simulation studies are prom oted for the design of ceramics coating materials. We study the computer simulation for m icrostructure of ceramics coating with pore, gap and particle with nano scale and the microstructure change during coating process and long time working at high temperature. Microstructure change, mechanical properties degradation and peeling are also predicted by the simulation.

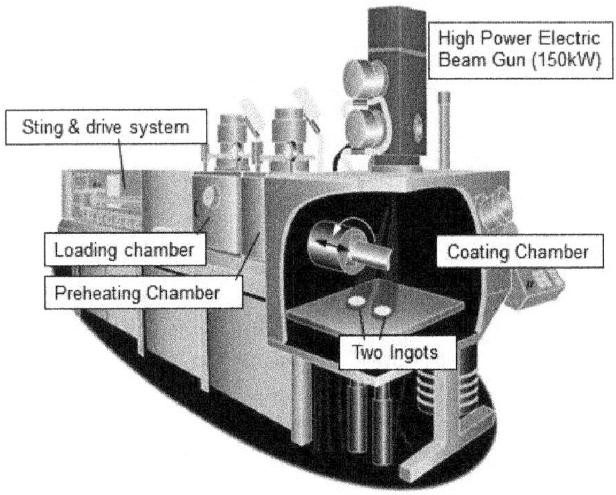

Fig.1 The EB-PVD equi pment at Japan Fine Ceram ics Center for the Nano-Coating Project (METI/NEDO) in Japan.

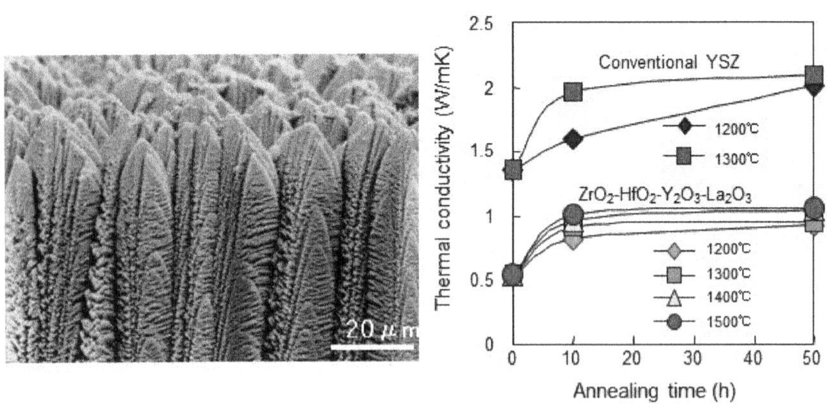

Fig.2 Microstructure and therm al conductivity of zirconia base ceramics layer fabricated by electron beam PVD process for thermal barrier coating in jet engine.

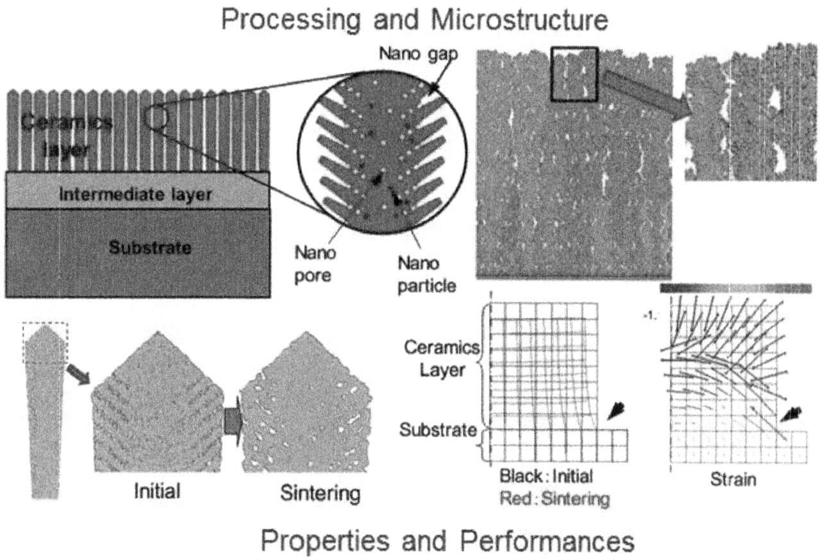

Fig.3 Images of simulation for ceramics coating in SIP project.

REFERENCES

[1]N. Yamaguchi, K. Wada and H. Matsubara, Microstructure modification of yttria-stabilized zirconia layers prepared by EB-PVD, J. Ceram. Soc. Japan, **111**, 883-889(2003).

[2]K. Wada, N. Yamaguchi and H. Matsubara, Crystallographic Texture Evolution in ZrO2-Y2O3 Layers Produced by Electron Beam Physical Vapor Deposition, Surface and Coatings Technology, Elsevier Science, **184**, 55-62(2004).

[3]B.K. Jang, M. Yoshiya, N. Yamaguchi and H. Matsubara, Evaluation of thermal conductivity of zirconia coating layers deposited by EB-PVD, J. Mater. Sci. Lett., **39**, 1823-1825 (2004).

[4]M. Matsumoto, N. Yamaguchi and H. Matsubara, Low thermal conductivity and high temperature stability of ZrO2-Y2O3-La2O3 coatings produced by electron beam PVD, Scripta Mater., **50**, 867-871(2004).

[5]K. Wada, N. Yamaguchi and H. Matsubara, Effect of Substrate Rotation on Texture Evolution in ZrO2-4mol% Y2O3 Layers Fabricated by EB-PVD, Surface & Coatings Technology, **191**, 367-374(2005).

[6]T.Kato, K. Matsumoto, H. Matsubara, Y. Ishiwata, H. Saka, T. Hirayama and Y. Ikuhara, Transmission Electron Microscopy Characterization of a Yttria Stabilized Zirconia Coating Fabricated by Electron Beam-Physical Vapor Deposition, Surface and Coating Technology, **194**, 16-23(2005).

[7]B. K. Jang, H. Matsubara, Influence of Rotation Speed on Microstructure and Thermal Conductivity of Nano-porous Zirconia Layers Fabricated by EB-PVD, Scripta Materialia, **52**, 553-558 '(2005).

[8]K. Wada, M. Yoshiya, N. Yamaguchi and H. Matsubara, Surface & Coatings Technology, Texture and microstructure of ZrO2-4mol% Y2O3 layers obliquely deposited by EB-PVD, Surface & Coatings Technology, **200**, 2725-2730(2006).

[9]B. K Jang and Hideaki Matsubara, Surface roughness and microstructure of yttria stabilized zirconia EB-PVD coatings, Surface and Coating Technology, **200**, 4594-4600 (2006).
[10]K. Suzuki, K. Wada, H. Matsubara, T. Shobu, M. Kawamura and K. Tanaka, Residual stress in zirconia coating by EB-PVD Method, Journal of the Society of Material Science, **56**, 588-593(2007).
[11]K. Suzuki, T. Shobu, K. W ada and H. Matsub ara, Residual stress and structure of EB-PVD thermal barrier coatings exposed at high temperature, Transactions of the Japan Society of Mechanical Engineers, **74**, 370-376(2008).
[12]K. Suzuki, T. Shobu, K. Wada, H. Matsubara and M. Kaw amura, Microstructure and residual stress of EB-PVD TB Cs grown under substrate rotation, Journal of the Society of Material Science, **57**, 674-680(2008).
[13]B. K Jang, Y. Sakka, H.Matsubara, H-T Kim, Thermal Conductivity and Microstructure of Nano-Porous Zirconia Co atings Fabricated by Electron Beam PVD, Materials Science and Technology, **47**, 32-39, (2010).
[14]H. Matsubara and R. J. Brook, Computational Modeling of Mass Transfer for Ceramic Microstructure", Ceramic Transactions, Am. Ceram. Soc., **71**, 403-418(1996).
[15]M. Tajika, H. Matsubara and W. Rafaniell, Experimental and Computational Study of Grain Growth in AlN Based Ceramics, J. Ceram. Soc. Japan, **105**, 928-933(1997).
Y. Okamoto, N. Hirosaki and H. Matsubara , Computational Modeling of Grain Growth in Self-reinforced Silicon Nitride, J. Ceram. Soc. Japan, **107**, 109-114 (1999).
[16]M. Tajika, H. Nomura, H. Matsubara and W. Rafaniello, Experimental and Computational study of Grain Growth and Microstructures in AlN Composite Ceramics, J. Ceram. Soc. Japan, **109**, 288-293 (2001).
[17]M. Yoshiya, K. W ada, B. K. Jang and H. Matsubara, Computer simulation of nano-pore formation in EB-PVD therm al barrier coating s, Surface & Coating s Technology, **191**, 367-374 (2004).
[18]K. Mori, H. Matsubara and N. Nogami, Micro-macro of sintering process by coupling Monte Carlo and finet elem ent methods, International Journal of Mechanical Science, **46**, 841-856(2004).
[19]M. Shimizu, H. Matsubara, H. Nomura, M. Okuhara and H. Tomioka, Computational modeling of phase connectivity in solid-liquid phase microstructures during grain growth, J. Ceram. Soc. Japan, **110**, 1067-1072(2002).
[20]M. Shimizu, H. Matsubara, H. Nomura and H. Tomioka, Computational modeling of phase connectivity in microstructures during sintering and grain growth, J. Ceram. Soc. Japan, **111**, 205-211(2003).
[21]H. ITAHARA, H. NOMURA, T. TANI and H. MATSUBARA, Design of Grain Oriented Microstructure by Using the Mont e Carlo Sim ulation of Sintering and Grain Growth; Isotropic Grain Growth, J. Ceram. Soc. Japan, **111**, 548-554 (2003).
[22]H. Matsubara, Computer Simulation Studies on Sintering and Grain Growth, J. Ceram. Soc. Japan, **113**, 263-268(2005).
[23]H. Itahara, T . Tani, H. Nomura and H. Matsubara, Com putational Design for Grain-Oriented Microstructure of F unctional Ceramics Prepared by T emplated Grain Growth, J. Am. Ceram. Soc., **89**, 1557-1662(2006).
[24]M. Shimizu, H. Nomura, H. Matsubara and S-G Shin, Analysis of porous structures of crystalline particles with anisotropic surface energy by the sintering simulation, J. Jpn. Society of Powder Powder Metallurgy, **55**, 3-9(2008).

Ceramic Optical Materials

NOVEL GLASS AND GLASS SCINTILLATORS FOR GAMMA-RAY AND NEUTRON DETECTION

Tapan K. Gupta[1], William Rhodes[1], Matthew M. Hall[2], Sean Breed[2], Urmila Shirwadkar[1], Michael R. Squillante[1], and Kanai S. Shah[1]

[1]Radiation Monitoring Devices, Inc.
Watertown, MA, USA

[2]Alfred University
Alfred, NY, USA

ABSTRACT

The ultimate goal of the research described here is to develop glass and glass-ceramic scintillators that are capable of efficient gamma-ray and thermal neutron detection, have high light yields, a fast response, moderately good energy resolution, and neutron/gamma discrimination. Such scintillators would provide the inherent advantages of glasses such as the fabrication into very large sizes and fibers at low cost. The scintillators will be based upon robust glasses (such as magnesium aluminosilicates) doped with both ^{6}Li as neutron absorber and Ce^{3+} as luminescent dopant. The ^{6}Li ion is incorporated directly into the glass matrix as network modifier. The role of the Ce^{3+} ion is more complex since in glass ceramics it can serve as a "nucleating agent", facilitating the precipitation of nanocrystals within the host glass matrix using engineered heat treatments. Consequently, the glass-ceramic matrix can retain the favorable properties of bulk inorganic single crystalline scintillators, while enabling more economical production into larger volumes.

The use of a glass as a scintilation detector was first reported in the late 1950s. They are widely used as neuton detectors with the incorporation of ^{6}Li or ^{10}B to increase neutron sensitivity. Non-crystalline glass materials offer low cost and improved flexibility in size and shape, including long fibers. They are generally not used as gamma-ray detectors because the glasses have low gamma-ray stopping power due to their low density and low atomic number constituents. Also, their light output is significicntly less than typical crystalline scintillators. For use in neutron detectors, it would be advantageous if the glasses were completely insensitive to gamma rays, and significant research has gone into reducing sensitivity. Because they are solid glasses, however, they will always have some undesireable level of sensitivity to gamma-rays that introduces background noise.

To make scintillating glass materials that are useful for detecting gamma-rays, the gamma ray stopping power must be improved by increasing the average atomic number and density, and the light output needs to be increased to provide sufficent signal for doing spectroscopy, and the materials must have the ability to discriminate gamma ray events from neutron events.

APPROACH

This paper describes the first step in our investigation of such glass scintillators, and examines gamma ray/neutron discrimination using pulse shape discrimination (PSD). PSD discriminates gamma rays from neutrons based on the shape of the light pulse emitted by the scintillator. Photoelectrons resulting from gamma-ray interactions travel further in materials than the charged particles generatd by neutron interactions. Thus, neutron interactions generate much denser charge clouds, which affects the decay time of the pulse. PSD uses this difference to identify the type of particle.

Glass scintillators can be easily and economically fabricated in the form of large structures (e.g., as large area plates, tubes, rods or bars). They can be directly cast into complex shapes or pulled into optical fibers.[1,2] At the same time the compositional parameter space for glass scintillators is much broader than that of a single crystal. Due to their thermal, mechanical and chemical durability, glass scintillators can also be used as radiation detectors under a variety of adverse conditions.[3] Glasses activated with rare earth ions are promising alternatives to single crystals. The cerium doped glass scintillators were first introduced by Ginther and Schulman[4,5,] and with the advancement of the technology these scintillators are being used in high energy physics, and in X-ray-CT.[6,7,8] The most common glass scintillators are terbium or cerium activated lithium silicates and boron silicates. Lithium and boron have high neutron cross sections, giving the detectors good sensitivity. Lithium is more widely used than boron since it creates a larger signal.

The main drawback of glass scintillators is low light yield, generally attributed to inefficient energy transfer from the glass matrix to the luminescent centers. In addition, the density of the glass scintillators (typically less than 3g/cc) is too low to provide good X-ray and gamma-ray stopping power.[9,10] The density of the glass scintillators can be increased by introducing some heavy metals like oxides of gadolinium or lutetium (e.g. Gd_2O_3 and Lu_2O_3) in the glass matrix.[11-14] These heavy metal oxides will increase the density of the glass and the concentration of the metal oxides, and at the same time have the potential to increase light yields. Adding gadolinium to Na-phosphate glass, for example, can increase the light yield by 20%.[15,16,17]

In this work, we examined magnesium aluminosilicate glass with composition on the cordorite area of the MgO-SiO_2-Al_2O_3 phase diagram (Figure 1). This was chosen for our initial investigation because its thermal expansion coefficient is close to that of silicon (used for scintillation readout), and has a high chemical durability, a high volume resistivity, a low dielectric constant, and a high mechanical strength.[18]

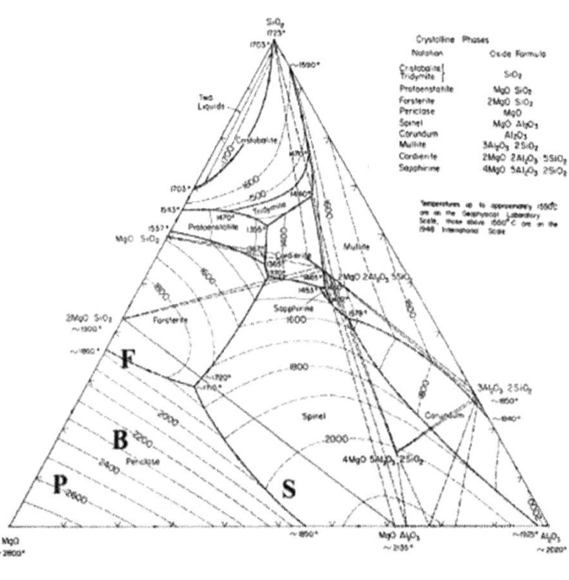

Figure 1 Phase diagram of MgO-Al_2O_3-SiO_2 System [From Muan[19]]. F indicates the forstetite castable, P the periclase, and S the spinel castable and B is the common band phase

^{6}Li and an activator $Ce^{3+}$was added to the cordorite composition. To address dispersion, deflocculants were used during processing. After filling the castable into a cone, the cone was removed and the castable allowed to spread by action of gravity alone. The castable was then subjected to vibration for 15 seconds.

EXPERIMENTAL PROCEDURE

A flow diagram of the glass synthesis process is shown in Figure 2.

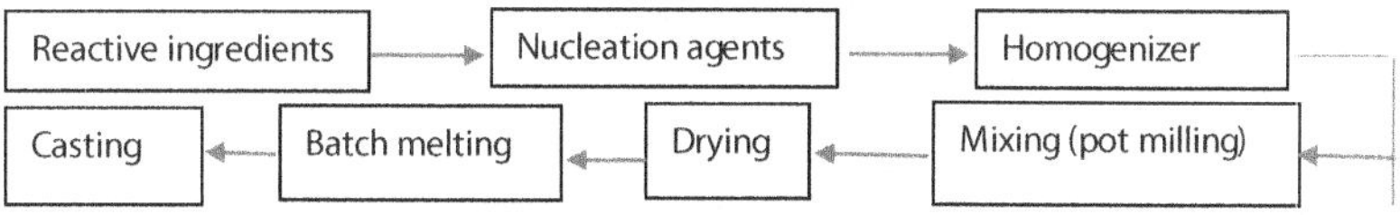

Figure 2. Flow diagram of the synthesis of magnesium aluminosilicate glass

The glass was synthesized as follows: 15 weight percent of magnesium oxide (MgO), 26.8 percent of alumina (Al_2O_3), and 39.5 percent of silica are mixed with 6.7 weight percent of cerium. The mixed chemicals are grounded in a mortar and pestle. The ground material is mixed with 8-10 weight percent of lithium carbonate. The charge material is loaded in a Pt-Rh crucible and heated ~1600C for 1 hour.

The molten mass is swirled to facilitate mixing then poured onto a hot ground Fe-Ni alloy surface. A second hot metal plate made of Fe-Ni alloy was used to press the glass immediately after pouring to prepare flat samples suitable for testing the scintillation properties. The preheated glass was then annealed in a furnace for 6h and cooled down to room temperature at 10C/min. Figure 3 shows the fabricated glass with activator cerium. Figure 4 shows activated glass doped with Li^{6}.

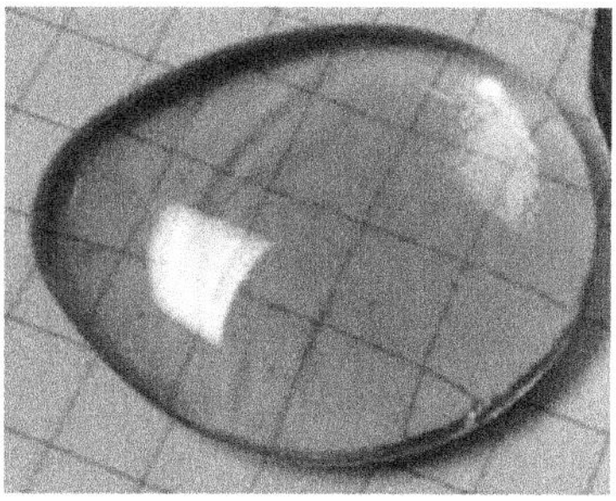

Figure 3. Transparent Ce-activated magnesium aluminosilicate glass

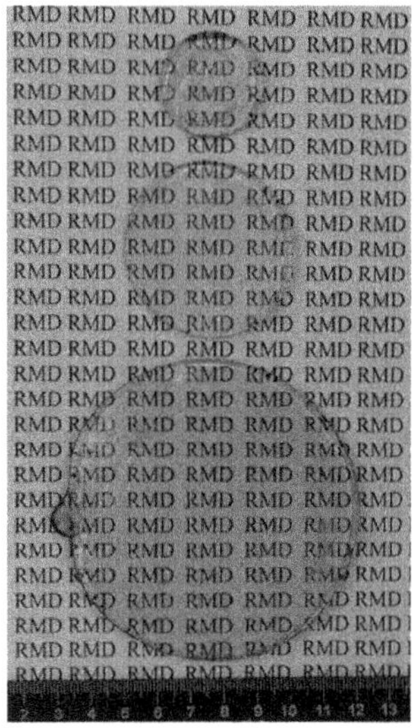

Figure 4 shows the Ce-activated glass with Li^6 dopant added

RESULTS AND DISCUSSIONS

The glass samples shown in Figure 4 were characterized for transparency and emission properties. The solid piece is placed on printed paper to show its clear transparency. The refractive index was measured to be ~1.58. The yellowish color of the material is due to the Ce^{3+} ion. Figure 5 shows the emission spectrum under X-ray excitation of the Ce^{3+} activated glass. The upper spectrum is for the test glass composition without lithium and the lower one is for the glass enriched with Li^6. The emission spectra show two broad bands, one ranging from 280-300nm region and the other one extending from 625-670nm region. The first one can be ascribed due to 4f-$5d_{3,4,5}$ transition, and the second one can be attributed to the $5d^1$-$^2F_{5/2,7/2}$ transition. The decay time was about 20 ns. The lithium does not significantly affect the transparency, or the emission spectrum.

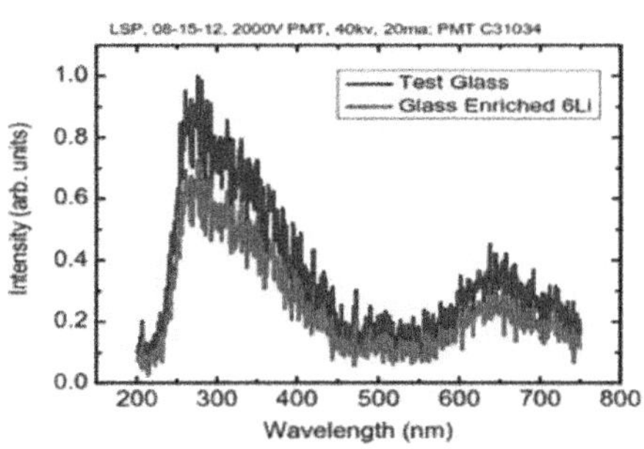

Figure 5. Emission spectrum of the cerium doped magnesium aluminosilicate glass without ^{6}Li (upper) and with ^{6}Li (lower).

We also characterized the response of the glass to gamma rays and neutrons ionizing radiation. One of the key requirements of a detector for use with both gamma rays and neutrons is the ability to distinguish gamma ray events from neutron events. A variety of approaches have been used in the past to discriminate gamma ray events from neutron events, with limited success. These include simply using pulse height discrimination, since the energy deposited by a neutron event is larger than most gamma rays encountered in many applications. The discrimination is not adequate for security applications and neutron imaging in the presence of high gamma ray backgrounds. Other approaches include two similar detectors: one enriched in ^{6}Li and one depleted in ^{6}Li[20,21], layered hybrid detectors of Li-glass and plastic scintillator[22], and post detection spectrum analysis[23]. None however, are suitable for producing practical gamma-ray/neutron detectors.

The scintillation response of the glass is tested with an Am/Be source. Figure 6 shows the energy spectra of magnesium aluminosilicate glass doped with lithium (Li^6) and activated with cerium (Ce^{3+}), and recorded during irradiation by Am/Be thermal neutrons and by ^{241}Am 60keV gamma rays. The counts in arbitrary units are plotted along the Y-axis and the channel numbers are plotted along the X-axis. The lower spectrum was collected under irradiation from a thermal neutron source (thermalized Am/Be). Upon interacting with Li^6 ions in the glass, the thermal neutrons generate conversion electrons and produce a scintillating pulse, which is recorded by the multichannel analyzer (MCA). The upper spectrum was collected with the thermal neutrons blocked by a boron rubber sheet (boron absorbs thermal neutrons), in order to record the spectrum attributable to ambient ionizing radiation.

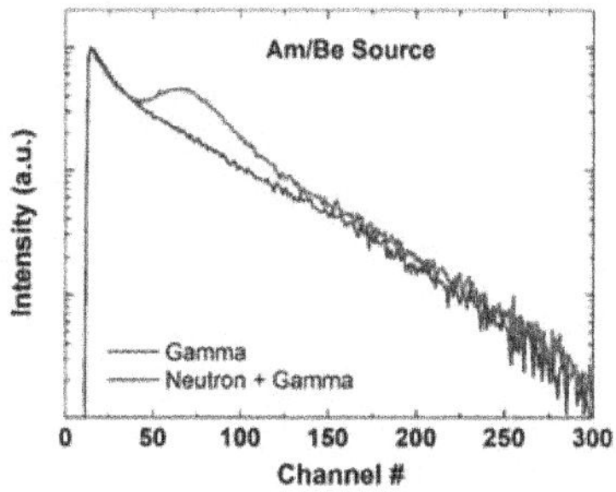

Figure 6. Pulse height spectra of a ^{6}Li-doped glass-ceramic scintillator under neutron and gamma-ray excitation. The neutron peak is clearly visible in the full spectrum (Neutron + Gamma) shown in the upper curve, and vanishes when the neutrons from the source are attenuated with boron-loaded rubber sheets.

Figure 7 shows the pulse shape obtained from gamma-ray and neutron events. These data were obtained by integrating a number of individual pulses. Clearly, there is a significant difference in the shape of the decay. This difference can be used for separating and identifying gamma-ray and neutron events in the detector. This is shown in Figure 8, which plots the ratio of two regions of interest in the pulse, one at a short time (less than 5 ns) and one at a longer time (tens of nanoseconds). While this is very encouraging and demonstrates the concept, better performance is needed to produce useful detectors. The performance of these devices is limited by the low light levels, which results in a poor signal-to-noise ratio. Higher light output would increase the S/N, and provide much better discrimination. These results do, however, demonstrate that glass-ceramic scintillators have the capability to discriminate the signals, and thus have the potential for use in gamma ray and neutron detectors.

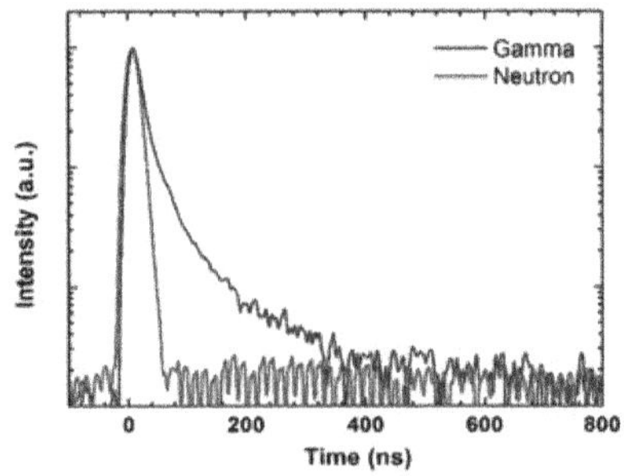

Figure 7. Scintillation pulse decay traces of a ^{6}Li-containing glass scintillator measured under gamma-ray (slow decay) and neutron (fast decay) excitation. Note the difference in the lifetimes for gamma-ray and neutron excitation, which is needed to provide n-γ pulse shape discrimination.

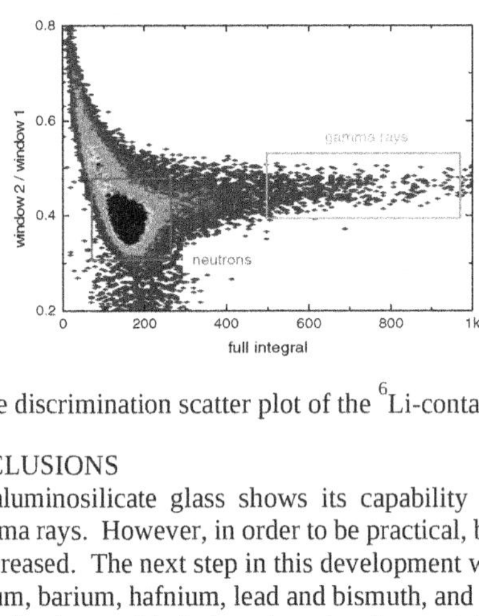

Figure 8. Pulse shape discrimination scatter plot of the ^{6}Li-containing glass scintillator

SUMMARY AND CONCLUSIONS

Our magnesium aluminosilicate glass shows its capability to detect and discriminate thermal neutrons and gamma rays. However, in order to be practical, both the stopping power and light output need to be increased. The next step in this development will be the addition of heavy metal ions such as strontium, barium, hafnium, lead and bismuth, and heavy metal rare earth ions, such as gadolinium praseodymium and holmium, into the glass matrix to improve the light yield and quantum efficiency. Polisadova, for example, studied Li_2O-B_2O_3-P_2O_5-CaF_2 system doped by Gd, Gd/Ce, Gd/Pr, Tb, and Tb/Ce.[15,24] In parallel, we will explore the process parameter space to improve transparency and uniformity. Beyond that, we plan to investigate glass ceramics in which nanocrystals (tens of nanometers) are incorporated into the glass to act as miniature single crystal scintillators. Transparent glass ceramics containing rare-earth doped halide nanocrystals have been investigated for applications in lasers and optical amplification.[6,25] Figure 9 is a micrograph of a glass ceramic developed for use as an LED host[26]. This approach has been recently investigated by a number of researchers, using a variety of materials containing (Pb,Cd)F_2, LaF_3, GdF_3, BaF_2, $BaCl_2$, and CaF_2 nanocrystals.[6,27,28] Kang, *et al.*, achieved light output of greater than 2000 photons/ Mev, with 20% FWHM energy resolution with a CdF_2 glass ceramic.[29] The use of glass scintillators optimized for gamma ray performance, and as the host for glass ceramics, holds significant potential to meet the need of large, inexpensive detectors.

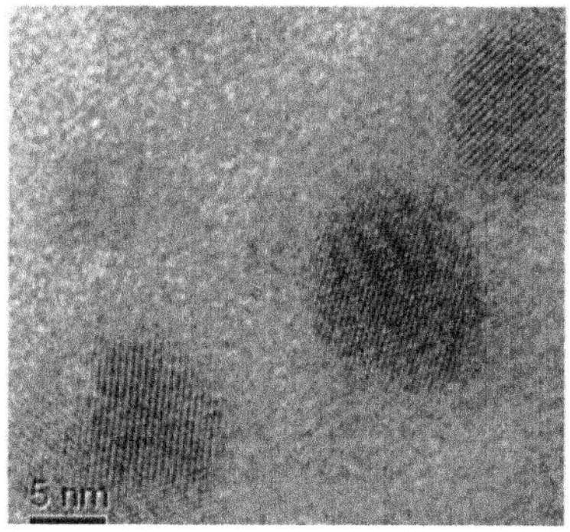

Figure 9. TEM image of CeF_2 glass ceramic [Chen[26]]

REFERENCES:

[1]K.H. Park, D.W. Shin, J. Ceram. Process. Res. 3, 153 (2002)

[2]L.H. Lao, H.P. Zhou, and C. Xu, J. Mater. Sci., Materials in Electronics, 13 (7), 381 (2002)

[3]G.C. Tyrrell, Phosphors and Scintillators in Radiation Imaging Detectors, Nuclr. Instrum. Methods.A, 546, 180 (2005)

[4]R.J. Ginther, J.H. Schulman, Glass Scintillators, IRE Trans. Nuclr. Sci. NS-5 (3), 92 (1958)

[5]R.J. Ginther, New Cerium Activated Scintillating Glass, IRE Trans. Nuclr. Sci. NS-7, 28 (1960)

[6] Leo, W. R. "Techniques for Nuclear and particle Physics Experiments", 2nd edition, Springer, ISBN 3-540-57280-5] (1994).

[7]D.G. Anderson, J. Dracass, and T.P. Flanagan, Luminescent Glasses for Use in the Detection of Nuclear Radiation, J. Electron. Control, 7, 463 (1959)

[8]M.S. West and W.P. Winfree, Temporal Response of Terbium Glass Scintillator Used for X-ray Tomography and Radiography, Rev. Prog. Quanti. ND Evaluation, 15, 457 (1996)

[9]W. Kononenko, Measurements for Muons in Scintillating Glass with Vacuum Photodiode Readout, IEEE Trans. Nuclr. Sci. 31 (1), 136 (2007)

[10]S.S. Patra, Synthesis and Characterization of Cordierite Based Glass Ceramic, B.S. Thesis, N.I.T. Rourkela, India, (2014)

[11]O.V. Mazurin, E.A. Porai-Koshits (eds.), Phase Separation in Glass, North Holland, NY., (1984)

[12]M. Tomozawa, in L.L. Hench, S.W. Freiman (eds.) Advances in Nucleation and Crystallization in Glasses, Am. Ceram. Soc. Westerville, OH, (1972), p-41

[13]L. Li, M.X. Qian, and D. M. Strachan, Gadolinium Solubility in Peraluminus Borosilicate Glasses. J. Non-Cryst. Solids, 281 (1-3), 189 (2001)

[14]M. Quin, L.Li, and D.M. Strachan, Partitioning of Gadolinium and its induced Phase Separation in Sodium-Aluminoborosilicate glasses, J. Non-Cryst. Solids, 333 1, (2004)

[15]E. F. Polisadova, D. T. Valiev, K. N. Belikov, N. L. Egorova, "Scintillation lithium-phosphate-borate glasses doped by REI", Glass Physics and Chemistry, Volume 41, pp 98-103 (2015),

[16]Nikl, M., Nitsch, K., Mihokova, E., Solovieva, N., Mares, J.A., Fabeni, P., Pazzi, G.P., Martini, M., Vedda, A., and Baccaro, S., Efficient radioluminescence of the Ce3+-doped Na-Gd phosphate glasses, Appl. Phys. Lett., vol. 77, p. 2159 (2000)

[17]Mares, J.A., Nikl, M., Nitsch, K., Solovieva, N., Krasnikov, A., and Zazubovich, S., A role of Gd3+ in scintillating processes in Tb-doped Na-Gd phosphate glasses, J. Lumin., 2001, vols. 94–95, pp. 321–324.

[18]Y. Demici and E. Gilnay, Crystallization Behavior and Properties of Cordierite Glass-Ceramics with added Boron Oxide, J. Ceram. Process. Res. 12 (3), 352 (2011)

[19]A. Muan and E.F. Osborn, Am. Ceram. Soc. 1960

[20]D.M. Schneider, C.T. Meriden, and B.G. Hubner, Neutron/gamma discrimination in a lithium-6 glass scintillator in MWD tool, IEEE Trans. Nuclr. Sci. Symp. & Med. Imag. Conf. 1991, Santa Fe, NM, 2nd Nov.-9th Nov., (1991)

[21]M. Foster, and D. Ramsden, A compact Neutron Detector-Based on the use of SiPM detector, IEEE Nuclr. Sci. N-26(3), 1882 (2008)

[22]R.A. Winyard, J.E. Lutkin, and G.W. McBeth, Pulse shape discrimination in inorganic and organic scintillators, Nuclr. Instrum. Meth., 95, 141 (1971)

[23]A. Lintereu, J. Ely, J. Stave, and B. Mcdonald, Neutron and Gamma ray pulse discrimination with polyvinyltoluene, Pacific North West Nat. Lab. Pub. March (2012)

[24]M. Brooke Barta, Jason H. Nadler, Zhitao Kang, Brent K. Wagner, Robert Rosson, Ye Cai, Kenneth H. Sandhage, and Bernd Kahn, "Composition optimization of scintillating rare-earth nanocrystals in oxide glass–ceramics for radiation spectroscopy"' Applied Optics Vol. 53, pp. D21-D28 (2014) , doi: 10.1364/AO.53.000D21

[25]Wang Y, Ohwaki J.: New transparent vitroceramics codoped with Er^{3+} and Yb^{3+} for efficient frequency upconversion" Appl Phys Lett. 1993; 63:3268

[26]D. Chen and Y. Yu and H. Lin and P. Huang and F. Weng and Z. Shan and Y. Wang, "CeF3 - based glass-ceramic: a potential luminescent host for white-light-emitting diode", Optics Letters 34(19), (2009), pp. 2882–2884.

[27]Chen G, Johnson J, Weber R, et al. Fluorozirconate - based nanophase glass ceramics for high-resolution medical X-ray imaging. Journal of Non-Crystalline Solids. 2006;352:610.

[28]Chen D, Wang Y, Yu Y, Huang P. Structure and Optical Spectroscopy of Eu-Doped Glass Ceramics Containing GdF_3 Nanocrystals. J Phys Chem C. 2008;112:18943–18947.

[29]. Z. Kang, B. K. Wagner, C. J. Summers, J. Nadler, R. Rosson, B. Kahn. "Polymer and glass matrix nanocomposites for scintillation applications", IEEE Nuclear Science Symposium and Medical Imaging Conference Record (NSS/MIC), pp. 1688–1691 (2012).

PRASEODYMIUM-DOPED SiAlON RED PHOSPHORS PREPARED BY POLYMER-DERIVED METHOD

Hui Yu [1], Quan Li [1], Xuan Cheng [1,2] Ying Zhang [1,2], Chaoyang Gong [1,2]
[1]Department of Materials Science and Engineering, College of Materials, Xiamen University
Xiamen, Fujian, P. R. China
[2]Fujian Key Laboratory of Advanced Materials
Xiamen, Fujian, P. R. China

ABSTRACT

The red emission praseodymium (Pr)-doped silicon based oxynitride phosphors were prepared by a novel polymer-derived synthesis route via nitridizing and calcining precursor formed by reacting polycarbosilane with Pr acetylacetonate (SiON:Pr) or Pr and Al acetylacetonates (SiAlON:Pr). The effects of sintering temperatures and nominal ratios of Si/Pr or (Si+Al)/Pr on photoluminescence properties were examined. It was shown that the α-Si_3N_4 and β-Si_3N_4 phases coexisted, and two emission peaks at 566 nm and 610 nm were observed for the SiON:Pr phosphor. The mixed β-SiAlON and Si_2N_2O phases were identified for the SiAlON:Pr phosphor sintered at 1550 °C, with two emission peaks at 566 nm and 610 nm. However, when sintered at 1600 °C and 1650 °C, the SiAlON:Pr phosphors exhibited only one emission peak at 623 nm with a predominant β-SiAlON phase. The double-exponential photoluminescence behaviors with the decay time of 30-45 μs were found for the SiAlON:Pr phosphors, which corresponded to the transitions of $^1D_2 \rightarrow {}^3H_4$.

INTRODUCTION

Rare-earth-activated silicon based oxynitride luminescence materials have attracted much attention in recent years due to their thermos-stability, non-toxicity and excellent optical properties. This novel class of inorganic phosphors exhibits a broad excitation and emission in ultraviolet to visible spectral region, and has potential applications in light emitting diodes. Silicon nitride (Si_3N_4) typically exhibits two phases, namely, α-Si_3N_4 with trigonal symmetry and β-Si_3N_4 with hexagonal symmetry [1]. Recently, Si_3N_4 and Si_3N_4-based derivatives have emerged as good candidates for rare-earth-doped phosphors because of their low thermal quenching, high chemical stability and encouraging luminescent properties (excitability by ultraviolet or blue light and the possibility of full color emission) [2]. The crystal structures of α-SiAlON and β-SiAlON are similar to those of α-Si_3N_4 and β-Si_3N_4, respectively. Little information is available for Pr-doped β-SiAlON phosphors. The only reported β-SiAlON:Pr^{3+} red phosphor had to start with the pure raw materials of α-Si_3N_4 and AlN nitrides, along with Al_2O_3 and Pr_2O_3 oxides. The conversion from α-Si_3N_4 to β-Si_3N_4 was first carried out under vacuum atmosphere (10^{-2} Pa) from room temperature to 1200°C then maintaining at 1200°C for 2h, and followed by the formation of β-SiAlON:Pr^{3+} through a complicated sintering procedure, i.e., 1200°C-1600°C, maintaining at 1600°C for 3h, 1600°C-1950°C, maintaining at 1950°C for 2h, under 0.92 MPa nitrogen atmosphere [3-5]. Apparently, the extremely high temperature and high pressure are required since the Al and Pr elements are introduced via solid state reaction.

Recently, we have reported a simple and facile polymer-derived method to synthesize Eu-doped

Si_3N_4 yellow phosphors [6]. Completely different from the traditional solid state reaction methods, a polymer-derived method can achieve not only a molecular level mixture of raw materials through polymer reaction, but also a uniform rare earth doping into the phosphor at mild conditions through an organic-inorganic transformation. This work was extended to prepare the polymer-derived Pr-doped Si-based oxynitrides and SiAlON red phosphors by using solid polycarbosilane, praseodymium acetylacetonate or aluminum acetylacetonate as raw materials. The effects of atomic ratios for Si/Pr or (Si+Al)/Pr and sintering temperatures without and with Al incorporation on crystal structures, photoluminescence properties and surface states of Pr and Al were preliminarily investigated.

MATERIALS AND METHODS

The empirical formula of home-made solid polycarbosilane (PCS) was $SiC_{1.78}H_{4.38}O_{0.03}$ with the softening point of 153 °C. The number and weight average molecular weights were 950 g/mol and 2157 g/mol, respectively, with the polydispersity index (PDI) of 2.27. Praseodymium acetylacetonate ($Pr(AcAc)_3$) and aluminum acetylacetonate ($Al(AcAc)_3$) hydrates of 99% purity were purchased from Shanghai Dibo Chemical Co., Ltd. China.

The 10 g of PCS powder was ground and well mixed with 5 g of $Al(AcAc)_3$ and 1 g of $Pr(AcAc)_3$ in a 500-ml three-neck flask which was connected to a closed system. Meanwhile, a heating jacket wrapped the flask to control the heating process. The flask was vacuum-pumped and imported N_2 (99.99%) repeatedly for three times, heated up to 320 °C in N_2 (99.99%) for 6 h, and then naturally cooled to room temperature. Finally, the precursor powder named PPACS was obtained. The 1.5 g PPACS was milled and put in an alumina crucible, and transferred into a tube furnace. The heating process was proceeded to raise temperature to 800 °C at 5 °C/min and held for 3 h in NH_3 (99.999%), then continued to the selected sintering temperatures of 1550 °C /1600 °C /1650 °C at 5 °C /min and held for 3 h in N_2 (99.999%). The polymer-derived Pr-doped SiAlON powders with the nominal ratios of (Si+Al)/Pr being (83+7)/1 were obtained after natural cooling to room temperature. Prior to the SiAlON preparations, the polymer-derived Pr-doped Si-based oxynitrides (SiON) powders were synthesized by the similar method using the Si/Pr ratios of 46/1, 62/1 and 90/1 at the selected sintering temperatures of 1550 °C, 1600 °C and 1650 °C without adding $Al(AcAc)_3$.

The crystal structures were characterized by Rigaku Ultima IV X-ray diffraction (XRD) analyzer using Cu K_α radiation. The chemical states of Pr and Al were analyzed by X-ray photoelectron spectroscope (XPS) with PHI Quantum 2000 scanning ESCA microprobe using Al K_α radiation. Photoluminescence and fluorescent lifetime spectra were collected by an FLS920 fluorescence spectrophotometer equipped with a Xenon lamp. The phosphorescence decay measurements were made with a pulsed Xenon microsecond flash lamp producing short microsecond pulses.

RESULTS AND DISCUSSION

Typical powder X-ray diffraction patterns of the Pr-doped SiON and SiAlON phosphors are shown in Fig. 1. For comparisons, the standard lines of α-Si_3N_4, β-Si_3N_4 and Si_2N_2O are included in the figure. In general, a mixed phase of α-Si_3N_4 and β-Si_3N_4 was identified from the Pr-doped SiON phosphors prepared with the nominal ratios of Si/Pr being 46/1, 62/1 and 90/1 when sintered at different temperatures, as evident in Fig. 1(a-c). However, the crystal phases formed were obviously changed with the introduction of Al (Fig. 1(d)). The formation of Si_2N_2O phase became evident, in particular at 1550 °C. The β-Si_3N_4 phase predominated at 1600 °C and 1650 °C.

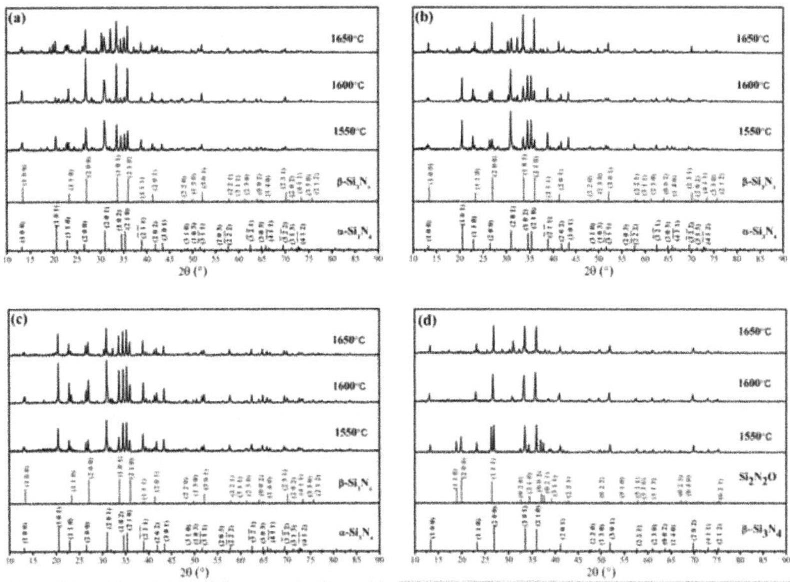

Figure 1. Typical powder X-ray diffraction patterns of Pr-doped SiON (a-c) and SiAlON (d) sintered at different temperatures. (a) Si/Pr=46/1; (b) Si/Pr=62/1; (c) Si/Pr=90/1; (d) (Si+Al)/Pr=(83+7)/1. The standard lines of α-Si_3N_4, β-Si_3N_4 and Si_2N_2O are included for comparisons.

To further recognize possible impure phases presented in the Pr-doped SiON and SiAlON phosphors prepared at different sintering temperatures, Figure 1(a) and Figure 1(d) are re-plotted at an expanded scale for 2θ=10°-50°. By carefully examining Fig. 2(a) and Fig. 2(b), minor impure phases such as SiO_2 (labeled as asterisk) and C_3N_4 (labeled as number symbol) were also presented when sintered at 1650 °C for the Pr-doped SiON (Fig. 2(a)). With the Al incorporation (Fig. 2(b)), a mixed phase of β-Si_3N_4 and Si_2N_2O was found at 1550 °C, while the main phase of β-Si_3N_4 sintered at 1600 °C and 1650 °C. Other impure phases of $PrSi_3N_5$ (labeled as asterisks) and $Pr_2(Al_{0.5}Si_{2.5}O_{3.5}N_{3.5})$ (labeled as numbers) were also detected.

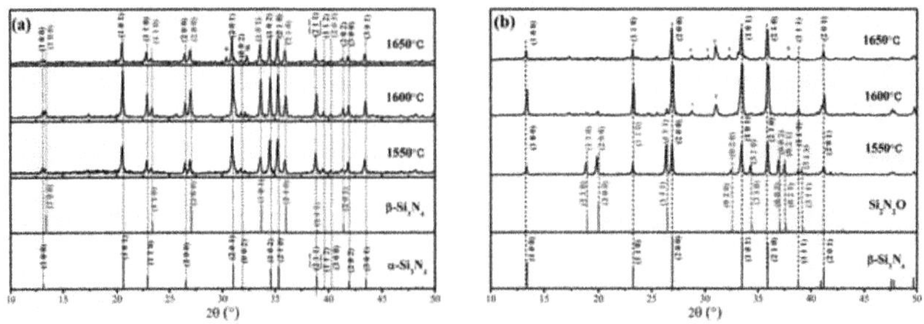

Figure 2. The X-ray diffraction patterns of Pr-doped SiON (a) and Pr-doped SiAlON (b) re-plotted at an expanded scale for 2θ=10°-50°.

The corresponding excitation and emission spectra of Pr-doped SiON and SiAlON phosphors are illustrated in Fig. 3. Typical Pr-activated photoluminescence responses were observed and the assignments to the origin of emission peaks [7] are indicated in the figures. It is evident from Fig. 3(a-c) that the Pr-doped SiON phosphors exhibited the emission peaks at 566 nm ($^3P_0 \rightarrow ^3H_5$) and 610 nm ($^1D_2 \rightarrow ^3H_4$) under 235 nm excitation, which are not significantly affected by the atomic ratios of Si/Pr and sintering temperatures. The relatively strong excitation bands observed at 235 nm could be assigned to the host lattice absorption [8], while the relatively weak excitation at 460-480nm corresponded to the f–f transitions of Pr^{3+} [9]. However, the Pr-doped SiAlON phosphors when sintered at 1600°C and 1650°C showed the very strong emission peak at 623 nm ($^1D_2 \rightarrow ^3H_4$) under 290 nm excitation (Fig. 3(d)). The excitation bands near 290 nm and 460-480 nm were enhanced by Al incorporation, which could be attributed to the ionization process of Pr^{3+} ions (transfer of an electron from the Pr^{3+} ion to the conduction band) and the transitions of $^3H_4 \rightarrow ^3P_0$, $^3P_1$ and $^3P_2$ (belonging to f-f transitions of Pr^{3+}), respectively [4,5]. The prominent red luminescence observed in Fig. 3(d) might be caused by relaxation of a low-lying Pr-to-metal intervalence charge transfer state (IVCT) that can lead to the population of the excited $^1D_2$ state and to the total or partial quenching of the $^3P_0$ emission [10,11]. This type of red-emitting behavior is similar to those reported previously by solid state reaction [3-5] since the predominant phase is β-SiAlON. On the other hand, the Pr-doped SiAlON phosphor sintered at 1550 °C displayed very similar optical properties to the Pr-doped SiON phosphors. The coexistence of β-Si_3N_4 and Si_2N_2O phases resulted in an additional excitation band at 260 nm, this band might belong to the absorption of Si_2N_2O phase. The excitation bands smaller than 235 nm induced by the absorption of Si_3N_4 host and the energy transformation from the host to Pr activator led to energy loss, therefore, a lower luminescence intensity is understandable without Al incorporation.

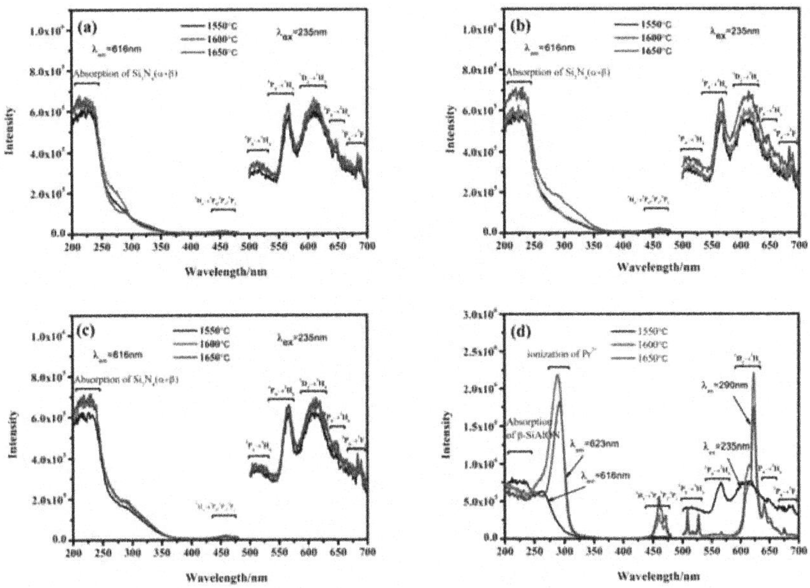

Figure 3. The excitation and emission spectra of Pr-doped SiON (a-c) and Pr-doped SiAlON (d) sintered at different temperatures. (a) Si/Pr=46/1; (b) Si/Pr=62/1; (c) Si/Pr=90/1; (d) (Si+Al)/Pr=(83+7)/1.

Apparently, the Pr-doped SiAlON phosphors exhibited stronger red emission, thus, their photoluminescence behaviors were further studied by measuring the photoluminescence decay curves under the excitation at 235 nm and the emission at 616 nm for 1500 °C, the excitation at 290 nm and the emission at 623 nm for 1600 °C and 1650 °C. The results are illustrated in Fig. 4. For the $^1D_2 \rightarrow {}^3H_4$ red emission, the decay curves in Fig. 4 were fitted double-exponentially, and the average decay time (τ) can be calculated using the following equation:

$$\tau = \frac{\int I(t)tdt}{\int I(t)dt} \tag{1}$$

The average τ values are evaluated to be 101 μs, 45 μs and 30 μs for 1550 °C, 1600 °C and 1650 °C, respectively. The average τ value for 1550 °C differed appreciably from those for 1600 °C and 1650 °C because of the different host structures. The former was a mix phase of Si_2N_2O and β-SiAlON, while the later were dominated by β-SiAlON, as confirmed with the XRD data. The double-exponential photoluminescence behaviors were observed since it could be caused by nonradiative depopulation of the $^1D_2$ state, especially the cross-relaxation energy transfer between two Pr^{3+} ions, or could be due to the presences of both internal and surface defects which may store the excitation energy feeding red emission and consequently the decay time of the $^1D_2$ level increases as well [4, 12, 13].

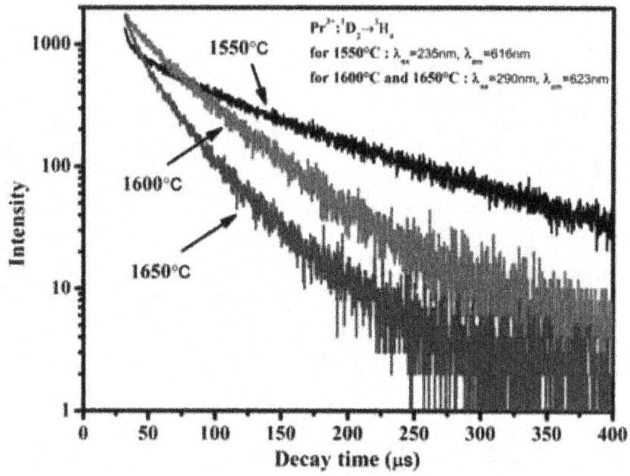

Figure 4. The photoluminescence decay curves of Pr-doped SiAlON phosphors sintered at different temperatures.

The polymer-derived method provided the Pr doping and Al incorporation into SiAlON at an atomic and a molecular level since the precursors containing Pr and Al elements were formed through the distinctive polymer reactions. The chemical states of Al and Pr were further studied by measuring the XPS spectra of Al 2p and Pr 3d for the Pr-doped SiAlON phosphors. The results are given in Fig. 5.

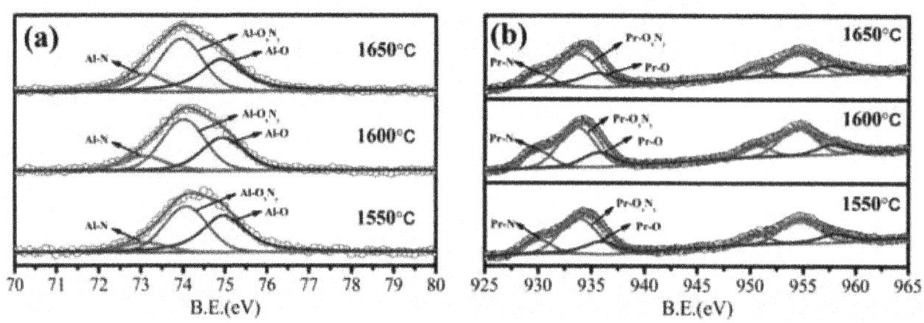

Figure 5. Deconvoluted XPS spectra of Al 2p (a) and Pr 3d (b) for SiAlON:Pr

The XPS spectra of Al 2p (Fig. 5(a)) could be deconvoluted into three peaks with the binding energies centered at 73 eV, 73.8 eV and 74.8 eV, which belonged to Al-N, Al-N_xO_y and Al-O, respectively. Similarly, the Pr 3d spectra can be decomposed and assigned to Pr-N, Pr-N_xO_y and Pr-O, as illustrated in Fig. 5(b). The relative amounts of every species evaluated based on the XPS analysis are summarized in Table I.

Table I. Fitting results of Al 2p and Pr 3d based on XPS data in Figure 5

XPS Spectrum	T (°C)	Relative Amount (%)		
		Al(Pr)-N	Al(Pr)-O_xN_y	Al(Pr)-O
Al 2p	1550	8.9	43.8	47.3
	1600	16.2	48.5	35.3
	1650	17.6	45.7	36.7
Pr 3d	1550	17.1	59.0	23.9
	1600	21.1	58.0	20.9
	1650	17.6	51.4	31.0

The Al-O species was the residuals majorly from the raw material of aluminum acetylacetonate ($Al(AcAc)_3$), while the Al-N and Al-N_xO_y species represented the complete and partial nitridations of $Al(AcAc)_3$, respectively. The relative amounts of Al-N increased from 8% at 1550 °C to 16% at 1600 °C and 17% at 1650 °C, implying the higher degrees of nitradation at 1600 °C and 1650 °C. The relative amount of Pr-N was significantly increased from 17% at 1550 °C to 21% at 1600 °C, but reduced to 18% when further rising to 1650 °C by the introduction of Al (Table I). This might be contributed to the stronger photoluminescence intensity observed with the phosphor sintered at 1600 °C as evident in Fig. 3(d).

CONCLUSIONS

The polymer-derived Pr-doped SiAlON red phosphors were obtained at mild conditions and the precursors containing Pr and Al elements were formed through polymer reactions before the organic-inorganic transformation and crystallization processes. The effects of nominal atomic ratios of Si/Pr and sintering temperatures on the crystal structures and optical properties of Pr-doped SiON were minimal. Without Al incorporation, a mixed phase of α-Si_3N_4 and β-Si_3N_4 was identified. Two emission peaks at 566 nm and 610 nm were observed under the excitation of 235 nm. With Al incorporation, the sintering temperature influenced the crystal structures and optical properties of the phosphors. A coexistence of β-SiAlON and Si_2N_2O phases was found when sintered at 1550 °C. A dominate phase of β-SiAlON was identified at both 1600 °C and 1650 °C, which led to a main emission peak at 623 nm as a red phosphor, similar to those prepared through solid-state reaction route. The enhanced photoluminescence behavior observed at 1600 °C might be contributed to the more complete reaction of Pr acetylacetonate, which increased the relative amount of Pr-N as indicated by the XPS analysis.

REFERENCES

[1]C.-M. Wang, X. Pan, M. Rühle, F. L. Riley, and M. Mitomo, Silicon nitride crystal structure and observations of lattice defects, *Journal of Materials Science*, **31** 5281-98 (1996).
[2]X. Rong-Jun and H. Naoto, Silicon-based oxynitride and nitride phosphors for white LEDs—A review, *Science and Technology of Advanced Materials*, **8** 588-600 (2007).

[3]T.-C. Liu, B.-M. Cheng, S.-F. Hu, and R.-S. Liu, Highly Stable Red Oxynitride β-SiAlON:Pr^{3+} Phosphor for Light-Emitting Diodes, *Chemistry of Materials*, **23** 3698-705 (2011).
[4]A. Lazarowska, S. Mahlik, M. Grinberg, T.-C. Liu, and R.-S. Liu, New Pr^{3+} site in β-SiAlON red phosphor, *Optical Materials*, **35** 2001-05 (2013).
[5]S. Mahlik, A. Lazarowska, M. Grinberg, T.-C. Liu, and R.-S. Liu, Luminescence Spectra of β-SiAlON/Pr^{3+} Under High Hydrostatic Pressure, *The Journal of Physical Chemistry C*, **117** 13181-86 (2013).
[6]Q. Li, C. Y. Gong, X. Cheng, and Y. Zhang, A novel polymer-derived method to prepare Eu-doped Si_3N_4 yellow phosphor, *Ceramics International*, **41** 4227-30 (2015).
[7] G Blasse and B. C. Grabmaier, Luminescent Materials, Springer Berlin Heidelberg (1994).
[8]Y. Q. Li, N. Hirosaki, R. J. Xie, T. Takeda, and M. Mitomo, Photoluminescence properties of rare earth doped α-Si_3N_4, *Journal of Luminescence*, **130** 1147-53 (2010).
[9]P. Dorenbos, A Review on How Lanthanide Impurity Levels Change with Chemistry and Structure of Inorganic Compounds, *ECS Journal of Solid State Science and Technology*, **2** R3001-R11 (2013).
[10]P. Boutinaud, E. Pinel, M. Oubaha, R. Mahiou, E. Cavalli, and M. Bettinelli, Making red emitting phosphors with Pr^{3+}, *Optical Materials*, **28** 9-13 (2006).
[11]P. Boutinaud, R. Mahiou, E. Cavalli, and M. Bettinelli, Red luminescence induced by intervalence charge transfer in Pr^{3+}-doped compounds, *Journal of Luminescence*, **122–123** 430-33 (2007).
[12]P. J. Deren, R. Pazik, W. Strek, P. Boutinaud, and R. Mahiou, Synthesis and spectroscopic properties of $CaTiO_3$ nanocrystals doped with Pr^{3+} ions, *Journal of Alloys and Compounds*, **451** 595-99 (2008).
[13]R. Naccache, F. Vetrone, A. Speghini, M. Bettinelli, and J. A. Capobianco, Cross-relaxation and upconversion processes in Pr^{3+} singly doped and Pr^{3+}/Yb^{3+} codoped nanocrystalline $Gd_3Ga_5O_{12}$: The sensitizer/activator relationship, *The Journal of Physical Chemistry C*, **112** 7750-56 (2008).

Alumina Materials

ALUMINA INSULATORS FOR HIGH VOLTAGE AUTOMOTIVE IGNITION SYSTEMS

William J. Walker, Jr.
Federal-Mogul Powertrain
Plymouth, MI, USA

ABSTRACT

Alumina insulators for spark plugs must reliably provide a high level of performance in the harsh environment of an internal combustion engine. These insulators provide the dielectric barrier to isolate the high voltage required to initiate combustion of the air-fuel mixture. With advances in engine technology, the insulators must withstand increasingly severe conditions including higher voltages and must deliver reliable performance over constantly increasing service lifetimes up to the lifetime of the engine. Dielectric breakdown of the alumina insulator results in misfiring of the engine, thus dielectric breakdown strength is a critical property of the ceramic and processing must be well-controlled in order to ensure a high degree of reliability.

INTRODUCTION

Worldwide, more than 4 million spark plug insulators are produced each day from high alumina ceramic. These are used in automobiles, aircraft, small engines and stationary industrial engines. The insulators serve as both a dielectric barrier to high voltage and a structural element. Figure 1 shows typical construction of an automotive spark plug. The insulator is mounted in a steel shell and supports the electrically conducting elements of the spark plug that transmit electrical current into the combustion chamber where a controlled spark between the center electrode and the ground electrode initiates combustion of the air-fuel mixture. The alumina insulators are a passive dielectric material, meaning that the ceramic acts as a barrier to the passage of electrical current, even at high applied electrical fields. Dielectric breakdown is the failure – in the form of a physical puncture of the ceramic – caused by an electrical field that exceeds the limit of the material. When a dielectric failure occurs, a secondary path to ground is formed which results in poor engine performance.

The ceramic used for spark plug insulators is comprised primarily of alumina. Alumina is produced in enormous quantities from bauxite using the Bayer process as an intermediate step in the production of aluminum metal. A small portion of that material is used for alumina chemicals and ceramics.

Alumina is a very hard, strong, chemically inert material that has been the basis for spark plug insulators since the 1930s[1]. Alumina is blended with small amounts of kaolin, talc and limestone, which react at high temperature to create a liquid phase that aids in consolidating the material during sintering and persists as a glass in the final ceramic material. Early alumina ceramics used for spark plug insulators contained about 85% alumina and 15% glass. The glass lowers the firing temperature of the ceramic, resulting in less costly processing, but also lowers the electrical and mechanical properties of the material. The trend over time has been to use more alumina and less glass. Today's spark plugs typically use ceramic compositions with 96% alumina. Recently a 99% alumina composition has been introduced to the market for premium automotive spark plug applications.

EVOLUTION OF INSULATOR MATERIALS FOR SPARK PLUGS.

A wide range of materials were used in early spark plugs, including porcelain, glass, rubber, and mica. There are even reports of wood being having been used for spark plug insulators. As the automotive industry began to grow in the early twentieth century, variants of triaxial porcelain (consisting of feldspar, kaolin and quartz) became the most commonly used materials in the United States. The shortcomings of triaxial porcelain became acute with the onset of the first World War and the development of the Liberty aircraft engine*. Thermal shock resistance was poor and the insulators would break due to the volumetric change associated with the large amount of crystalline silica in the body, which undergoes displacive phase transformations at 117°, 230° and 573°C – temperatures well within the operating range of spark plugs. High temperature electrical resistance was also inadequate because monovalent potassium and sodium ions from the feldspar are fairly mobile charge carriers. Work conducted from 1917 to 1919 by the National Bureau of Standards Laboratory in Pittsburgh by Bleininger and Riddle[2] resulted in a mullite-based porcelain. Mullite, formed by calcining kaolin was substituted for quartz . Feldspars were replaced by alkaline earth silicates or beryl in order to reduce the alkalis. Alumina was recognized as a potential substitute for quartz, but was not available with sufficient purity and in sufficient quantities at the time. Champion Spark Plugs adapted the results of the work of Bleininger and Riddle to develop a sillimanite-based ceramic. Sillimanite is an aluminosilicate mineral in the mullite family. A natural deposit of sillimanite was discovered in the California mountains and shipped by rail to Detroit to be processed into spark plug insulators.

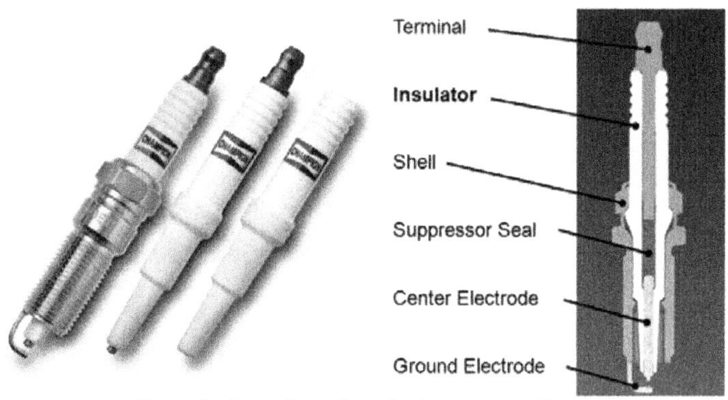

Figure 1. Overview of spark plug construction.

With the advent of leaded gasoline, the shortcomings of mullite porcelain became evident. Tetra-ethyl lead was added to gasoline to boost the octane rating and prevent engine knocking in the high compression internal combustion engines that were developed prior to World War II. The lead readily attacked the silicate based ceramic. The first alumina ceramics were developed in Germany by Siemens and Halske in 1932. The material, known as "sinterkorund" provided the needed resistance to leaded fuels, but was difficult to process. Sinterkorund was 99.3 – 99.7% alumina, and was produced by casting or pressing an acidified suspension of alumina particles and firing to about 1800°C, which is about 250°C below the melting point of pure alumina[3]. Performance of sinterkorund was somewhat unreliable, in part because the high firing temperature was difficult to achieve consistently. An alternate technology for high alumina ceramics was developed in the United States and became widely used in the 1940s and 50s. This ceramic consisted of alumina grains bonded by alkaline earth silicate glass. Early compositions used between 80 and 95% alumina, with minerals such as talc, limestone, beryl and kaolin to form the glass phase. The development of tabular alumina by Thomas S. Curtis of the Aluminum Company of America in 1934 provided a consistent raw material that was available in large quantities at a reasonable price, and contributed to the success of alumina ceramic spark plug insulators[2].

Advances in the technology for producing calcined alumina enabled further advances in spark plug insulators. Champion Spark Plugs initially used tabular alumina as the primary raw material. The tabular alumina was ground and chemically treated to reduce the soda content. A suitable low-soda calcined alumina to replace the chemically treated tabular alumina became commercially available in the 1960s and was used in 90 – 92% alumina compositions. Reactive grades of low soda alumina were adopted in the 1990s, which facilitated higher alumina content. Today, most spark plugs insulators are made from 96% alumina compositions using low soda alumina with primary crystal size in the 1.8 to 2.7 micron size range and specific surface area in the range of 0.5 to 1.0 m^2/g. Typical specifications for alumina powder used for spark plug insulators are shown in Table I. Moderate grain growth occurs during sintering resulting in grain size of about 10 microns. Typical fired specific gravity is >3.70 which corresponds to about 97% density. Typical microstructure is shown in Figure 2.

Table II shows how the alumina content of spark plug insulators has changed over time. Alumina was present in early compositions, but in the form of mullite crystals. Since the introduction of alumina ceramics, the alumina content has increased to meet more severe engine conditions.

Table I: Typical Specifications for Low Soda Alumina used for Spark Plug Insulators

	A	B	C	D
Specific Surface Area (m^2/g)	0.60 – 0.75	0.7 – 0.9	0.5 – 1.0	0.6 – 1.3
Primary crystal Size (μm)	2.0 – 2.7	1.8 – 2.4	~2.0	~2.5
Na_2O	<0.08	<0.06	<0.07	<0.10
Fe_2O_3	<0.04	<0.04	<0.03	<0.03
SiO_2	<0.04	<0.10	<0.04	<0.05
CaO	<0.06	<0.03	--	<0.03
B_2O_3	<0.05	--	--	--

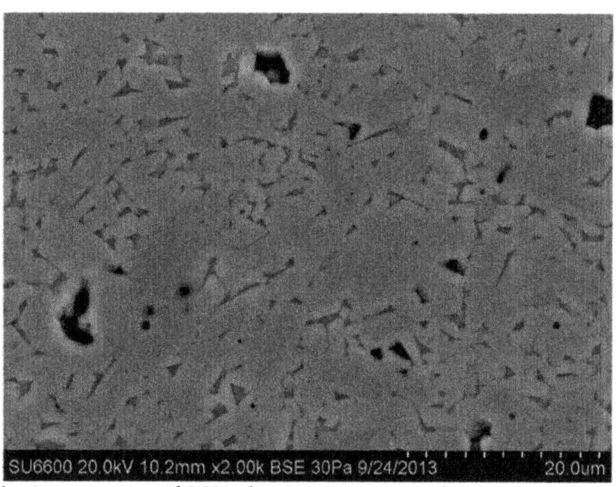

Figure 2. Typical microstructure of 96% alumina ceramic used to produce spark plug insulators.

Table II: Alumina Content of Spark Plug Insulators

	Triaxial Porcelain	Mullite Compositions	High Alumina		
Years in Use	1900-1920	1920-1940	1940 - 1965	1965 – 1990	1990 – Present
Alumina Source	Kaolin, Feldspar	Calcined Kaolin, Sillimanite	Tabular Alumina	Calcined Low-Soda Alumina	Reactive Low-Soda Alumina
Al_2O_3	30%	53%	90%	90%	96%
SiO_2	63%	44%	7%	7%	2%
Firing Temp	1345°C	1470°C	1540°C	1540°C	1600°C
MOR	70 MPa	150 MPa	340 MPa	340 MPa	400 MPa
Voltage Requirement	6 kV	10 kV	20 kV	20 kV	35 kV

SPARK PLUG INSULATOR REQUIREMENTS

Each new generation of engine technology results in increased operating demands on spark plugs. In general, today's automotive engines operate at higher temperatures and require higher voltages in order to deliver increased performance and economy. These factors, in combination with the demand for more compact plug designs and ever-longer service life, continually push the capabilities of the materials. At the same time, the available "footprint" for the spark plug is shrinking due to the increased sizes of cooling passages and valves. Just a few years ago, automotive spark plugs typically used a 14 mm thread to mount in the engine. Today, 12 mm has becoming the norm and 10 mm is expected in the next few years, with the diameter of the insulator reducing by corresponding amounts (see Figure 3). Moreover, the utilization of leaner fuel mixtures, increased compression and/or alternative fuels depends on available energy to ignite the mixture. In spite of these rising demands, vehicle manufacturers and other customers expect longer plug life, with the ultimate target being the lifetime of the engine.

Spark plug manufacturers have addressed these demands through design changes and the use of longer-lasting precious-metal electrodes, such as iridium. One of the remaining barriers to increased plug performance and durability, however, is the ceramic used in the plug insulator. Being the heart of the spark plug, the ceramic insulator isolates the high voltage applied to the terminal of the spark plug from the grounded shell. With increased applied voltages, higher operating temperatures and thinner insulator cross-sections, existing ceramic materials are reaching their practical limits. Improvements in ceramic technology are critical to developing higher-performance ignition systems. As engine technology progresses, the role of the ceramic spark plug insulator in controlling voltage becomes increasingly important.

Consumers clearly are demanding increased power and overall performance and durability, as well as increased fuel efficiency from automotive engines. These needs have helped drive the downsizing of under-hood compartments and engine packages, thereby affecting spark plug geometries. These new size and shape restrictions redouble the need for improved ceramic materials.

MANUFACTURING SPARK PLUG INSULATORS

Spark plug insulators are almost universally formed by a dry-bag isostatic pressing process. A flow chart of the manufacturing sequence is shown schematically in Figure 4. The starting materials, alumina powder and minor amounts of kaolin, talc and limestone, are finely ground in an aqueous suspension or slurry. A small amount of a binder-lubricant is added and the slurry is spray-dried by atomization in a heated chamber, forming a free-flowing, granulated feed material for pressing. The spray-dried powder is fed into rubber molds and formed into insulator blanks by using isostatic pressure to compact the powder around steel arbors which form the bores of the insulators. The outer surfaces of the insulators are formed by profile grinding in the green state using shaped abrasive wheels. The parts are then fired to about 1600°C to sinter the powder into a dense ceramic. Subsequent processing adds decoration and glaze, which are fired to a lower temperature.

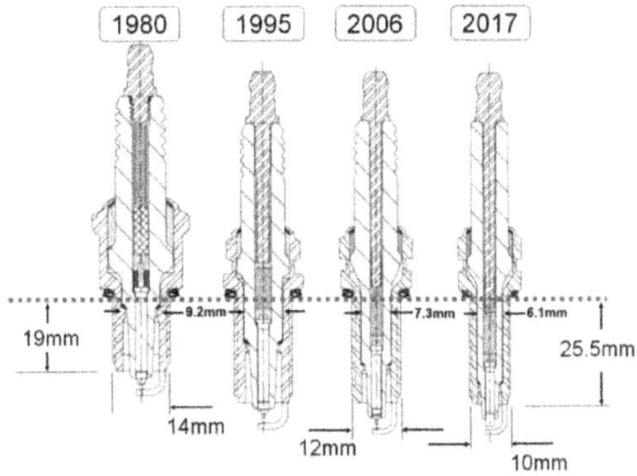

Figure 3. Evolution of spark plug design. Increases in the size of valves and cooling passages in engines limits the available space for spark plugs, necessitating smaller diameter insulators with reduced thickness.

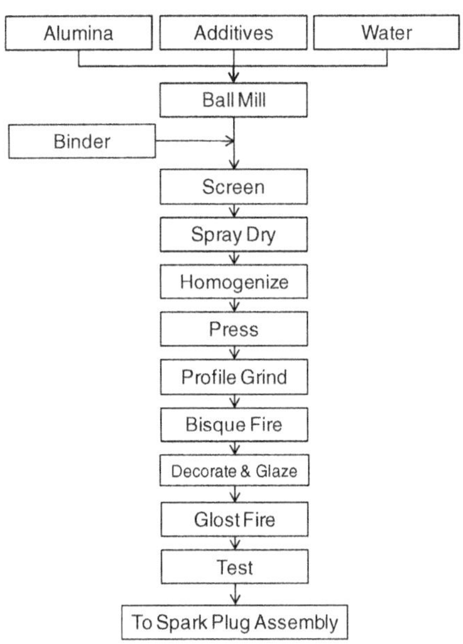

Figure 4. Flow chart of spark plug insulator manufacturing process.

PROPERTY REQUIREMENTS FOR SPARK PLUG INSULATORS

The insulator for a spark plug is generally cylindrical with an axial bore, and is mounted in a metal shell which threads into the engine block. The high-voltage lead is connected to a metal terminal projecting from the end of the bore away from the combustion chamber. An electrode projects from the other end of the bore into the combustion chamber. This electrode is typically a nickel alloy with a copper core, and may be tipped with platinum, iridium or some other precious metal alloy. The electrode and terminal are connected by means of an electrically resistive glass seal, which has the combined functions of forming a gas-tight seal to prevent passage of combustion gasses and providing sufficient electrical resistance to suppress electromagnetic interference from the spark. The outer profile of the insulator is contoured to have an enlarged section to lock the insulator in the shell, and a tapered tip or "nose" that extends beyond the end of the shell into the engine. The design of the nose controls the heat range of the spark plug and prevents tracking of the spark along the surface of the insulator to the shell. The portion of the insulator that extends out of the engine to the terminal is generally cylindrical and is coated with a glaze that prevents buildup of dirt that can cause flashover to occur.

In order to withstand the severe environment of a modern engine, the ceramic must have the following properties:

(1) High Dielectric Strength – Ignition voltages for the most demanding engine applications can exceed 42,000 volts.
(2) High Electrical Resistance – Leakage currents reduce the power to the spark, and degrade the long-term performance of the insulator.
(3) High Mechanical Strength – The insulator is mechanically locked in a metal shell, and is subjected to vibration and cyclical high pressure loading of as much as 20 MPa.
(4) Resistance to High Temperatures – The tip of the insulator can reach 900° – 1000°C.
(5) High Thermal Conductivity – The spark plug insulator contributes to the conduction of heat away from the firing tip in order to avoid becoming a source of pre-ignition, while retaining enough heat to burn off conductive deposits.
(6) Good Thermal Shock Resistance – The ceramic must withstand the rapid change from a cold start to the operating temperature of an engine, and withstand the constant thermal cycling between impingement of the cool fuel-air mixture and combustion temperatures.
(7) Chemical Inertness – The ceramic cannot degrade due to exposure to corrosive combustion gasses and fuel additives.

Very few ceramic materials combine the necessary properties for spark plug insulators, and of those, only alumina is cost effective.

DIELECTRIC BREAKDOWN

Dielectric breakdown strength is of particular importance in today's engines. Figure 5 shows the dielectric loading on an insulator. The maximum field strength occurs where the insulator seats in the shell. A secondary maximum occurs at the head of the center electrode.

Dielectric breakdown occurs when an insulation material fails to insulate due to a rapid increase in charge carriers passing through the material once the applied electric field exceeds some critical level. In the case of gasses, this occurs when an arc forms between electrodes. A spark plug is designed to apply electric fields that exceed the dielectric strength of the gaseous mixture of air and fuel in the cylinder of an engine in a reliable and consistent manner in order to control ignition. The insulator, however, must be capable of withstanding these same electric fields.

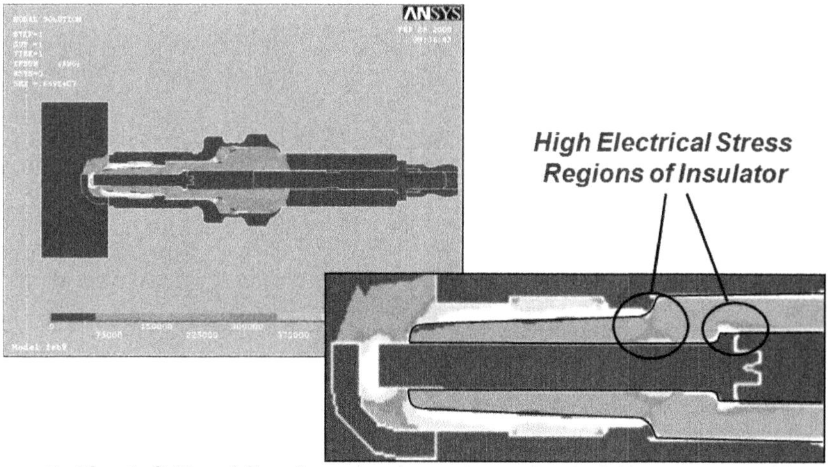

Figure 5. Electric field modeling shows that the maximum electric field occurs at transitions in the internal and exterior profile of the insulator in close proximity to contact points with the shell and center electrode.

Dielectric breakdown of the insulator results in a physical puncture through the material (see Figure 6). Breakdown may occur due to one of three mechanisms[4].

- Intrinsic or electronic breakdown occurs at very high electric fields where electrons may be ejected from the electrode into the insulating material, or electrons may be promoted across the band gap from the valence band to the conduction band. Some of these electrons collide with atoms or electrons, generating more electrons, resulting in an avalanche of conducting electrons concentrated in a small area. Intrinsic breakdown is rarely observed in solids because of the high electric fields that are required.
- Thermal breakdown is the most common mechanism of dielectric breakdown in ceramic insulators, occurring at electric fields about two orders of magnitude lower than those required for intrinsic breakdown. Movement of charge carriers within the material in response to the applied electric field results in localized heating, which in turn results in additional movement of charge carriers and more localized heating. The resulting thermal runaway reduces the voltage required to initiate the avalanche of conducting electrons that cause dielectric breakdown.
- Ionization breakdown occurs as a result of the ionization of gasses within pores and cracks of insulators with a non-homogeneous structure. The high-energy ionized gas generates heat locally, which promotes thermal breakdown.

The factors affecting dielectric breakdown of a material are those that contribute to the thermal and ionization breakdown mechanisms: the number and mobility of charge carriers and microstructural inhomogeneities such as cracks and pores. Spark plug manufacturers have long recognized the importance of minimizing alkali ions by using low-soda alumina and high purity minor additives. The overall composition of the glass phase also influences susceptibility to thermal breakdown. Improvements in processing methods can minimize the effects of ionization breakdown by reducing the size and number of pores within the microstructure of the ceramic.

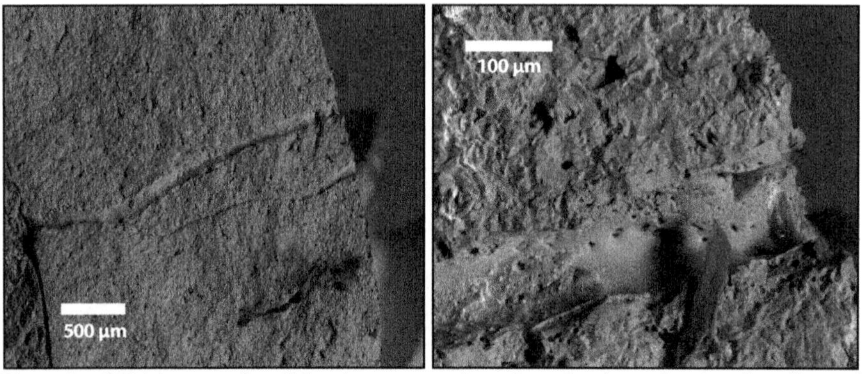

Figure 6. SEM images of a dielectric puncture through an insulator showing the effects of thermal breakdown. The puncture propagated radially from the center electrode to the shell and generated sufficient heat to melt the ceramic. The insulator was carefully broken to reveal the path of the puncture.

ELECTRICAL RESISTIVITY

Electrical conduction through the ceramic insulator is low; however, there is a small amount of leakage current which increases with temperature. At 550°C, the electrical resistance through the ceramic, between the internal conducting path and the shell of a spark plug is typically in the 100 M-ohm to 2.5 G-ohm range, and is dependent on the composition of the glass phase of the ceramic. Figure 7 shows the electrical resistivity of two alumina compositions used for spark plug insulators. As temperature is increased, the electrical resistivity decreases as the number and mobility of charge carriers increases. The temperature dependence of the two compositions has a similar slope, but the resistance at any given temperature differs by a factor of 10. Alumina 1 is typical of today's insulator compositions, while Alumina 2 is an older composition that was developed in the 1970s but is still used for lower performance applications. Alumina 2 is prone to dielectric failure due to the thermal breakdown mechanism.

MEASUREMENT OF DIELECTRIC BREAKDOWN STRENGTH

Dielectric breakdown strength of solids can be measured according to ASTM D149[5] and IEC 60243-1[6] by placing a planar test specimen between two electrodes and increasing the applied electric field until failure occurs. Failure is detected by measuring an increase in the current between the electrodes, and verified by visually locating a puncture. It is usually necessary to immerse the test fixture in a dielectric fluid to prevent flashover between the electrodes. A 60 hz AC electric field is increased at a rate of 500 V(rms)/sec until failure occurs. Automotive companies generally prefer dielectric testing performed on assembled spark plugs using an automotive ignition source in order to test the performance of the device rather than the property of the material alone. The spark plug is threaded into a pressure vessel which is pressurized with air or some other gas in order to suppress the spark between the electrodes and force puncture through the ceramic[7]. The high-voltage lead from an automotive ignition coil is attached to the terminal of the spark plug and the pressure vessel is grounded. The voltage signal is a series of DC pulses at 50 hz, and is increased in 1 kV steps in 1 sec. intervals until puncture is detected by a change in the waveform of the voltage and verified by inspection. Sets of 30 or more spark plugs ensure reliable statistics.

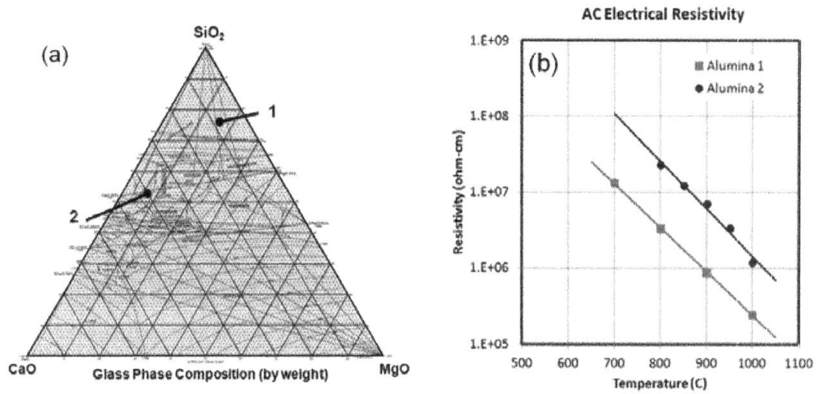

Figure 7. Electrical resistivity of alumina ceramic is dependent on the composition of the glassy phase. Ternary diagram (a) shows the glass phase composition of two typical spark plug insulator compositions. Electrical resistivity (b) decreases with increasing temperature and differs by a factor of 10 for these two compositions.

Figure 8 shows Weibull plots of the dielectric breakdown voltage of two sets of spark plugs processed under different conditions. Set A was produced under less than optimal conditions. At the low end of the distribution, two points deviate from the best-fit line for the Weibull distribution. The lowest point is outside the 95% confidence interval range for the Weibull distribution, suggesting that this low flier was caused by a defect that is due to a different type of defect than those in the rest of the population. The poor performance of this one specimen severely limits the dielectric reliability of spark plugs made with those insulators. In contrast, Set B was produced under improved conditions. The Weibull modulus was increased from 26.5 to 31.0 and the characteristic puncture voltage was increased from 41.8 kV to 43.3 kV, but more importantly, the low fliers were eliminated. The reliability of the insulators is significantly improved.

SUMMARY

Alumina ceramics provide the necessary combination of properties for spark plug insulators. Advances in automotive engine technology push the demands on spark plugs, necessitating more compact plug designs that operate under more severe thermal and electrical stresses over an ever-longer service life. Ceramic insulators with improved properties are needed to meet these demands, and need to be manufactured on a scale of millions of parts per day in order to supply the automobile industry. Insulators for spark plugs are almost universally made from aluminum oxide ceramic and formed by a dry bag compaction process. The ceramic insulator provides the dielectric barrier to isolate the high voltage required for ignition, and acts as the mechanical support around which the spark plug is built. Dielectric failure of an insulator results in poor engine performance so the insulators must be manufactured with a high degree of reliability. The microstructural defects that cause low dielectric breakdown strength in a small number of insulators can be eliminated by carefully controlling the manufacturing process. Weibull statistics are a useful tool for ensuring reliability of dielectric performance.

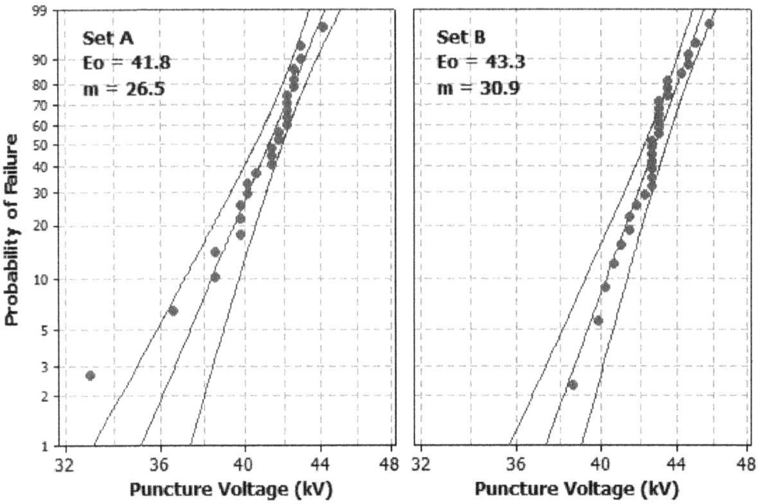

Figure 8. Weibull plots with 95% confidence interval showing dielectric performance of two sets of spark plugs using insulators produced under different conditions. Testing was conducted on complete spark plugs using an automotive ignition coil to provide the voltage, using pressurized gas to suppress normal sparking at the firing tip in order to force puncture through the ceramic insulator.

FOOTNOTE

*The Liberty engine program was initiated in May 1917, one month after the United States entered the First World War, with the goal of developing, as rapidly as possible, an engine with high power-to-weight ratio that was easy to manufacture. The program resulted in a 27-liter, V-12 engine that produced 400 hp, which was one of the most powerful engines of its time.

REFERENCES

[1]J.S. Owens, J.W. Hinton, R.H. Insley and M.E. Poland, "Development of Ceramic Insulators for Spark Plugs," American Ceramic Society Bulletin, 56 [4] 437-440 (1977).

[2]F.H. Riddle, "Ceramic Spark Plug Insulators," Journal of the American Ceramic Society, 11 [32] 333-346 (1949).

[3]"Sinterkorund, a New Ceramic Material," Ceramic Age, July 1932, 28-37.

[4]R.C. Buchanan, "Ceramic Insulators," pp. 1-86 in Ceramic Materials for Electronics, 3rd Edition, R.C. Buchanan, editor. Marcel Dekker, Inc., New York, 2004.

[5]ASTM D149-97a (2004) Standard Test Method for Dielectric Breakdown Voltage and Dielectric Strength of Solid Electrical Insulating Materials at Commercial Power Frequencies.

[6]IEC 60243-1: Electrical Strength of Insulating Materials - Test Methods - Part 1: Tests at Power Frequencies.

[7]JIS B8031-2006 Japanese Industrial Standard for Internal Combustion Engines-Spark Plugs.

Thermal Protection Materials and Systems

PHOTOGRAMMETRIC SURFACE RECESSION MEASUREMENTS ON ABLATIVE SAMPLES OF VARIOUS SHAPE

Thomas Reimer[1], Stefan Löhle[2], Rainer Öfele[3]

[1]DLR, German Aerospace Center, Institute of Structures and Design
Pfaffenwaldring 38-40, 70569 Stuttgart, Germany

[2]University of Stuttgart, Institute of Space Systems, Head of High Enthalpy Flow Diagnostics Group, Pfaffenwaldring , 70569 Stuttgart, Germany

[3]University of Stuttgart, Institute of Space Systems, Pfaffenwaldring , 70569 Stuttgart, Germany

ABSTRACT

Results from plasma wind tunnel testing of a lightweight carbon phenolic ablative material are presented. The tests were carried out in a high-enthalpy subsonic flow at a condition corresponding to the Hayabusa re-entry condition in 78 km altitude. The samples were produced from ZURAM which is a carbon-fiber structure impregnated with phenolic resin. The focus of the investigation was on the measurement method of the recession. Stereoscopic photogrammetry was applied to reconstruct the sample surface in three-dimensional space over the testing time. Two cameras were used to capture images simultaneously. Using a three-step process the sample surface for each image pair was reconstructed and the recession between the image pairs was calculated. The recession resolution is in the range of 45 μm. The advantage of the method is that the time-resolved in-situ surface recession measurement gives insight into the local recession distribution over the sample surface. This means that recession differences due to the sample geometry or due to material inhomogeneities can be identified. In order to verify this, samples were prepared that include distinct geometric features which influence the recession. Starting with a flat surface cylinder, samples with a rounded surface were prepared as well as a cone-sphere sample and one with a defined indentation in the form of a half-sphere. The temporally and locally resolved recession was successfully measured and related to the sample geometric features.

INTRODUCTION

A spacecraft entering into the atmosphere of Earth or another planet has to withstand severe heat loads which depend on factors like entry speed, vehicle geometry and mass and also the type of atmosphere. A heat shield is used to protect the vehicle from the resulting high heat loads[1]. There are different concepts of heat shields that are used to fulfill the task. Re-usable systems like the heat shield of the Space Shuttle bear the heat load without changing their shape or losing mass which is a good thing with respect to aerodynamics. The disadvantage of these systems is that they are partially made from CMC which is relatively expensive to manufacture and maintain and that current materials are limited in their maximum service temperature.

Non-reusable systems like ablators rely on a multitude of effects to deal with the enormous amount of heat[2]. They are good insulators, so the surface temperature is very high which favors radiation and energy consumption takes place due to chemical processes and physical change. In addition there is film cooling due to outgassing and also radiation blocking because of particles transported into the boundary layer.

Ablators are currently the thermal protection systems of choice when the heat loads become very high as e.g. in the case of a superorbital entry like in the case of a moon or mars return[3-5]. A complex set of processes takes place when ablative materials are heated up. Part of the incoming heat is rejected via surface radiation due to the high surface temperature as is the case with reusable systems, but there are also a lot of processes that consume heat as a result of the material transformation[6]. The materials undergo massive changes and lose a considerable share of their initial thickness. The decrease of the ablator thickness, or in other words, the recession is a critical issue for the design of an ablative heat shield. For a given mission the remaining thickness has to be sufficient to guarantee the effective function as heat shield. The investigation of ablative materials with regard to their thickness loss or recession is usually carried out with measurements before and after a test, most often done in arc jet facilities. Since missions become more demanding and the design of the space vehicles increasingly sophisticated, the understanding of the TPS materials needs to be improved to be able to reduce margins to meet the mission mass targets. Therefore, the aim of this work was to get more detailed information out of tests, specifically the local time-resolved recession rate of ablative materials.

MATERIAL AND SAMPLES

For this study, two types of material were tested. For an initial test, a carbon preform material was used. After that, dedicated samples made from the carbon phenolic lightweight ablative material Zuram were manufactured. Although there is a wide range of ablators with different compositions and properties used for different kinds of entry missions, these materials are hard to get by. Therefore, DLR recently has begun to develop its own ablative material in order to enlarge the range of available heat shield materials.

Carbon felt and ablative material

The carbon preform material is CALCARB®, a commercial thermal insulation product from the Mersen company which was chosen to have an uncomplex material with a relatively high recession rate and known properties. It was used before in similar investigations[7].

The ablative material that was used for the investigations is ZURAM® which is a lightweight ablative material with a density of 0.36 g/cm^3, designed for high enthalpy entry missions. Employing a specific manufacturing process, a nanoscale phenolic foam is created in the carbon-phenolic-based material. The microstructure of the phenolic foam results in high damage tolerance of the composite with a compression strength of 6.9MPa at a compression strain of 0.39. In arc jet tests at enthalpies of 13.5MW/m^2 and a pressure of 675 hPa, average values for the mass loss and recession rate were 0.323 g/s and 0.308 mm/s respectively. The thermal conductivity of the uncharred material is 0.18W/mK at ambient conditions.

Samples

The carbon felt sample was cylindrical with a diameter of 40 mm, a flat face and a center borehole for additional spectroscopic measurements on the stagnation streamline[8].

As the intention of the work was to demonstrate the potential of the measurement method to resolve the sample recession over time and space, sample shapes were chosen that would show an unevenly distributed recession over the surface. All samples were of a cylindrical shape with different front surface geometries. As the basic case a sample with a flat face was selected. Additional surface geometries included a conical shape with a rounded tip and a spherical face. The spherical front was chosen along the considerations for a sample shape adapted to the flowfield in a supersonic test facility in order to have a resulting uniform heat load over the surface. The shape that has been suggested from these considerations is called Iso-q shape and was proposed by Milos and Chen[9]. As a variation of that shape, a sample with an elliptical front surface cross section was also included. This shape was proposed as an improvement with regard to the heat load distribution in supersonic flow. The semi-major of the ellipse is defined as the cylinder radius R, the semi-minor as $R(1-\sqrt{3}/2)$. Finally, as a shape to contrast the conical sample, an Iso-q sample was produced that featured a spherical indentation in the stagnation region. Table 1 gives an overview of the samples used in the test campaign and Fig. 1 shows the geometries of the tested samples.

Table I: Overview of sample shapes.

Material	Surface Geometry	Diameter	Height	Details
		mm	mm	
Carbon felt	Flat	50	40	
ZURAM	Flat	50	40	Edge radius 3.125 mm
ZURAM	Spherical	50	40	Sphere radius 50 mm
ZURAM	Spherical, indentation	50	40	Indentation depth 5 mm
ZURAM	Elliptical	50	40	Semi-minor: $R(1-\sqrt{3}/2)$
ZURAM	Conical, round tip	50	40	Cone angle 135°, tip radius 5 mm

PHOTOGRAMMETRY

Photogrammetry is applied to retrieve non-intrusively three-dimensional geometric data from two-dimensional images from image pairs. Nevertheless, more than two images can be used and do in fact increase accuracy[10]. The absolute sizes of the objects are no limitation to the method. Photogrammetry is applied in air and space imagery as well as in near-field applications. In order to be able to calculate three dimensional objects from two-dimensional images, the images have to be central projections. A central projection is the two-dimensional representation of three-dimensional objects. In contrast to a parallel projection, the projection rays that connect object points with the projection plane meet in the same point, the center of projection. In that way each object point is related to an image point. To be able to describe the projection rays in three-dimensional space, the internal and external orientation of all cameras must be known. The system is calibrated with an appropriate test object. The calibration in the present work was realised with the commercial product AGISOFT PHOTOSCAN. The measured data is analysed using SURE[11]. The theory of photogrammetry shall not be detailed further in this work, the process has been described in a previous publication [7]. Photogrammetric recession measurements were previously done by another group using a different technique[12].

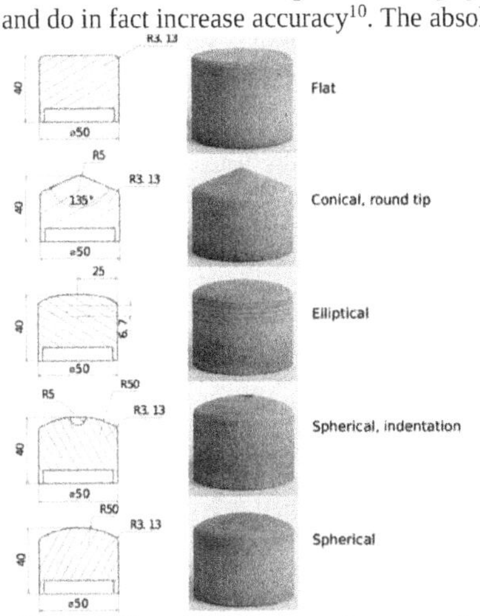

Figure 1: Tested ZURAM sample geometries.

EXPERIMENTAL SETUP

The experiments were performed in the plasma wind tunnel facility PWK 1 at the Institute of Space Systems of the University of Stuttgart. This facility is used to simulate the aerothermal situation that materials are facing during an atmospheric entry manoeuver[13, 14]. A high enthalpy flowfield is established using an appropriate generator and the samples are positioned in this flowfield at positions that correspond to the heat flux and total pressure of the flight condition. Diagnostic methods can be applied outside the vessel when sufficient and useful windows are available. The camera system for the photogrammetry is mounted on tripods to observe the sample surface through windows in the front lid[6].

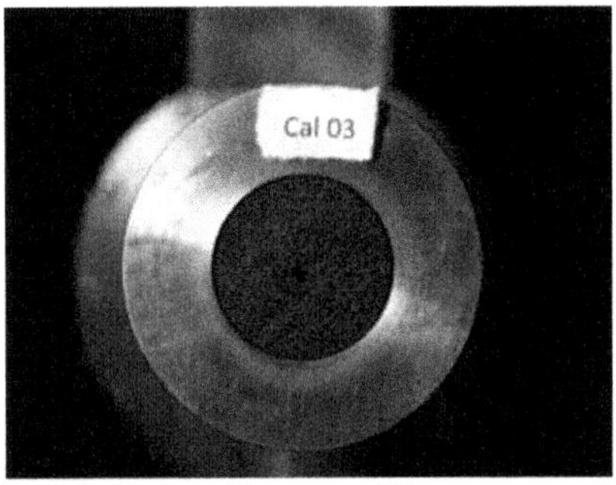

Figure 2: Probe with mounted CALCARB sample.

The present investigation was conducted with different geometries tested in this way for the first time in the facility. The conventional probe technique which is used in the present investigation for the carbon preform is used regularly for ablation material testing (see e.g.[8, 15, 16]). Fig. 2 shows a photo of the sample as mounted in the sample holder. Further details of this setup are described in[7].

The chosen flow condition was developed for ground test analysis of the observed Hayabusa re-entry in 2010[13, 17]. The parameters of the chosen flow condition are summarized in Table 2. Heat flux and total pressure were measured in separate tests using the same probe geometry. Heat flux is derived from a calorimetric cold wall heat flux measurement on a pre-oxidized copper surface.

Table 2: Plasma wind tunnel condition corresponding to Hayabusa at 78.8 km.

Parameter	Value
mass flow $\dot{m}$	18.0 g/s
ambient pressure p_∞	16.6 hPa
total pressure p_{tot}	24.3 hPa
arc current I	1220 A
arc voltage U	133 V
electric power P	162 kW
probe position	x = 270 mm, y = 0 mm
heat flux $\dot{Q}$	4100 kW/m²

There are only small optical windows to observe the plasma flow and the probe. Fig. 3 shows a photograph and a schematic of the setup using two cameras. A possible influence of the windows with respect to the photogrammetric measurements was analysed during previous work[19]. In this configuration the angle between the imaging axis of the two cameras was between 95° and 100°. Depending on other instruments used in the experiment, the distance between probe and camera varies between 2.8 m and 4.0 m. A closer position is preferable, but is restricted by the space required to open the front lid of the vacuum vessel.

Figure 3: Plasma wind tunnel PWK 1 (left)[7] and geometry for the photogrammetry (right)[18].

CAMERA AND SAMPLE SETUP

The tests were carried out in the plasma wind tunnel PWK 1 of the Institute of Space Systems at the University of Stuttgart. The setup is a triangle with two cameras symmetrically positioned at an angle of approximately 50° to the axis where the sample is positioned as depicted in Fig. 3.

To acquire the images, two identical Canon EOS 60D DSLR cameras were used in combination with Canon EF 300mm 1:4,0 L IS USM telephoto lenses shown in Fig. 4. In some cases an additional Extender EF 1.4x III was used to increase the focal length by a factor of 1.4. The cameras were mounted on sturdy tripods to minimize vibration during exposure. A three-axis geared tripod head was very helpful in adjusting the exact pointing tojwards the sample. The camera has a CMOS sensor with a size of 22.3 by 14.9 mm corresponding to a resolution in pixels of 5184 by 3456. It was desired to take images at a high frame rate. The camera supports a maximum frame rate of 5 images per second, however, when the image quality is set to the raw-format with a large file size, that rate cannot be kept for long. Therefore, the images were saved as jpeg-files at a rate of 3 frames per second which allowed for continuous operation. Both cameras were triggered at the same time via a pulse generator. The brightness of the heated sample front during testing was known roughly from previous tests at comparable conditions, so neutral density gray filters ND 1.2 were used which reduce the intensity of the transmitted light to 6.3 % to arrive at feasible exposure settings. The cameras were controlled via a computer interface to avoid disturbing their position after the calibration.

Figure 4: Camera with lens on tripod head.

The calibration procedure required a total of 20 image pairs of a calibration object taken under different angles. The calibration plate that was used included a scale to have the size information available.

DATA PROCESSING

In a first step the commercial software AGISOFT PHOTOSCAN was used to calibrate the setup. PHOTOSCAN can actually also generate three-dimensional object representations, however, for this purpose it was only used to establish the camera calibration data with the orientations which was then used by another tool in a following step. The images were cut to have the sample fill the complete image. It was important to cut all the images in the same way since that has an influence on the photogrammetric processing. Then the local contrast of the images was increased using a standard image editing tool. Experience had shown that a moderate contrast increase resulted in a higher success rate for the photogrammetry software to find corresponding points in the image pairs.

Using the calibration data from PHOTOSCAN, SURE calculated high-resolution point clouds from the images. SURE is a software developed by the Institute of Photogrammetry of the University of Stuttgart. The software processes every image pair separately. First the images are rectified and orientated. Then corresponding image points are identified on the basis of epipolargeometric calculations. For every image pair a point cloud is calculated. Thereby, a consecutive set of point clouds is generated which represent ablation of the sample over time. The point clouds were processed further with CLOUDCOMPARE to bring all the point clouds in relation to each other and to do distance calculations. The first point cloud is used as the reference for all the following time steps. The actual data evaluation in terms of recession and recession rate and their visualization was calculated with MATLAB.

In the following, data is presented for the total recession of the sample and the corresponding recession rate over time. For the analysis in this work, the simplifying assumption was made that the recession would be rotationally symmetric over the sample axis. So, the surface of the sample was divided into circular areas of even spacing and each circular area was averaged. Each averaged value for one circular area is given in the discussion of the recession rate over the sample radius. For the evaluation of the total recession the averaged values of each circular area were taken and averaged with a weighting regarding the area of each, arriving at one value for the total recession of the sample over time. The assumption of a symmetric condition was verified with the test data from the carbon felt sample. One of the circular areas was further divided into sub-sections which were compared with regard to the recession data. The result indicated that the assumption was justified.

RESULTS

The experiments were successful with the carbon felt sample, the flat and spherical samples and the indented spherical sample. In the case of the sample with elliptical surface the data could not be evaluated because of a glitch during the setup of the cameras; the images of one camera did not capture the sample fully. In the case of the conical sample it was realized during the data processing that the cone angle was apparently still too small - i.e. the cone was too pointed - and the recession data could not be calculated.

CARBON PREFORM

The carbon felt sample was tested for 60 s. The measured recession rate increased from 56 µm/s to 59 µm/s which is almost constant over time and a mean value of 58 µm/s. Corresponding to that, the total recession increased in a linear way to 3.36 mm as shown in Fig. 5.

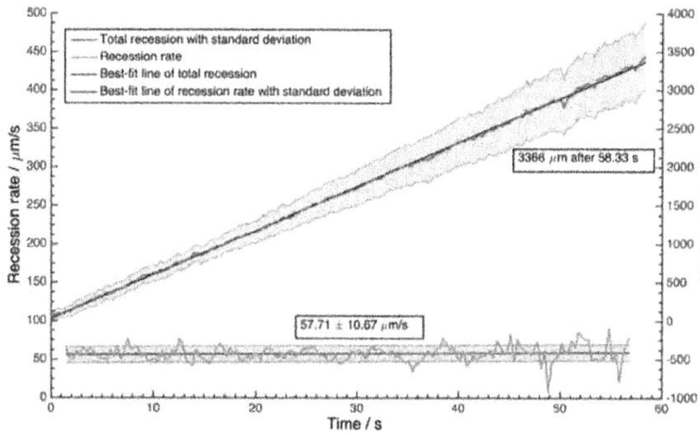

Figure 5: Total recession and recession rate of the carbon felt sample over time.

The distribution of the recession and the recession rate over the radius of the sample is shown in Fig. 6. From Fig. 6 it can be noted that the recession was strongest in the center of the sample leading to a distinctive washout.

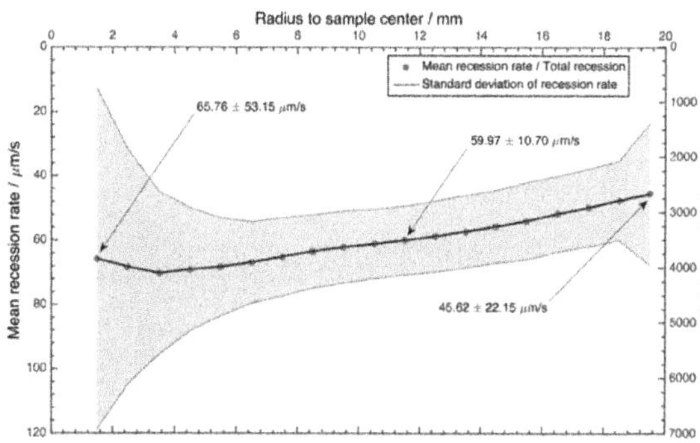

Figure 6: Total recession and recession rate over the radius of the carbon preform sample.

FLAT SAMPLE

The flat sample was the first ZURAM sample that was tested. The test duration was 60 s but after 30 s the fixation of the sample to the model holder failed so that only the first 25 s can be regarded as useful for the analysis. Therefore, for all remaining samples the test time was reduced to 25 s. When this time span is considered, the recession rate increases slightly from 42 μm/s to 44 μm/s, with a mean value of 43 μm/s. The total recession is 1.07 mm as presented in Fig. 7.

When a regression line is fitted to the recession rate curve, it is almost at a constant value and has only a theoretical slope of 0.06 μm/s over the test time which is negligible, so the recession rate is practically constant over the first 25 s of the test. Also, the standard deviation of the recession rate is smaller when compared to the carbon felt sample.

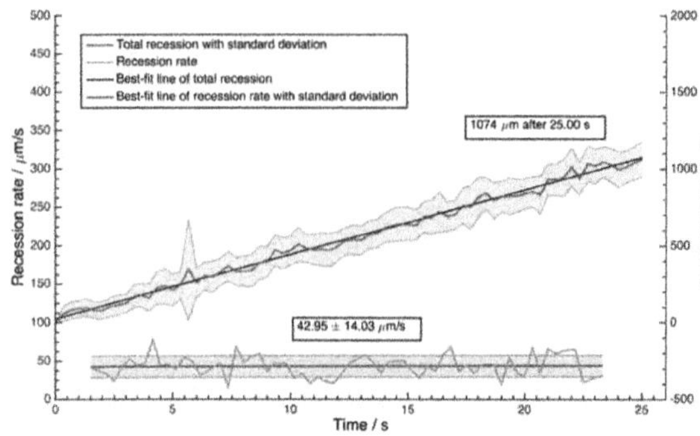

Figure 7: Total recession and recession rate over time for the flat sample.

The recession rate and the total recession are almost constant over the radius of the sample. The recession rate in the center is 44 μm/s, falling slightly to 40 μm/s at the sample edge, shown in Fig. 8. This is explained with the fact that the sample geometry and the free-standing position in the flow lead to more homogeneous ablation conditions on the sample surface when compared to the setup with the metallic holder.

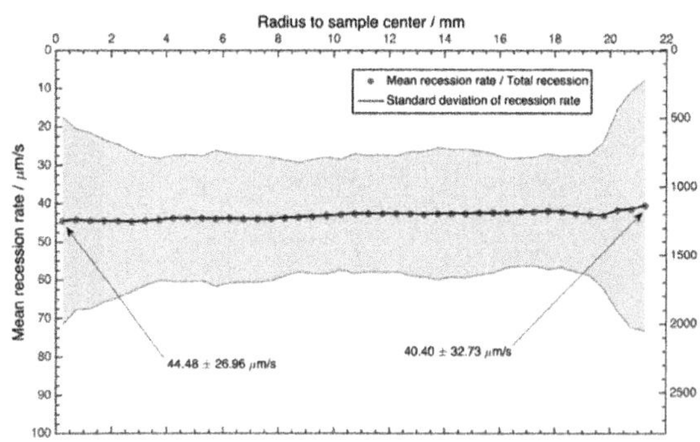

Figure 8: Total recession and recession rate over the radius of the flat sample.

SPHERICAL SURFACE SAMPLE

In the case of the sample with spherical surface the recession rate decreased over time from 53 μm/s to 33 μm/s as shown in Fig. 9. The total recession at the end of the test was 1.28 mm. Fig. 10 shows that the average recession rate over the sample radius is not constant. In the sample center it is 70 μm/s, whereas it is only 32 μm/s at the edge of the sample. This observation can be explained with an uneven distribution of the heat load over the sample surface. Obviously the spherical shape is not well suited for subsonic flows if a constant heat load over the surface is desired.

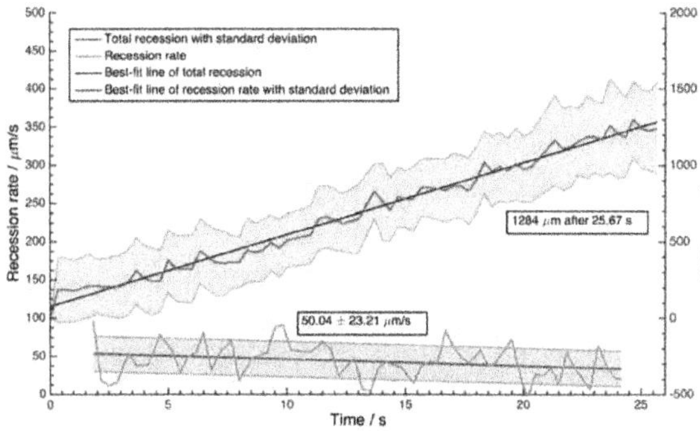

Figure 9: Total recession and recession rate over time for the spherical surface sample

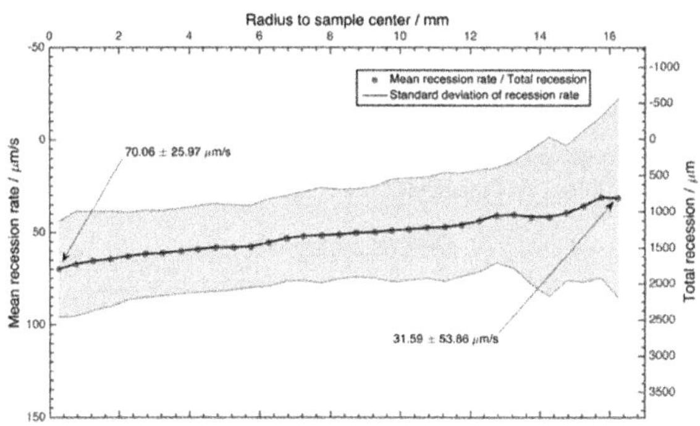

Figure 10: Total recession and recession rate over the radius of the spherical sample.

SPHERICAL SURFACE SAMPLE WITH CENTER INDENTATION

The indentation was introduced in the sample to have a shape which is certainly experiencing non-constant heat load and thus greater differences in recession over the surface. The indentation had the shape of a half-sphere with 5 mm radius which was also the depth measured from the theoretical tip of the non-indented sample. In terms of the recession measurements it was expected that the inner surface of the indentation could not be evaluated completely, which was indeed the case due to the fact that each camera can only see a certain part of the inner indentation surface and vice versa. In addition it has to be considered that the indentation remains relatively cool and thus dark at the beginning of the test so that SURE has trouble in finding points in the beginning and those that are calculated have a greater uncertainty about them. The average recession rate in the regions that can be evaluated is 62 μm/s, which seems to be due to the influence of the indentation. The slope of the regression curve through the rate data is -0.91 μm/s, which is on the same level as that for the same shape without indentation. However, the standard deviation of the recession rate is higher at 44 μm/s, which is attributed to the complex contour shown in Fig. 11.

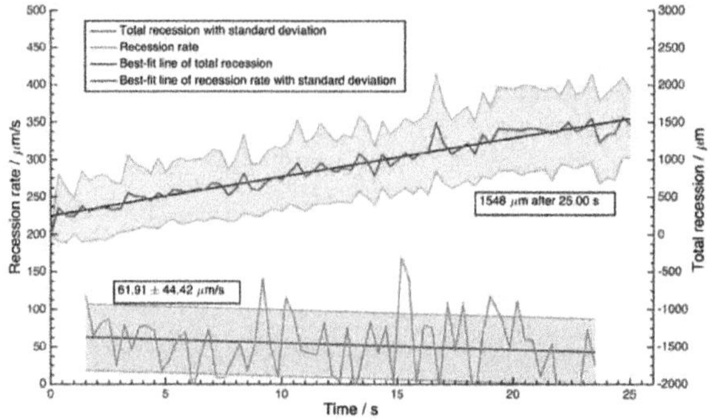

Figure 11: Total recession and recession rate over time for the indented spherical sample.

The distribution of the recession rate over the sample radius as given in Fig. 12 shows the influence of the indentation. In the sample center, on the bottom of the indentation, a recession rate of almost 150 µm/s is calculated, however, the large error values of +/- 200 µm/s indicate that this value is unreliable. At the edge of the indentation, the recession rate is 99 µm/s, with a standard deviation that is down to 66 µm/s which is still comparatively high but in the order of values obtained for the other samples. To the edge of the sample the rate decreases to 44 µm/s.

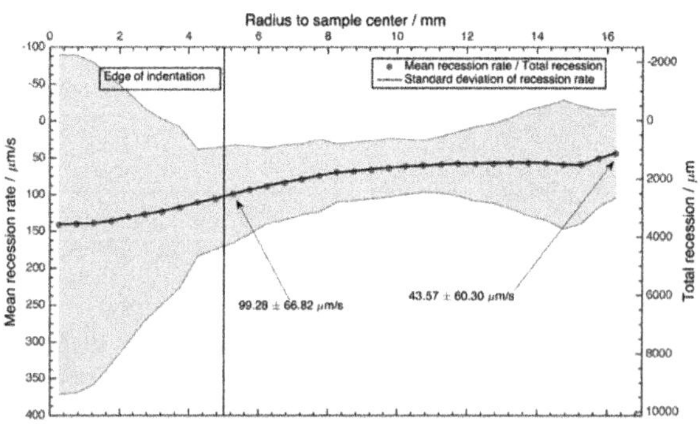

Figure 12: Total recession and recession rate over the radius of the indented spherical sample

It was assumed that the recession is quite transient for this sample, so the analysis was repeated with two different evaluation time intervals. The first one comprised the time from the beginning of the test up to 12.3 s, given in Fig. 13, the second one the remaining time up to the test end at 25 s, given in Fig. 14. There is indeed a difference when the recession rate in the same location is compared. At the indentation edge the recession rate for the first evaluation interval was 115 µm/s whereas it is only 94 µm/s for the second interval, compared to the average of 99 µm/s for the whole test.

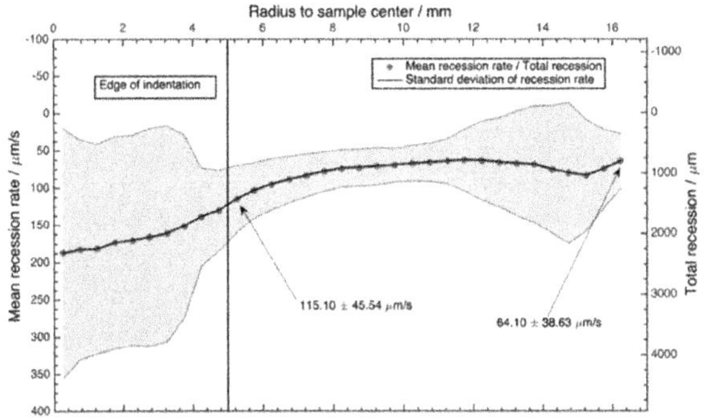

Figure 13: Total recession and recession rate over the radius of the indented sample evaluated over the first 12.3 s of the test.

A similar observation can be made for the sample edge where the rate for the first interval is 64 µm/s and 43 µm/s for the second interval, compared with 44 µm/s for the whole test, see Fig. 13 and Fig. 14. In addition to the points mentioned above, it was observed that the average recession rate during the first half of the test was approximately 20 µm/s higher than during the second half of the test.

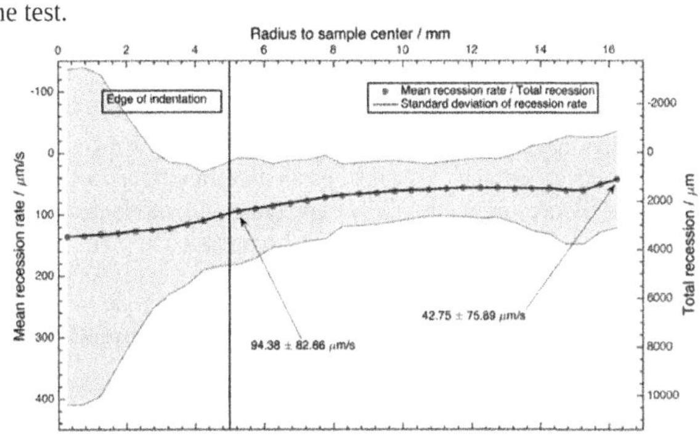

Figure 14: Total recession and recession rate over the radius of the indented sample evaluated over the time from 12.3 s to 25 s.

Since the edge of the indentation was the region with the highest recession values, this region was analyzed in more detail. A circular region between the radii 5.5 and 6.0 mm was considered – the indentation radius itself was 5.0 mm. The intention was to select an evaluation region with a slightly larger radius to avoid any problems with areas not visible by one of the cameras. The results given in Fig. 15 show that the recession rate at the indentation edge is 119 µm/s in the beginning and then drops to a value of 64 µm/s towards the end of the test. This is a change of -2.5 µm/s^2 which is the highest change in the recession rate measured.

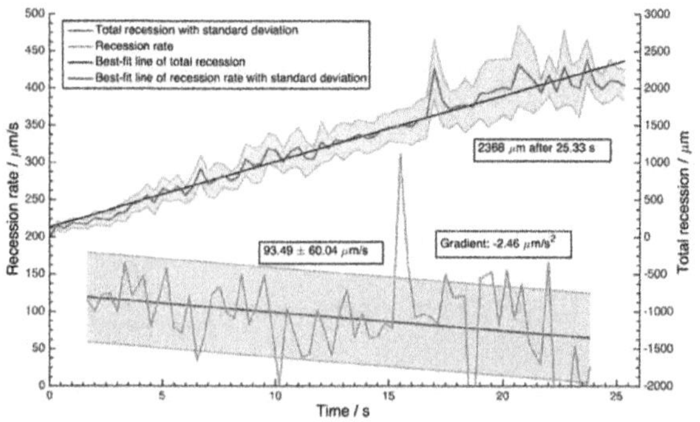

Figure 15: Total recession and recession rate over time for the indented spherical sample evaluated at the edge of the indentation.

CONCLUSION

An investigation of the recession of different geometries of a lightweight carbon phenolic ablator material was conducted. Using a photogrammetric setup, the recession behavior was investigated to a high level of detail. The selected geometries proved to yield interesting results. As for the surface shape, it was evident that the Iso-q geometry does indeed only lead to constant surface heat load in a supersonic flow. Of the investigated shapes, the flat faced sample was closest to a constant surface heat load for the subsonic flow conditions in the tests. The indented sample showed that very localized phenomena can be investigated.

It can be concluded that the method allows to identify surface effects arising from surface inconsistencies and geometrical defects. This is of particular importance when materials are tested for future flight scenarios. For future material investigations with the focus on material properties, a flat faced sample should be considered to allow for homogeneous recession and thus a distinct experimental situation.

The goal for further development of the measurement method is to calculate the recession rate for smaller time intervals, ideally the momentary rate as the derivative of the recession and to do this with a higher degree of discretization over space.

ACKNOWLEDGEMENTS

The helpfulness of Christian Zuber from DLR is acknowledged for providing the ZURAM material batch.

NOMENCLATURE

CMC	Ceramic Matrix Composite
CMOS	Compementary Metal Oxide Semiconductor
PWK	Plasmawindkanal
SURE	Photogrammetric Surface Reconstruction from Imagery Software
ZURAM	Trade name of ablative DLR material

REFERENCES

[1]Lu, F. K. and Marren, D. E., editors, Advanced Hypersonic Test Facilities, *Progress in Astronautics and Aeronautics*, **198**, AIAA, 2002.

[2]Lachaud, J., Cozmuta, I., and Mansou, N. N., Multiscale Approach to Ablation Modeling of Phenolic Impregnated Carbon Ablators, *Journal of Spacecrafts and Rockets*, **47**, No. 6, 2010.

[3]Johnston, C. O., Gnoffo, P. A., and Sutton, K., Influence of Ablation on Radiative Heating for Earth Entry, *Journal of Spacecrafts and Rockets*, **46**, 2009, pp. 481–491.

[4]Wright, M. J., Grinstead, J. H., and Bose, D., A Risk-Based Approach for Aerothermal/TPS Analysis and Testing, *Experiment, Modeling and Simulation of Gas-Surface Interactions for Reactive Flows in Hypersonic Flights*, **17**, RTO, 2007.

[5]Wright, M. J., Tang, C. Y., Edquist, K. T., Hollis, B. R., Krasa, P., and Campbell, C. A., A Review of Aerothermal Modeling for Mars Entry Missions, 2010.

[6]Löhle, S., Hermann, T., Zander, F., Fulge, H., and Marynowski, T., Ablation Radiation Coupling Investigation in Earth Re-entry Using Plasma Wind Tunnel Experiments, *30th Aerodynamic Measurement Technology and Ground Testing Conference*, AIAA, 2014.

[7]Löhle, S., Staebler, T., Reimer, T., and Cefalu, A., Photogrammetric Surface Analysis of Ablation Processes in High Enthalpy Air Plasma Flow, *30th Aerodynamic Measurement Technology and Ground Testing Conference*, AIAA, 2014, accepted for publication.

[8]Hermann, T., Zander, F., Fulge, H., Löhle, S., and Fasoulas, S., Characterization of a Re-entry Plasma Wind Tunnel Flow with Vacuum Ultraviolet to Near Infrared Spectroscopy, *Journal of Thermophysics and Heat Transfer*, 2015, submitted.

[9]Milos, F. S. and Chen, Y.-K., Ablation and Thermal Response Porperty Model Validation for Phenolic Impregnated Ablator, *AIAA Journal of Spacecrafts and Rockets*, **47**, No. 5, 2010.

[10]Luhmann, T., *Nahbereichsphotogrammetrie: Grundlagen, Methoden und Anwendungen*, Wichmann, Berlin, 3rd ed., 2010.

[11]Rothermel, M., Wenzel, K., Fritsch, D., and Haala, N., SURE: Photogrammetric Surface Reconstruction from Imagery, Berlin, Dec. 2012.

[12]Schairer, E. T. and Heineck, J. T., Photogrammetric Recession Measurements of Ablative Materials in Arcjets, *Measurement Science and Technology*, **21**, 2010.

[13]Löhle, S., Brandis, A., Hermann, T., and Peter, J., Numerical Investigation of the Re-entry Flight of Hayabusa and Comparison to Flight and Ground Testing Data, *43rd AIAA Thermophysics Conference*, AIAA, LA, 2012.

[14]Auweter-Kurtz, M. andWegmann, T., Overview of IRS PlasmaWind Tunnel Facilities, *RTO Educational Notes*, **8**, Oct. 1999.

[15]Eswein, N., Herdrich, G., Fasoulas, S., and Röser, H.-P., Investigation of Graphite Ablation at IRS, *42nd Thermophysics Conference*, AIAA, 2011.

[16]Wernitz, R., Eichhorn, C., Herdrich, G., Löhle, S., Fasoulas, S., and Röser, H.-P., Plasma Wind Tunnel Investigation of European Ablators in Air Using Emission Spectroscopy, *42nd AIAA Thermophysics Conference*, 2011.

[17]Löhle, S. and Jenniskens, P., High Resolution Spectroscopy of the Hayabusa Re-entry using a Fabry-Perot Interferometer, *Journal of Spacecrafts and Rockets*, **51**, No. 6, 2014.

[18]Marr, J., Untersuchung von photogrammetrischen Methoden zu in-situ Oberflächenanalyse von Ablatoren, Diplomarbeit, Universität Stuttgart, 2011, IRS-11-S18.

[19]Pitzer, P., Einfluss refraktiver und reflektiver Elemente auf die photogrammetrische Kamerakalibrierung, Bachelorarbeit, Universität Stuttgart, 2012.

Author Index

Lightning Source UK Ltd.
Milton Keynes UK
UKOW06n1539051116
286901UK00001B/72/P

9 781119 323648